MICROFINANCE:
Model, Policies and Experiences

Susanta Kumar Barik (b.1972) is Master of Arts (Economics), Master of Business Administration (Human Resource Management), M. Phil in (Resource Management) and Ph.D. He has authored five books on MFIs, Tribal Development, Non-Timber Forest Product (NTFP) Management, Contract Farming (CF) etc which are published nationally and internationally. Besides, he has penned many research articles published in national and international journals.

During his Two Decades of professional life, he has been associated with various national and international institutions. As a researcher he has been associated with institutions like DANLEP, ORG-MARG, ODI-Uk, and CDR-Denmark.

As development professional to execute rural development projects in coordination with Government, he has worked with Institutions like CARE-India and Terre des hommes. During this stint, he had worked with 30 Non-Governmental Organisations (NGOs) to execute EU supported Sustainable Tribal Empowerment Project (STEP).

As a corporate professional, he has worked with Ashley Bio-fuels, Ashok Leyland (Hinduja Group) to steer the bio-fuel fleet at Visakhapatnam and currently, working as professional Director for Biomass Development Cooperation Pvt Ltd (BDCPL) and Vice-President for Agrigold Foods and Farm Products Ltd (Agri Division) of Agrigold Group of Industries.

About the Book

Microfinance has been viewed as institutions to bridge the rural-urban development gap. These institutions are spreading across villages of Indian states and addressing the existing financial gaps of rural people, thereby provisioning effective means to cater livelihood activities of choice to survive. The syndrome of economic-development-through-institutional means is amply presented in this book, while questioning the very objective of formation of the institution.

Historic perspective of microfinance journey is thoroughly captured. Broad models of MFI's being in practice by different institutions across the states of the country have been presented. Broad MFI regulation and control policies of state governments and the central government are succinctly captured.

The operational modalities of SHG-mode of MFI operation are thoroughly gauged in a tribal dominated ITDA of Paderu, Visakhapatnam District of Andhra Pradesh. The book is giving a close development view of MFI in its operating regions.

This book would be a base book for learners, practitioners of MFIs, NGOs, Donning agencies and various Government departments dealing with issues of rural development. The comprehensive information of MFIs i.e. formation, models, regulating policies and practical implications, provided in this book would certainly be a capsule information kit on MFIs.

MICROFINANCE: Model, Policies and Experiences

Susanta Kumar Barik

2015

Regency Publications

A Division of

Astral International (P) Ltd

New Delhi 110 002

ISBN 9789351302971

Published by : **Regency Publications**
A Division of
Astral International Pvt. Ltd.
– ISO 9001:2008 Certified Company –
House No. 96, Gali No. 6,
Block-C, 30ft Road, Tomar Colony, Burari
New Delhi-110 084
E-mail: info@astralint.com
Website: www.astralint.com

Sales Office : 4760-61/23, Ansari Road, Darya Ganj
New Delhi-110 002 Ph. 011-23245578, 23244987

Laser Typesetting : **Rajender Vashist**
Delhi - 110 059

Printed at : **Replika Press Pvt. Ltd.**

PRINTED IN INDIA

Preface

Finance is the backbone of the economy. It is often related with the power of directing the economy using all available resources and manpower. Available natural resources would be low or little meaning without sufficient availability of finance to best utilize the resources of the economy, let alone be the case of tribal economy.

The *built-in* ability of finance gets all the institutions and systems activated irrespective of its area and place of operation. Indeed systemic change *inter alia* could be better and faster in a financially better economy and vice versa. The flow of finance steers the activities to be undertaken in an economy, involvement of institutions, extend of market reach and involvement of agents and/or players.

The current document is a small initiative to present the overall gamut of functioning of microfinance finance, financial institutions, experiment of IKP model of financing and its impact on the tribal communities living on a hinterland. The document has attempted to collect and collate all rules, regulations, and legal provisions of Government of India (GoI) and Government of Andhra Pradesh and present in the document. This has been shared with the view to provide understanding to the readers on the complicacies as well as legal operating environments set in for the financial institutions in operation in the country.

There are several models of finance and especially microfinance being under trial and also tried in different parts of India and the functional efficiency is tested over the period of time. This has enabled to modify, re-modify and complete alteration of modality of operations. The evolution through learning with different setup communities and areas such as Scheduled Tribe, Scheduled Cast, Other Back Ward Caste, Literate and

Illiterates, male and female, rural and urban invited the competitive and complementing models into operation.

This document is an attempt to portray the economic, institutional, legal and environmental functioning of the microfinance institutions and/ or systems in place with tribal areas. It is a sincere effort to examine the IKP model of microfinance in execution in a tribal agency area of the state of Andhra Pradesh.

Though birth of SHGs often correlated with DWCRA Scheme, the popularization and reaping of benefits to the women in a tribal agency area is visualized only after intervention of Indira kranthi Patham (IKP). Tribal women's across the different clans get united to wipe out economic poverty of the fellow community, find out means for effective engagement of herself, family members and show direction to the youths.

The meaning of living, livelihood, social acceptance and above all social recognition has been broadened. The socio-economic-institutional revolution in tribal agency area through IKP is aptly comprehended in this book.

I gratefully acknowledge several individuals who contributed in various ways to comprehend my work. Without their co-operation & encouragement, I could not have completed my work. I acknowledge my irredeemable debt to them.

I owe deep debt of gratitude to Dr. Balla Appa Rao, Rector, Rajiv Gandhi Group of Educational Institutions, Rajahmundry, East Godavari District for his keen interest, encouragement and discussions which has enriched my knowledge and understanding on the subject.

I express sincere gratitude and applaud for the comprehensive support offered by Prof B. Mohan Venkata Ram, Professor of Management and Placement Chair, A.U Collage of Arts and Commerce, Andhra University in completing my research work. He has stood as a source of inspiration and encouragement throughout the research in a friendly manner. He has given the freedom of sharing ideas and getting his suggestions-a novel deed from a novel professor.

I am expressing my heartfelt thanks to our beloved Chairman Sri Venkat Rama Rao Avvas, Agrigold Group of Industries for his great vision on micro-activities for macro development which inturn enabled me to conceptualize the document framework to its present form.

I express my sincere thanks to Dr. A. Sarath I.A.S, Joint Collector & Addl. District Magistrate, Sanga Reddy, Medak District and Dr. T. Babu Rao Naidu I.A.S, Joint Collector & Addl. District Magistrate, Eluru, West Godavari District, for their guidance and support to conceptualizing the IKP operation in the tribal areas of the state of Andhra Pradesh.

I am thankful to my wife Anasuya Barik, and 7-year's old son Swastik Kumar Barik. No amount of expression of thankfulness would compensate the cost of keeping me aloof from him at the time when he needs my attention the best. Finally, I am dedicating this document to my father Sj. Pitabas Barik and mother Smt. Marua Barik whose nurturance, care, confidence and trust has made me to bring this document to the public.

Susanta Kumar Barik

Contents

Chapter 1
Outlook of Microfinance

"To the people of poor nations, we pledge to work alongside you to make your farms flourish and let clean waters flow; to nourish starved bodies and feed hungry minds. And to those nations like ours that enjoy relative plenty, we say we can no longer afford indifference to the suffering outside our borders; nor can we consume the world's resources without regard to effect. For the world has changed, and we must change with it."

–Oath taking speech of Barrack Obama, President U.S.A.

1.0. Introduction

The World Bank puts the number of poor around the world at 1.4 billion. It says this is larger than previously thought. But, it notes new estimates confirm that between 1990 and 2005, the number of people living in extreme poverty has fallen from 1.8 to 1.4 billion. This indicates the 1990 global poverty rate is likely to be halved by 2015, which would put the world on track to achieving the Millennium Development Goals.

Global poverty can be cut in half by 2015 and eliminated by 2025 if the World's richest countries including the United States, Japan and Germany more than double aid to the poorest countries (Koffi Annan). The investments needed to meet the U.N. goals adopted by world leaders at the Millennium Summit in 2000 to tackle poverty, hunger and disease and promote education and development, mainly in African and Asian countries. "What we're proposing is a strategy of investment to help empower the lives of very poor people that lack the tools and sometimes even the basic means to stay alive, much less be productive members of a fast-paced world economy," said Professor Jeffrey Sachs, head of the U.N. anti-poverty effort (M. Lederer ©MMV, The Associated Press, January 17, 2005).

However, the UN report says these figures mask large disparities among regions. It says poverty is declining most in Eastern Asia, particularly China. Other regions are seeing only a modest fall in poverty rates. And, UN officials say the number of poor has increased in sub-Saharan Africa and the Commonwealth of Independent States. "One hundred and five more million people will be falling into extreme poverty according to the World Bank estimates as a direct consequence of these increases. The peak is behind us and the prices have been declining since early July, 2008. But, the prospects are that we will remain with comparatively higher prices in the future," De Schutter said.

Voice of America reported that more than half a million mothers in developing countries still die in childbirth or from pregnancy complications every year. It says about one-quarter of developing world children are undernourished, and almost half of the developing world still lacks good sanitation (Lisa Schlein Geneva,*11 September 2008, Voice of America*).

The scenario of development is still worse in the third world countries. Riddled with extreme poverty and coupled with the political imbalances, sectoral development is lagging behind adequately to help enable a balanced development of sectors and causing for in-appropriate regional development. The debate on poverty is not for the developed and developing nations of the world, rather poverty in general. This has been getting reflected in many forms across the globe while the developing nations are home to it. Recent studies shows that during the 1990's, the poor in the developing countries were mostly concentrated in South Asia, typically located in rural areas, landless and participant in the rural labour markets, living in disadvantaged and resource-poor areas, and generally women and members of backward and economically disadvantage communities.

Poverty therefore causing for a lopsided development of the world economy in general and economies of the third world countries in particular. Low level of literacy, poor health facilities, poor infrastructural facilities, declining environment and ruined natural resources, abundant supply of manual labourer, etc. is thereby becoming the characteristics of the nations riddled with poverty. Further analysis of the poverty scenario across the different sections of the society in many rural areas of the developing countries suggests a very discernible pattern among the poverty groups, the aged often rely on the extended family-a structure that tend to be stable over time. Studies further confirm that poverty is highly correlated with landless in South Asia, South Africa, and much of Latin America.

The discernible pattern of development is becoming the global concern and is attempted to address through a strategic movement through the "Millennium Development Goal Approach". Developing nations are shifting a part of their resources in the form of charity, grants, grants-in-aid, subsidized grants, towards brings up sustainable development of the people and their institutions. The history of fund flow, from the developed nations to the third world countries, shows that till the end of 1980's it is in the nature of provisions of consumptives for the poor, which has later on changed towards development and promotion of self-propelling institutions to take care of long term measures of development.

The provision and effort of developed nations with vigorous internal changes in governments developmental policies to look into the *in situ* development devising different need based strategies on the basis of region, gender, natural resource base, manpower etc are contributing in a way to address issues of poverty (urban and rural).

1.1. Rural Development Strategies in India- At a Glance

Development of India is *prima facie* seen in terms of the development of rural villages. The common saying of India lives in villages and "India shines" when the villages of the country prospers. Through the five years plan, special attention is being given on the development of rural villages. Time to time necessary changes had been brought up in the development strategies to deal with diversified and complex issues peeping in with the passage of time. A close analysis of the poverty alleviation programmes in India clearly falls under two categories *i.e.* (1) interventions where the poor households/persons are direct beneficiaries, and (2) area level interventions where geographical areas/locations with high incidence of unemployment and poverty are targeted for development. This classification excludes programmes oriented towards social assistance and relief, which are of transient in nature.

Since the inception of economic planning in India, efforts have been made in successive plans to mitigate the incidence of poverty. Depending on the dominant development paradigms of the age, India has tried various strategies for addressing the issue of poverty. Some of them could be summarized as the Community Development, Trickle Down, Basic Needs, Human Resource Development, Labour-Intensive growth with targeted programmes and empowerment and enhancing security, to name a few. Despite all these efforts, there has been no conspicuous change at the poverty front: though the relative poverty has fallen marginally, the absolute poverty is still alarmingly high.

The growth strategy followed in 1960s and 1970s had presumed that a higher rate of economic growth, through "Trickle Down" effect, would enhance the standard of living of the poor. But the "Trickle Down" concept has failed to precipitate. Therefore, during the 1970s the Government of India had initiated Anti-Poverty Programmes. Since the Sixth Five Year Plan (1980-85), a more distinct and direct approach was adopted. "The 'direct' approach to poverty reduction emphasized that it is essential to directly provide the poor with adequate purchasing power, other assets or access to food grains at subsidized prices to meet their minimum consumption requirement" (Nayyar, 2005). "Bypassing the traditional growth approach, special Poverty Alleviation Programmes (PAPs) were to be implemented in order to reduce poverty level to 30 per cent by 1985. The schemes involved income generation for the poor, meeting their minimum basic needs (like rural drinking water supply, primary education, primary health care facilities, rural infrastructure and electrification, low cost housing and other social services), and provide specific support for the backward areas" (Stuijevenberg, 1996).

Programmes such as these were considered an acute necessity because there had been a gradual decline in the incidence of poverty, in absolute terms 277 million persons were still living below the poverty line, facing conditions of ill health and short life expectancy (Planning Commission, 1996-97). Lack of basic educational skills and access to the means of production, prevented the masses to participate in, and derive benefit from, economic growth. Besides, households in India often suffered from transient rather than chronic poverty. Their economic position also varied from year to year depending on a good or bad harvest, and within a year due to the seasonality of employment and wage earnings. The Anti-Poverty Programmes (APP henceforth) could be broadly classified into two groups: Rural Wage Employment Schemes and Rural Self- Employment Schemes. We have focused here on Self-employment programmes only.

The *first* category comprises credit-based self-employment programmes, wage employment and training programmes for the rural poor. Programmes under the first category are (i) Integrated Rural Development Programme (IRDP), (ii) Jawahar Rozgar Yojana (JRY), (iii) Employment Assurance Schemes (EAS), and (iv) Training of Rural Youth for Self Employment (TRYSM). Development of Women and Children in Rural Areas is a specific scheme that targets the upliftment of rural poor women and children. The Government of India coinage all self-employment programmes in to one and termed them as Swarnajayanti Gram Swarojgar Yojana (SGSY) and JRY is rechristened as Jawahar Gram Samridhi Yojana (JGSY).

The *Second* categories of schemes executed towards development of areas are (i) Drought Prone Area Programme (DPAP), (ii) Desert Development Programme (DDP), and (iii) Integrated Water Development Projects (IWDP). All these schemes are being executed for mitigating the effects of droughts, controlling the process of desertification and restoring ecological balance in the affected areas.

In recent times, special emphasis is laid on social mobilization through Self-help Groups (SHG's), especially DWCRA groups for creating access to development schemes and programmes and bringing together the hitherto socially marginalized, economically weaker and politically excluded sections of the rural poor. This approach is built on the fundamental principles of promoting the cause of the poor as partners in development and thereby ensuring the benefits of participation to the poor because the market oriented growth will not provide opportunities to the poor to participate in it and drive the benefits.

It is being realized by the planner and policy maker as well as the grass-root field executives of the plans and programmes that response from the public as well as their participation is more in the people involvement execution of schemes and programmes. The dent towards people involvement in project formulation and execution is seen, though of late, in all the poverty alleviation programmes by the centre and the state Government schemes and programmes. The recently launched participatory poverty alleviation programmes in the states of Andhra Pradesh with support of the World Bank *i.e.* Indira Kranthi Patham bears the testimony.

1.2. Microfinance- A Historic Look

The microfinance movement emerged in the 1970s[1]. Pioneers of microfinance, like Grameen and Opportunity International, were driven by a view that poor people are full of inherent potential: capable of pulling themselves out of poverty, if only they could get access to capital that banks would not provide. Small loans (micro credit) were devised to meet this need. Microfinance institutions at present serve an estimated 120 million clients in the world.

1.2.1. Microfinance- The Global Outreach

The microfinance sector has grown over time with more and different types of actors getting involved, more and more geographic regions around the world being serviced, new types of products and services being developed, and with new ideas and technologies evolved to support it. The global picture regarding microfinance outreach is quite impressive. From a mere 7.6 million poorest families in 1997, the Microcredit Summit Campaign reported an outreach of more than 113 million clients by

December 31, 2005. It included about 82 million poorest families when they started with the program. Of the 3,133 institutions that had reported to the Microcredit Summit Campaign by December 31, 2005, 1652 were in Asia, 959 in Sub-Saharan Africa, 439 in Latin America and Caribbean, 35 in North America and Western Europe, 30 in the Middle East and North Africa, 18 in Eastern Europe and Central Asia. The increase in the number of institutions reporting from 618 in 1997 to 3,133 in 2005 is definitely an indication of an impressive growth in the field of microfinance.

Of the over 113 million people reached in 2005, 96.7 million were in Asia, 7.4 million in Africa and 4.4 million in Latin America and the Caribbean. Only 5.7 million of the 61.5 million poorest families in Africa and the Middle East were covered by microfinance programs by the end of 2005. Asia, which is the home to some 67 per cent of the world's people living on less than US$ 1 a day, can, therefore, rightfully boast of a vibrant microfinance sector. In Asia, Bangladesh distinguishes itself by reaching more than 75 per cent of poor families with microfinance. It is the home for 31 per cent of the largest programs in the world who have individually reached more than 100,000 clients including some reaching more than 5 million poor families each. Microfinance Programmes in Bangladesh reached over 22.5 million poorest clients by the end of 2005. The intensity and density of microfinance is greater in Bangladesh than in any other country.

The pioneering role of Grameen Bank, the bold initiative of NGO-MFIs, the participation of banks, the implementation of government programs like BRDB, PDBF etc., the operation of PKSF, the strong commitment and competitive spirit of the major players in the field have largely contributed to such a development of microfinance in Bangladesh.

1.2.2. Microcredit Summit Campaign: Phase II Goals

In April 2005, the Microcredit Summit Campaign announced the extension of the Campaign up to 2015 at the Summit in Santiago, Chile. Phase II of the campaign from 2006-2015 has two goals. The first goal is to ensure that 175 million of the world's poorest families, especially the women receive credit for self-employment and other financial and business services by the end of 2015.

The second goal of the campaign is to ensure that 100 million of the world's poorest families move from below US$ 1 a day adjusted for Purchasing Power Parity (PPP) to above US $ 1 a day adjusted for PPP, by the end of 2015. With an average of 5 people per family, it can be estimated that 500 million poor would rise above US$ 1 a day, which would imply the near accomplishment of the Millennium Development Goal of halving absolute poverty.

This is exactly the kind of progress needed to reach the Millennium Development Goals (MDG). The task at hand is to make adequate institutional and policy preparations to ensure that the MDG and the two new goals set by the summit campaign are achieved on time. Based on the lessons learnt and successes enjoyed so far, this is the time to intensify the efforts to get to the goals as planned. It will therefore, be very important to see who, what, when, where, why and how of microfinance to project its possible expansion over the next ten years.

1.3. Development of Microfinance

In envisioning the future of microfinance it is important to know how the microfinance movement has come so far, who took the initiatives, who supported the cause and what opportunities and constraints it faced during the course of its development. It may be mentioned that the poverty-focused microfinance has come into existence as a private initiative and has grown almost unnoticed through a process of learning by doing. It is the inner urge that worked as a driving force for Muhammad Yunus. He took the initiative and then succeeded in developing a sustainable microfinance system that brought financial services to the doorsteps of the poor, especially to the poorest women who were always considered un-bankable. It came into being as a result of feeling for suffering human being, continuous thinking and innovative efforts. So is the case with many other MFIs that evolved as results of private initiative and that have grown overtime.

1.3.1. Global Issues of Microfinance

In the 30 years since its inception, the microfinance industry has received mixed reviews. There have been inspirational stories of households and even entire villages that have lifted themselves from the depths of poverty with the help of Microcredit loans. However, there have also been disappointing accounts of the inability of microfinance institutions (MFIs) to sustain their operations or to reach the hundreds of millions of poor who still lack access to basic financial services.

The world of microfinance, including that in Asia, is undergoing a major shift. Once the realm of donor supported, grassroots nonprofit organizations, the microfinance sector is forming partnerships with other stakeholders to create a broad based approach to the provision of financial services to the poor. By joining forces with governments and commercial banks, the microfinance sector is gaining recognition as a key part of the world's mainstream financial system. This shift is enhancing the sector's ability to reach out to larger numbers of clients with increased efficiency.

The Asia Microfinance Forum, held in Beijing on March 21-24, 2006 reflected the movement of microfinance into the mainstream. Organized by the Banking With The Poor (BWTP) Network and the Foundation for Development Cooperation (FDC), with lead sponsorship from the Citigroup Foundation, the forum focused on how partnerships between MFIs, governments and commercial financial institutions can create a better environment for microfinance services. Closer collaboration between stakeholders encourages innovations such as new product development and the use of information technology, and galvanizes resources for capacity building, including financial education training for MFI staff and their clients.

Many examples of partnerships and innovations in Asia that are contributing to international best practice in the microfinance field are showcased. Some of these are the cases of Philippines, Srilanka and Bangladesh. In the Philippines, a telecom company has transformed mobile phones into electronic wallets that allow lower income Filipinos to handle remittances, donations, loan settlements, disbursement of salaries and payment of bills through simple text messaging. Hatton National Bank in Sri Lanka has formed a partnership with the postal service so that post offices can provide electronic banking services for microfinance clients in remote areas. Central to discussions of the forum was the need for nonprofit organizations that provide microfinance services to conduct operations on a more businesslike basis, and for commercial banks to recognize the expertise that MFIs bring to the table in working closely with very poor clients in slums or scattered across vast rural areas.

With regard to the tensions between social mission and sustainability, there was general agreement at the Asia Microfinance Forum that the sector needs to achieve both these purposes. However, some Asian MFIs expressed concern that commercialization might lead to 'mission drift' – a dilution of the commitment to poverty reduction. Commercial style competition for resources could also disadvantage MFIs working in poor and more remote areas by diverting scarce resources to bigger MFIs that can show more stellar achievements in less challenging environments.

Several participants at the forum pointed to the activities of some larger and more established MFIs to support smaller and less developed organizations, especially in countries where microfinance is not so advanced. Other participants underlined the need for the private sector's role in microfinance to move beyond philanthropy and work with MFIs as partners to accommodate small scale savings, lending and insurance coverage, and to overcome the limitations that conventional banks have regarding access to poor and remote areas. 'At Citigroup, we support the industry through a combination of philanthropy and commercial initiatives,' says Robert Morse, CEO, corporate and investment banking, Citigroup Asia Pacific. 'In addition to Citigroup Foundation grants worth nearly $27 million that increase the capacity of large numbers of MFIs around the world, Citigroup businesses have commercial relationships with many MFIs, microfinance networks and investors to develop products and services that promote greater financial inclusion.

Further discussions at the forum reinforced the vital role played by government in creating a flourishing microfinance environment, but expressed caution about the overregulation of MFI activities, pointing to interest rate caps, inappropriate subsidies and political interference that hinder MFI efficiency and expansion. 'The microfinance industry has suffered from the tendency of its advocates to talk up its successes and of its detractors to sell it short,' says Beris Gwynne, executive director of FDC. 'Only through rigorous information gathering and evaluation can we truly appreciate the potential of microfinance to reduce poverty and accelerate sustainable development[2].'

1.4. Microfinance in India

Indian history withstands to the fact of rule of kings and their kingdom. Reference of Chanakya is being quoted as a role model of management of economics and management. Through ages, it is being mentioned that poor people use to develop their own mechanism of survival and manage their crises. The multiple form of coping is seen in the formation of local level organizations/institutions to extend help and assistance in all forms to waive off the difficulties and grain banks, seed banks, etc are the symbols. However, the small groups were always not been able to mitigate the credit and/or financial requirements of the entire community. The end result is often seen with creation of a disparity of wealth between and among the community members- a group of community member always depends on others in long-term perpetually.

The emergence of banking sector and its liberalization as well as nationalization of banking sector was able to manage the credit requirement of certain section of the community, while a major section being unable to access the service. The rich and wealthier people use to access the banking facilities and services extended while the poor remain away of access of the services offered by the banking sector. The concern for poor gets revealed in national as well as international forums and mandated to extend financial support to poor and marginalized to earn their bread and butter with dignity.

The Fifth General Assembly of the Asia Pacific Regional Agricultural Credit Association (APRACA) held at Bangkok in December 1984 exhorted the agricultural and rural development finance institutions in the region to mobilize savings from the rural areas with the objective of providing loanable funds for agriculture and rural development. The experience in some of the countries where informal self help groups (SHG's) of rural people which promoted savings among members and used these resources for meeting their credit needs, were considered useful innovations.

Further, the Third Consultation on the Scheme for Agricultural Credit Development (SACRED) held at Rome in 1985 called for active promotion of linkages between banking institutions and SHGs as a means of improving the access of low income group to banking services. The Executive Committee Session of APRACA held at Seoul in October 1985 approved the holding of a South-East Asian sub-regional workshop to devise ways and means of improving such linkages. The APRACA Regional Workshop held at Nanjing, China in May 1986 recommended national level consultation and organization of national surveys of SHG's in collaboration with APRACA and other agencies.

The Sixth General Assembly of APRACA held at Kathmandu, Nepal in December 1986 considered a project proposal on 'promotion of linkages between banking institutions and SHGs in rural savings mobilization and credit delivery to the rural poor'. It was decided that each member country would form a Task Force to conduct a survey of SHGs and thereafter, formulate suitable national level programmes. Consequent upon this, a Task Force was set up in India in the Ministry of Agriculture, to identify the existing SHGs, undertake a survey of the groups and draw a plan of action for channeling the flow of savings and credit between the rural poor and banks through SHGs and identify concrete projects for action research in this field.

Accordingly, in February 1987, it was decided that a study team led by NABARD and comprising of representatives from various financial institutions, should be constituted to undertake the survey. The survey was undertaken in September 1987 and the report discussed at the 18th Executive Committee Session and 10th Foundation Anniversary of APRACA held at New Delhi in November 1987. This survey report laid the foundation of the SHG-bank linkage programme in India launched as a pilot project in 1992. The pilot project was able to link 225 SHGs to the banks as against the target of 500 by the end of March 1993. The Self-help group and bank linkage programme operation in the country gave birth to three different unique models of financial delivery such as;

Model-I : SHGs promoted, guided and financed by banks

Model-II: SHGs promoted by NGOs/Government agencies and financed by banks

Model-III: SHGs promoted by NGOs and financed by banks using NGOs/formal agencies as financial intermediaries

All the three models are found executed across the country and are at different level of execution. Invariably all the models have been accepted as conduits of micro-credits liquidation. Interestingly, all the three up-surging models have been gaining momentum and popularity across the years, and have depicted an upward increasing trend of expanding the micro-credit net as well as the base. More numbers of self-help groups are being encouraged and so also the amount of credit disbursement and linkage with the banks (see table: 1.1).

However, a comparative look at the self-help group coverage as well as linkage with banks reveals that model-II is more vibrant and efficient. The model-II has emerged as the most popular model of microfinance delivery in India with a persistent increasing in SHG-bank linkage portfolio (see table: 1.2). The reason of being more popular and convenient mechanism is the involvement of local people and the institutions in the process of financial transactions, which has created auto most faith and belief on the agencies dealing with finance as well as the representatives. This has, in fact, resulted to formation of many Self-help groups (SHGs) in the country to expand the roots the micro credit. Needless to mention, Self-help groups formed in the country positioned women as the king-pin, who looks after management as well as decision maker for fund utilization as well as debt recovery. All formal lending agencies and informal agencies are mostly adopting the mechanism for financial liquidation.

Table 1.1: Model-wise Bank linkage Position

Models	2004		2005		2006	
	No. of SHGs	Bank Loans (Rs. crores)	No. of SHGs	Bank Loans (Rs. crores)	No. of SHGs	Bank Loans (Rs. crores)
Model-I	218	550	343	1013	449	**1673**
Model-II	777	3165	1158	5529	1646	**9200**
Model-III	84	189	117	356	143	**561**
Total	1079	3904	1618	6898	2238	**11434**

Note: Due to change in data formatting of NABARD, Model-wise figures after 2006 is not available
Source: NABARD, Annual Report, 2006.

A closer look at the Central Governments' initiative to confer economic reliability and sufficiency to women folk through extension of micro credit support reveals an uninterrupted increase in fund support to the target clientele through multifarious conduits of NGOs, SHGs and others. Grounded in 1993-94 and with an initial outreach of 37,000 women across the country, the programme has extended 140.82 lakhs loan support through 3707 self-help groups. The scenario has entirely been changed in terms of serving the destitute women of the country over the one and half decade and with an out reaches of nearly 11.4 lakh women (see table 1.3).

The microcredit support through different conduits has able to identify and utilize the entrepreneurial potentiality of the women of all races and castes across the country. Some of the notable investments of micro credit is seen around the use of natural resources available around the village community, while in the erstwhile cases instances of risk-ventured investments for income generation is also noted in terms of establishment of small scale industries in selected states.

The NGOs conduit of microcredit support extension is enabling the community to tap multiples sources of finance to the same group of people and in different doses. This is enabling to think bigger and act better as well as earn a smarter income for self-reliance and sustainability. Central Government's effort of women empowerment is in turn backed by the state governments in different forms of activities. Inter related departmental help is working a long way to entail and capacitate the women run institutions for sustainable income generation to reach at the final decision making point.

The wholesome picture of the country in terms of serving the women for conferring economic empowerment and self-sufficiency through extension of micro credit for undertaking entrepreneurial activities is very fragile. Moreover, more advanced states, as has been revealed from the statistics, have increased accessed to the designated fund for women empowerment of the country. The cumulative fund devolvement for women empowerment and reliability of the Rastriya Mahila Kosh of Government of India during 1993-2012 reveals that the state of Andhra Pradesh has comparatively greater access to the fund followed by Tamil Nadu. Invariably, it is the fact that the South-Indian states have greater to the kosh (See table 1.4) than the North-Indian states of the country.

Table 1.2: SHGs linked to Banks in India (1992-93 to 2011-12)

Year	No. of SHGs	Bank loan (₹ Cr)
1992-93	255	0.29
1993-94	365	0.65
1994-95	1502	2.45
1995-96	2635	6.06
1996-97	3841	11.84
1997-98	5719	23.76
1998-99	18678	57.07
1999-2000	81780	192.98
2000-01	149050	480.90
2001-02	197653	1026.30
2002-03	255882	2048.70
2003-04	361731	3904.20
2004-05	549385	6898.46
2005-06	610089	11398.00
2006-07	686408	18041.00
2007-08	1227770	8849.26
2008-09	1609586	12253.51
2009-10	1586822	14453.30
2010-11	1196000	14547.73
2011-12	1148000	16534.77

Source: Various Reports of MCID, and NABARD

Table 1.3: Cumulative and Year-wise achievement

Year	Loan Sanctioned (in Lakhs)	Loan Disbursed (in Lakhs)	No. of New NGOs	No. of SHGs	No. of Women beneficiaries
1993-94	439.33	140.82	24	3707	37066
1994-95	512.45	429.53	28	2506	25059
1995-96	861.66	563.83	48	3830	38302
1996-97	1700.55	916.76	70	9026	90264
1997-98	1271.40	1395.39	85	5962	59621
1998-99	1484.62	1148.5	131	5198	51982
1999-00	1958.75	1389.45	264	5746	57457
2000-01	2118.50	1675.8	151	4380	43799
2001-02	739.11	633.76	154	1561	15613
2002-03	1600.23	1058.37	89	4417	44172

contd...

Table 1.3: Contd...

Year	Loan Sanctioned (in Lakhs)	Loan Disbursed (in Lakhs)	No. of New NGOs	No. of SHGs	No. of Women beneficiaries
2003-04	2505.60	1446.07	69	3906	39056
2004-05	1549.74	1886.87	40	2236	22365
2005-06	1930.90	2068.42	82	2731	27305
2006-07	3071.00	2410.60	88	3469	34692
2007-08	3228.80	2557.74	52	3583	35827
2008-09	3030.45	2647.70	44	3617	36166
2009-10	1471.00	1563.03	23	1540	15404
2010-11	1278.00	1249.15	22	1336	13362
2011-12	1985.00	1631	22	1818	18182
2012-13	2258.00	1849	15	2147	21465
Total	**34995.09**	**28661.8**	**1501**	**72716**	**727159**

Source: Rastriya Mahila Kosh, Delhi

Table 1.4 : Cumulative and State-wise Achievement of fund accessibility and utilization from Rastriya Mahila Kosh

Name of States	Amount Sanct-ioned	Amount Disbursed	No. of New NGOs	No. of SHGs	No. of Women benefi-ciaries
1. Andaman Nicobar	40.00	40.00	1.00	67.00	667.00
2. Andhra Pradesh	9850.81	8596.18	379.00	23031.00	230313.00
3. Arunanchal Pradesh	10.00	5.00	1.00	10.00	100.00
4. Assam	485.90	453.90	13.00	660.00	6605.00
5. Bihar	577.25	408.83	65.00	1680.00	16806.00
6. Chhatisgarh	14.50	8.50	6.00	37.00	370.00
7. Delhi	954.72	758.62	27.00	746.00	7458.00
8. Gujarat	223.50	183.50	9.00	717.00	7173.00
9. Haryana	310.01	264.16	11.00	439.00	4392.00
10. Himachal Pradesh	676.48	655.63	22.00	1292.00	12925.00
11. Jammu & Kashmir	112.00	72.00	5.00	177.00	1768.00
12. Jharkhand	133.43	91.20	20.00	169.00	1688.00
13. Karnataka	1019.25	559.00	47.00	1889.00	18885.00
14. Kerala	866.22	682.52	55.00	3426.00	34257.00
15. Madhya Pradesh	2207.30	1932.40	27.00	2187.00	21866.00
16. Maharashtra	1671.90	1350.05	23.00	3567.00	35672.00
17. Manipur	284.05	216.04	36.00	842.00	8421

Contd...

Table 1.4: Contd...

Name of States	Amount Sanct-ioned	Amount Disbursed	No. of New NGOs	No. of SHGs	No. of Women benefi-ciaries
18. Mizoram	4.00	2.00	1.00	7.00	70.00
19. Nagaland	74.00	45.50	8.00	86.00	859.00
20. Odisha	2292.15	1972.40	128.00	3897.00	38971.00
21. Puducherry	51.00	16.00	1.00	30.00	300.00
22. Punjab	141.00	103.50	2.00	78.00	784.00
23. Rajasthan	1943.90	1584.74	40.00	3011.00	30108.00
24. Tamil Nadu	7491.34	6119.52	268.00	16586.00	165857.00
25. Uttar Pradesh	917.85	600.42	139.00	2657.00	26569.00
26. Uttarakhand	361.24	331.89	20.00	293.00	2928.00
27. West Bengal	2281.29	1608.29	147.00	5135.00	51347.00
TOTAL	**34995.09**	**28661.78**	**1501.00**	**72716.00**	**727159**

Source: Rastriya Mahila Kosh, Delhi.

1.4.1. Microfinance in Andhra Pradesh: A Synoptic View

The genesis of SHGs has been initiated by NGOs in late 1980s in Andhra Pradesh. However, the movement for women's savings and credit in the state have really began in 1995, in Nellore district as "*podupulakshmi*" movement. This has led to large-scale promotion of SHGs in every district of AP. The *podupulakshmi* movement started with the slogan of "save a rupee a day". Almost all the groups across the State began monthly individual savings of ₹ 30 with each DWCRA group consisting of around 15 women.

In the year 2000, DWCRA became a flagship program of Government of AP to implement poverty reduction programs. State Government introduced schemes like Deepam (subsidized gas connections for DWCRA women) and State funded Revolving Fund, in addition to the Central Government Schemes to channelize fund for poverty reduction. Several thousands of groups were formed instantly to access various Government Schemes. In order to strengthen the operation of DWCRA movement, the State Government has established a separate department called "Women Empowerment and Self Employment" under the Rural Development Department to focus on promoting DWCRA groups, increase bank linkage and to promote microenterprises. NABARD played a proactive role in sensitizing NGOs, Government Officials and Bankers on SHG – Bank linkage. Greater Coordination was witnessed among DRDAs, Banks and NGOs during 1999 – 2002.

NGOs that had internal capacities supported capacity building of DWCRA groups. During 1996-2001, UNDP funded Poverty Alleviation Program was implemented in Mahabubnagar, Kurnool and Anantapur districts with a total outreach of 3,200 SHGs. In 1999, CARE began the implementation of a DFID funded project called Credit and Savings for Household Enterprise (CASHE) to support the emergence of quality SHGs through NGOs and Government. Instances of competition among NGOs, DRDA and UNDP were also noticed. In June 2000, World Bank funded District Poverty Initiatives Project (DPIP) was initiated in six poorest districts of AP based on the UNDP experience. In 2002, the project was scaled up to all the 22 rural districts of AP with funding support from World Bank and the project is called AP Rural Poverty Reduction Project.

Irrespective of the fact of government run micro credit programme, there are many non-governmental institutions in to micro financing activities in the state. Agencies like Sa-dhan foundation, Grameen Bank, SHARE microfinance Ltd. BASIX, SKS Microfinance, The Bellwether Microfinance Fund, SPANDANA, ASMITA, UMDUMA PODDU PEDATHA and MACs and many more public sector banks are also engaged in the activities. The multiple agencies in microfinance in the state of AP thereby follow different models for financial transactions. However, of all the agencies extending microfinance support, the World Bank supported Velugu programme is working in the entire state. There are 8752540 members organized into 699056 SHGs in 33700 Village Organizations (VOs) and 1083 Mandal Samakhyas (MSs). In addition, there are 126 Mandal Vikalangula Sangams, 17 Chenchu Mandal Samakhyas and 15 Fishermen Mandal Samakhyas. Zilla Samakhyas have come into existence in all the 22 districts. The SHGs in A.P. constitute around 25 per cent of all 2799850 SHGs in India. Following the approach of SHG-BL, ₹ 5300090 have been mobilized for 4105309 SHGs across the districts (see table-1.5). Self-help Group statistics of the state reveals that 74.58 lakh rural poor women are actively covered under the scheme.

Table 1.5: SHG-Bank Linkage in Andhra Pradesh

Year	Achievement (in Lakhs)	No. of Groups	Per Group Finance
2000-01	17300	101765	0.17
2001-02	26200	119091	0.22
2002-03	47700	170357	0.28
2003-04	89800	249444	0.36
2004-05	123800	294762	0.42

contd,,,

Table 1.5: Contd...

Year	Achievement (in Lakhs)	No of Groups	Per Group Finance
2005-06	200100	290000	0.69
2006-07	306400	369157	0.83
2007-08	588300	432574	1.36
2008-09	663700	477482	1.39
2009-10	650100	414076	1.57
2010-11	709300	389725	1.82
2011-12	807600	351130	2.3
2012-13	1069790	445746	2.4
Total	**5300090**	**4105309**	**13.81**

Source: Compiled from various reports of Indira Kranti Pratham, Hyderabad

1.4.2. Scenario of Microfinance in Visakha Agency Area: Synoptic View

The PoP study in the Visakha agency area shows that almost 48.49 percentage tribal households in the Paderu Agency Area falls under the category of 'Poorest of the Poor' and similarly 45.23 per cent household are Poor, and a just 6.27 per cent household are 'Above the Poverty Line'(see table-1.7). With the objective of ensuring micro income, employment and food, it has been attempted by MACs as well as by Indira Kranti Patham (IKP) to ensure 100 per cent coverage of POP household as well as poor under the SHG led institutional coverage. The study also reveals that Mandals having less POP families are supported with more number of Self-help Groups and vice-versa. Interestingly, Mandal like Paderu, which is the ITDA head quarter and home to many 'APL House hold', is provided with a significant number of Self-help group coverage in comparison to the most interior Mandals like Munchingput. Indeed, all POP household in the agency are given with Self-help group coverage.

Indira Kranthi Pratham (IKP) is executed with involvement of active 8325 women Self-Help Groups, 582 Village Organizations, 11 Mandal Mahila Samkshya and One Zilla Mahila Samkshya. Self-help groups are generally promoted by the IKP and some of them are also formed by the Non-Governmental Organisations (NGOs) working in the agency area. The microfinance following the SHG conduits are generally executed following the hierarchy *i.e.* ZMS, MMS, VO and then SHGs.

Notes

1. What is microfinance, Opportunity International, Australia, 2007.
2. Extract from Advertising Supplement, *International Herald Tribune*, April 3, 2006.

Chapter 2

MFI's in India–
The Past, Present and Future

2.0 Microfinance Institutions: The Past

2.1 History of MFI's

The concept of microfinance is not new. Savings and credit groups that have operated for centuries include the "susus" of Ghana, "chit funds" in India, "tandas" in Mexico, "arisan" in Indonesia, "cheetu" in Sri Lanka, "tontines" in West Africa, and "pasanaku" in Bolivia, as well as numerous savings clubs and burial societies found all over the world. Micro Finance is always an integral part and parcel of human development and sustenance. It has acclaimed significant bearing in the economies of human society and dated as old as the existence of human society on the earth and of course not in the era of stone-age and new-stone age. History withstand to the fact that microfinance did gain momentum and necessity of it was massively recognized at macro-level in the mediaeval and modern era.

In 15th century, the Catholic Church funded pawn shops served as an alternative to usurious money-lenders in Europe. Formal credit and savings institutions for the poor have also been around for decades, providing customers who were traditionally neglected by commercial banks a way to obtain financial services through cooperatives and development finance institutions. One of the earlier and longer-lived microcredit organizations providing small loans to rural poor with no collateral was the Irish Loan Fund system, initiated in the early 1700s by the author and nationalist Jonathan Swift. Swift's idea began slowly but by the 1840s had become a widespread institution of about 300 funds all over Ireland. Their principal purpose was making small loans with interest for short periods. At their peak they were making loans to 20% of all Irish households annually[1].

In 18th century Europe saw the emergence of formal savings and credit institutions. All these institutions focused on savings and credit needs of rural and urban poor. The financial cooperative was developed in Germany. It aimed to help the rural population break dependency on moneylenders and to improve Welfare. The movement emerged in France in 1865 and Quebec in 1900. Many of today's financial cooperatives in Africa, Latin America, and Asia find their roots in this European movement. Another early example during the century is the Indonesian People's Credit Banks (BPRs) opened in 1895 and became the largest microfinance system in Indonesia, with close to 9,000 branches.

Early 19th century witnessed dawn of a new era of microfinance and wide variation in functioning of micro financing operation started creeping into the system. A new definition on the functioning of microfinance emerged, and style of functioning, priorities and policies changed. The rural finance interventions broaden the vision and aimed at modernizing the agricultural sector, mobilize idle savings, increase investment through credit, and reduce oppressive feudal relations enforced through indebtedness.

Between the 1950s and 1970s, governments and donors focused on providing agricultural credit to small and marginalized farmers under the plea that financial assistance would improve farm production and in turn fetch adequate income. The effort resulted to expansion of state-owned development finance institutions, and/or farmers' cooperatives. All these institutions sanction agricultural loan at below-market interest rates in massive scale.

Meanwhile, the 1970s saw the birth of microcredit Programmes in Bangladesh, Brazil, and a few other countries began lending to poor women entrepreneurs. Early micro enterprise credit was based on solidarity group lending in which every member of a group guaranteed the repayment of all members. Examples of early, pioneers include Grameen Bank in Bangladesh, which started out as an experiment by Prof. Muhammad Yunus; ACCION International, which began in Latin America and then spread to the United States and Africa; and the Self -Employed Women's Association Bank in India, which is a bank owned by a women's trade union. These institutions continue to thrive today and have inspired countless others to replicate their success.

In the 1980s, micro-credit programmes throughout the world improved on the original methodologies and defied conventional wisdom about financing for the poor. First, well-managed programs showed that poor people, especially women, paid their loans more reliably than better-off people with loans from commercial banks. Second, they demonstrated that poor people are willing and able to pay interest rates that allow microfinance institutions (MFIs) to cov-er their costs. MFIs that cover their costs can become viable businesses that attract deposits, commercial loans, and investment capital. They can reach huge numbers of poor clients without being limited by a scarce and uncertain supply of subsidized funds from governments and donor agencies. Bank Rakayat Indonesia (BRI) is a dramatic example of what can happen when MFIs focus on collecting loans and covering costs. BRI's village-level branch system now serves more than 30 million low-income savers and borrowers.

The 1990s saw growing enthusiasm among international development agen-cies and networks for promoting microfinance as a strategy to alleviate poverty. Microfinance blossomed in many countries where multiple MFIs serve the needs of micro-entrepreneurs and poor households. These gains, however, tended to concentrate in urban and densely populated rural areas. In the early 1990s, the term "microfinance" rather than "micro-credit" began to be used to refer to a range of financial services for the poor, including credit, savings, insurance, and money transfers.

To reach ever larger numbers of poor clients, MFIs and their networks in-creasingly began to pursue a strategy of commercialization, thus transforming themselves into for-profit corporations that could attract more capital and be-come more permanent features of the financial system. An emphasis on creat-ing and growing strong institutions (as opposed to channeling credit to specific groups) is a core element of this recent history.

2.2 Development of Microfinance in India: The Major Milestones

Microfinance in India left its foot print in 1904 with the enactment of Co-operative Credit Societies Act. The primary objective of enactment of co-operative credit system was to combat the problems of usury and indebtedness of the farmers to moneylenders and to bring together people of small means for promoting thrift and mutual help for development. The Co-operative credit agencies included the primary agricultural credit society (PACS) at the village level, which was the basic unit in the short-term credit structure and were federated into the Central Co-operative Banks (CCBs) at the district level. The CCB were, in turn, federated at the state level as apex co-operative banks. The Land Development Banks (LDBs) usually meets the long-term credit needs of the farming community. Unlike -the short-term credit societies, the LDBs issued loans only against the mortgage of land and, therefore, the landless poor were excluded from the purview of these banks.

After Independence, the Government of India (GOI) appointed a number of Committees to review the arrangements for institutional credit delivery. The recommendations of these Committees, dealt more with the need for streamlining the flow of credit to the rural sectors, mainly agriculture sector, through appropriate institutional arrangements than with providing access to institutional credit for the asset less rural poor. The All-India Rural Credit Review Committee appointed in 1966 suggested that credit must be made accessible to the small farmers. As a follow up to this recommendation Small Farmers Development Agency (SFDA's) were established in 45 selected districts during the Fourth Five-Year Plan period (1969-74) to extend financial assistance to farmers with land holding size of up to 2 hectares. The Government of India had approved 40 projects under the Marginal Farmers and Agricultural Labourers (MFAL) Development Agency with a view to provide supplementary employment opportunities for marginal farmers, agricultural and landless labourers.

2.2.1 Nationalization of Banks

In 1969 there are 14 major banks nationalized and operated under the single-door administration of Reserve Bank of India. In the post nationalization era, the country witnessed not only rapid expansion of the banking network in the rural and semi-urban areas but also an increased flow of credit to the priority sectors like agriculture, small scale, tiny, cottage and village industries, etc. The induction of the commercial banks into the realm of rural credit also ensured that a larger proportion of the rural poor, especially marginal and small farmers, rural artisans, had access to banking facilities. Nevertheless, it was increasingly felt at all level of Government administration that the rural poor needed exclusive attention and specialized services, especially in remote and hitherto unbaked areas.

The bank nationalization has impacted positively towards growth of bank branches in rural areas. The number of bank branches increased from 8,262 in 1969 to 68,682 as on 2007, of which rural area branches are 30,572 constituting of 45 per cent of the branches. The proportion of credit disbursed to rural areas tripled in the 1970s, and continued until 1980s. After 1980, however, the share of total bank credit that went to rural areas declined, from 15 per cent in 1987 to 10 per cent by March 2000, and further declined to 8 per cent in 2007. This is due to overall increase of industrial credit in the country. Rural deposits also grew rapidly after nationalization. Their share of aggregate deposits doubled in the 1970s, from 6 per cent in 1969 to 13 per cent by 1980 and become 16 per cent by 1990 and later on declined to 11 per cent in 2007.

2.2.2 Regional Rural Banks

India lives in villages and more than 70 per cent of working community depends on agriculture and allied activities. With a view to serve the growing demand of the rural poor, artisans and others rural banks have been established and are functioning in the country. The Government of India appointed a working group on Rural Banks in 1975 and the group recommended the setting up of State sponsored, regionally based and rural oriented banks called Regional Rural Banks (RRBs) through out the country. The RRBs would grossly aim at developing the rural economy by providing credit and other facilities to agriculture, trade, industry and other productive activities in the rural areas, particularly to the small and marginal farmers, agricultural laborers, artisans and small entrepreneurs. The coverage of the weaker sections and the envisaged "low-cost" structure of the RRBs characterize them as the "small man's bank". The GOI enacted the RRBs Act in 1976 and thus, the RRBs formed the third pillar of the multi-agency credit system for agriculture and rural development.

2.2.3 Differential Rate of Interest (DRI) Scheme

Government of India introduced the differential rate of interest (DRI) scheme to benefit the poor in April 1972 across 162 districts. Experiencing the benefit, the scheme was later on extended to the entire country. Under the scheme, banks use to earmark one per cent of their previous year's total outstanding advances for giving loans at a concessional interest rate of 4 per cent to the weaker sections of the society. The scheme was targeting to disburse almost 40 per cent of DRI fund to SC/ST. The amount disbursed under the scheme during 1997-98 was ₹ 598.92 crore to 14.07 lakh borrowers while the amount disbursed (₹ 349'.96 crore) to SC/ST constituted slightly more than 58 per cent of the amount disbursed under DRI. During the three-year period from 1995-96 to 1997-98, the number of accounts and the amount disbursed under the scheme has shown a declining trend.

2.2.4 Establishment of NABARD

The Reserve Bank of India constituted a Committee to Review Arrangements for Agriculture and Rural Development (CRAF1CARD) under the Chairmanship of Shri. B. Sivaraman in 1979. The committee had recommended for opening up of a formal institution to cater the need of the poor and agriculturists and their by create enough resource potential so that rural development would take its own course. On the recommendations of the committee, National Bank for Agriculture and Rural Development (NABARD) was founded in June 1982 to provide undivided attention, forceful direction and pointed focus to the credit problems arising out of integrated rural development". Prior to the formation of National Bank, the emphasis in financing the weaker sections was largely confined to the agricultural sector due to the national priority of attaining self-sufficiency in food production. Since its inception in 1982, National Bank's efforts have been directed towards extending the coverage of weaker sections in other sub-sectors of the rural economy. In 1985-86, NABARD has adopted twin strategy of (i) promoting the development of rural non-farm sector through various promotional schemes for imparting training through Training and Production Centres, Training by Master Craftsmen, Rural Entrepreneurship Development Programmes in collaboration with NGOS, and (ii) extending refinance support on liberalized terms to the credit institutions for financing the setting up of Mother Units, Common Facility Centres. Further, refinance support is also available for financing individuals under the Composite/integrated Loan Scheme for NFS. Emphasis was also laid on empowering women in rural areas by constituting a Women Cell in the bank in 1992 to address issues relating to credit and other supportive needs of rural women through refinance. The institutionalization of the rural credit delivery system had facilitated the building up of a vast network of credit infrastructure and helped in weaning away a substantial proportion of the borrowing households from the usurious moneylenders. The share of rural household borrowings from formal institutional sources increased from a meager 3 per cent in 1951 to 64 per cent by 1991, while 41 per cent of the non-cultivator households were dependent on informal sources of credit in 1991.

2.2.5 Integrated Rural Development Programme (IRDP)

Integrated Rural Development Programme (IRDP) Scheme of Government of India came in to force in 1978-79. Various schemes like the Command Area Development, Drought Prone Areas Programme, SFDA and MFAL were dovetailed into Integrated Rural Development Programme (IRDP) was initially executed in 2300 blocks and was later on extended to all the blocks of the country in1980. The main thrust of the programme was to help the poorest of the poor in rural areas. The target groups under the programme are the rural poor belonging to the families below poverty line (BPL) and comprise of small and marginal farmers, agricultural labourers, rural artisans, etc., whose annual family income does not exceed the poverty line of ₹ 11,000 at 1991-92 prices. The programme aimed at providing income generating assets and self-employment opportunities to the target group through providing financial assistance in the form of subsidy by the Government and credit by banks. Adequate safeguards have been taken to ensure maximum coverage of the disadvantaged sections of the society like

SCs/STs, physically handicapped and women were also provided for in the programme. These included stipulations that at least 50 per cent of the families assisted belonged to SC/ST families, 40 per cent of the total beneficiaries were women and 3 per cent physically handicapped. A separate agency called the District Rural Development Agency (DRDA) was created to co-ordinate with various agencies, including banks, to implement the programme.

The concurrent evaluation of IRDP suggested for fine-tuning the current policy as well as practices adopted. The important policy changes effected in the programme include, (a) giving supplementary dose of assistance to the families who were provided financial assistance but could not cross the poverty line due to variety of exogenous factors, (b) cash disbursement of assistance in selected blocks, (c) issue and updating of pass books to the beneficiaries, (d) providing back ended subsidy to prevent misuse of subsidy and (e) increased emphasis on group loans.

Basing on the recommendation, the concept of Family Credit Plan (FCP) was introduced in the year 1994-95. Under the scheme, more than a single member of the family would be assisted for acquiring multiple assets, subject to the overall ceiling of subsidy for the family and banks were expected to mobilize higher amounts of credit for the families selected. The scheme was operationalized in 316 selected districts of the country. A special scheme to provide basic technical and managerial skills to the rural youth (in the age group 18-35) from families below the poverty line and to facilitate them to take up self-employment, the scheme of Training of Rural Youth for Self Employment (TRYSEM) came in to operation in 1979 as part of the IRDP. Looking at the potentiality and commitment of rural women, a special sub-scheme (part of IRDP) called Development of Women and Child in Rural Areas '(DWCRA), was launched in 1982-83 in 50 selected districts of the country. The scheme was subsequently extended 1994-95 to cover all the districts in the country. The scheme was aiming at economic empowerment of women in rural areas through providing training, employment management skills and additional channel of funds in the form of working capital and investment credit. Such an intervention not only seeks at raising the incomes of women but also facilitates organized 'participation of women, who can not take up economic activities individually on their own, and in groups.

2.2.6 Swarnjayanti Gram Swarozgar Yojana (SGSY)

Several Evaluation and impact studies institutionalized by Government of India from time-to-time on all individual rural development programmes established some fundamental inefficiency. Some of these inefficiencies included (i) poor quality of assets procured by the beneficiaries, (ii) inadequate financial assistance including subsidy, (iii) lack of inter-agency co-ordination in ensuring backward and forward linkages, (iv) concentration on a few activities, (v) misutilisation of loans for meeting the consumption requirements and (vi) Inappropriate identification of beneficiaries and activities. In order to overcome the operational inefficiencies and yield maximum benefit to the maximum number, the Hashim Committee in 1999 recommended one umbrella scheme "Swarnjayanti Gram Swarozgar Yojana" (SGSY) of Poverty alleviation. The primary objective of the scheme was to bring the assisted poor families above the poverty line in three years, by providing them with income generating assets (IGA) through a mix of bank credit and government subsidy to ensure that the family has a monthly income of at least ₹ 2,000 following the Self-Help Groups (SHGs) approach.

2.2.7 Priority Sector Lending: Focus on Poor

As per the directions of RBI, Commercial banks are directed to provide loans to the extent of 40 per cent of their credit portfolio to the priority sector comprising agriculture and allied activities, small, tiny, cottage and village industries, rural artisans, etc. It is also specified that 18 per cent of loan would be exclusively extended to agriculture and allied activities and10 per cent is required to be provided to weaker section of the society covering scheduled castes, scheduled tribes, small and marginal farmers, agricultural labourers, rural artisans, etc. These loans are essentially of the nature of microfinance. The share of total advances to priority sector in the post nationalization period rose from 14 per cent in 1969 to 21 per cent in 1972 to 33 per cent in 1980 and 42 per cent in 2007.

2.3.0 Development of Informal Sector

Existence of voluntary organization dated back to the pre-independence period of the country. Multiple factors ensured for formation and growth of voluntary sector. Some of the noted reasons are illiteracy, geographic cut-off, lack of awareness, poor communication, poor health etc. Most of the Voluntary Organizations following the developmental activity root later on sticking to micro credit operation in the present era. NGOs like SHARAN in Delhi, Federation of Thrift and Credit Association (FTCA) in Hyderabad or SPARC in Mumbai have adopted the first model of SHG-BL, where they initiate the groups and provide the necessary management support. Others like SEWA in Ahmedabad or Baroda Citizen's Council in Baroda pertain to the second model.

The experience of these informal intermediaries shows that although the savings of group members, small in nature do not attract high returns, it is still practiced due to security reasons and for getting loans at lower rates compared to money lenders. These are short term loans meant for crisis, consumption and income generation needs of the members. The Wholesalers of formal and non-formal micro credit players will include agencies like NABARD, Rashtriya Mahila Kosh-New Delhi and the Friends of Women's World Banking in Ahmedabad. Few of the NGOs supporting SHG Federations include MYRADA in Bangalore, SEWA in Ahmedabad, PRADAN in Tamil Nadu and Bihar, ADITHI in Patna, SPARC in Mumbai, ASSEFA in Madras etc. While few of the NGOs directly retailing credit to Borrowers are SHARE in Hyderabad, ASA in Trichy, RDO Loyalam Bank in Manipur.

2.4.0 Indian Experience with various schemes: A snapshot view

Year	Events
1952	Community Development Programme to initiate process of transformation of socio economic life of villagers
1955	Establishment of State Bank of India
1969	1st phase of Nationalization of Banks (14 banks) Introduction of Lead Bank Scheme
1971	Small Farmers Development Agency (SFDA)
1972	Differential Rate of Interest Scheme (DRI) at 4 per cent rate of interest
1974	Introduced Priority Sector Lending to lend 1/3rd of PSBs Net Banking Credit (NBC) – Later on increased to 40 per cent in 1980
1975	Introduction of Regional Rural Banks
1979	Integrated Rural Development Programme (IRDP)
1980	2nd phase of Nationalization of Banks Introduction of REGP (Rural employment Guarantee programme)
1982	Introduction of National Bank for Agriculture and Rural Development for refinance support to banks
1988-89	Public Sector Banks to Finance 17 per cent advances to Agriculture to their total outstanding of Advances later on increased to 18 per cent
1989	Service area approach (SAA)
1992	Financial Sector Reforms by Narasimham Committee and Basel Committee

1992	Introduction of microfinance (SHGs)
1999	Introduction of SGSY- Deregulation of interests
2004	Doubling of agricultural credit
2005	Financial Inclusion- Removal of Service area approach

2.5.0 MFI in India: The Present Scenario

Microfinance emerged as a noble substitute for informal credit and an effective and powerful instrument for poverty reduction among people who are economically active but financially constrained and vulnerable in various countries (Morduch and Haley, 2002). It covers a broad range of financial services including loans, deposits and payment services, and insurance to the poor and low-income households and their microenterprises. Convincing research evidence exists showing significant role of MFIs in improving the lives of the deprived communities in various countries. Persuaded with the potential role of micro financing in alleviating poverty, the South Asian countries especially India and Bangladesh have been actively pursuing the policy of setting up formal network of microfinance institutions.

Microfinance institutions of India has acclaimed international stance in terms of performance, functionality and coverage. The international rating positioned 10 MFIs of India in the top-10 list; exhibiting the relevance and competitive functionality advantages of microfinance institutions across the globe putting behind the pioneering MFIs of Bangladesh. There are roughly some 800 MFIs of different forms is in operation across the country and CRISIL rating for 50 top Indian MFIs reveals that MFIs of southern India rated high than rest of the country[2]. According to State of the Sector Report on Microfinance, over 800 microfinance institutions (MFIs) now operate in India, reaching out to over 140 lakh clients, with an outstanding cumulative loan of ₹ 5,900 crores in small loans. Over 54 million clients are accessing microfinance services from the Self Help Group (SHG)-Bank linkage programme and MFIs today. The Indian microfinance sector has seen phenomenal growth in the last few years. In 2007-08 alone, MFIs have had a growth rate of 40 percent[3].

2.5.1 Value attributes of Microfinance

First, microfinance is done with a laudable intention

Second, microfinance is something done exclusively or predominantly with the poor. Banks typically will have to be classified as promoters, rather than providers.

Third, microfinance grows out of developmental roots. This is what can be termed as "alternative commercial sector." This encompasses the first two points – the organizations are promoted by the alternative sector, and targeting the poor. However the new organizations growing out of these roots need not necessarily be "developmental" in the form of incorporation. There are MFOs that have been offshoots of NGOs run on commercial lines. There are also instances where new MFOs are promoted on commercial lines. People who have worked with NGOs in the past and have "developmental" credentials usually promote such MFOs. The lending and related activities done by such MFOs are usually understood to be microfinance.

Lastly, the Reserve Bank of India (RBI) has defined microfinance by specifying criteria for MFOs to seek exemption from registration under the Non-Banking Finance Company (NBFC) guidelines. This definition is limited to not-for-profit companies and as of date only two MFOs in India – Sanghamithra Rural Financial Services (SRFS), and Indian Association of Savings and Credit (IASC) – qualify to be classified as microfinance companies.

2.5.2: Scenario of Microfinance Institutions across the Indian states

Micro finance in India has been viewed as a means of socio-economic development of the poor and specially targeting active involvement of the women section of the community. Despite all attempts by formal and non-formal microfinance institutions, recent the CRISIL estimate in 2009 reveals that, around 120 million households in India continue to face financial exclusion; this translate into a credit demand of around ₹ 1.2 trillion[4]. It has been observed that almost all states of the country are being covered by some form of Microfinance institutions (see table 2.1).There are 215 microfinance institutions having Headquarters of their operation in all states of the country and are extending financial outreach through 349 branches

The microfinance market in India is expected to grow rapidly, supported by the Government of India's initiatives to achieve greater financial inclusion, and growth in the country's retail sector. The Micro finance sector in India has passed its evolutionary phase, when the profit oriented working model of MFIs was perceived by the market as exceptionable; acceptance of the same has increased gradually. Also investors now have a wider choice of MFI's with scalable process. NGO-MFIs have been acquiring dormant NBFCs for regulatory, financial and operational reasons. Many large microfinance institutions are now started extending financial operation in urban areas of the country and have begun extending loan to individuals.

However, the realm of growth and development of microfinance institutions started with financing the poors' in the rural area of the country and the base of their operation is from the rural villages of India. The following table reveals that there are six states not being extended financial services to the rural community and are operating with a strong urban-base. However, the clientele groups in the urban centers are the bellow poverty households residing in urban slums and leading a life in cities. Interestingly, microfinance institutions in states like are Bihar, Orissa and Jharkhand are primarily working with the poor districts and poor households (see table-2.2 and Fig.1).

2.5.3 Trend of Loan outstanding of MFIs : An Analysis

The microfinance sector in India is fragmented. There are more than 3000 microfinance institutions, NGOs, and Government-MFIs are working in the country and of which about 400 have active lending programmes. The top 10 MFIs are estimated to account for around 74 per cent of the total loans outstanding for MFIs. The loan disbursement is skewed in nature. States like Andhra Pradesh (31.75%), Karnataka (18.02%) and Tamil Nadu (9.39%) accounted for 59.17 per cent of the total loan disbursement of the country (see table.2.3). The detail financial disbursement and performance of 241 leading MFIs in India is annexed to this chapter (see annex-1). It is worth noticing that 51.95 per cent of clients is accessing 59.17 per cent of resource of three states; accounting to 40 per cent fund availability for rest 22 states of the country. The density of fund devolvement by microfinance institutions is confined to southern states of the country. Statistics reveals that as of March 2009, 17 MFIs had outstanding loan for morethan ₹ 1 billion.

The per district loan disbursement average is in line with the loan fund distribution; exception is being notices in states like West Bengal, Delhi and Orissa where proportionately larger district coverage is accompanied by a significantly higher loan disbursement (See table.2.4). However, more fund devolvement in a small geographic location increases the per capita fund availability and hence seemingly a better prospect of taking up of Income Generation Programmes; which would likely multiply the chance of faster acceleration of socio-economic development.

Table 2.1 : State-wise Major of MFIs and MFI-Branches as on 2008

Sl. No.	Name of the state	No. of MFI (HO) in the state	No. of MFI (HO& Branches) in the state
1	Andaman Niccobar	0	1
2	Andhra Pradesh	32	38
3	Assam	7	8
4	Bihar	18	26
5	Chhattisgarh	0	7
6	Delhi	2	7
7	Gujarat	2	14
8	Haryana	0	2
9	Jharkhand	13	21
10	Karnataka	14	25
11	Kerala	7	11
12	Maharashtra	18	28
13	Manipur	1	2
14	Madhya Pradesh	9	17
15	Meghalaya	0	1
16	Nagaland	0	1
17	Odisha	26	35
18	Puducherry	0	3
19	Rajasthan	6	13
20	Tamil Nadu	21	32
21	Tripura	2	3
22	Uttar Pradesh	11	18
23	Uttarakhand	3	8
24	West Bengal	23	28
25	Punjab	0	0
	Total	215	349

Source: Data Compiled from various reports of Bharat Micro Finance, Sa-dhan, SKS, Access Development Services

Table 2.2: State-wise MFI and District Coverage in India

Sl. No.	Name of the State	No. of MFI (HO& Branches) in the State	No. of District Served	Out of which No. of poor Districts	% of poor districts to total districts covered by MFIs
1	Andaman Niccobar	1	1	0	0.00
2	Andhra Pradesh	38	23	19	82.61
3	Assam	8	20	7	35.00
4	Bihar	26	35	35	100.00
5	Chhattisgarh	7	16	4	25.00
6	Delhi	7	1	0	0.00
7	Gujarat	14	16	2	12.50
8	Haryana	2	11	1	9.09
9	Jharkhand	21	20	19	95.00
10	Karnataka	25	28	11	39.29
11	Kerala	11	11	3	27.27
12	Maharashtra	28	38	18	47.37
13	Manipur	2	2	0	0.00
14	Madhya Pradesh	17	32	15	46.88
15	Meghalaya	1	0	0	0.00
16	Nagaland	1	1	0	0.00
17	Odisha	35	30	24	80.00
18	Puducherry	3	2	1	50.00
19	Rajasthan	13	20	3	15.00
20	Tamil Nadu	32	30	10	33.33
21	Tripura	3	2	1	50.00
22	Uttar Pradesh	18	36	15	41.67
23	Uttarakhand	8	7	2	28.57
24	West Bengal	28	19	17	89.47
25	Punjab	N/A	N/A	N/A	0.00

Source: Data Compiled from various reports of Bharat Micro Finance, Sa-Dhan, SKS, Access Development Services

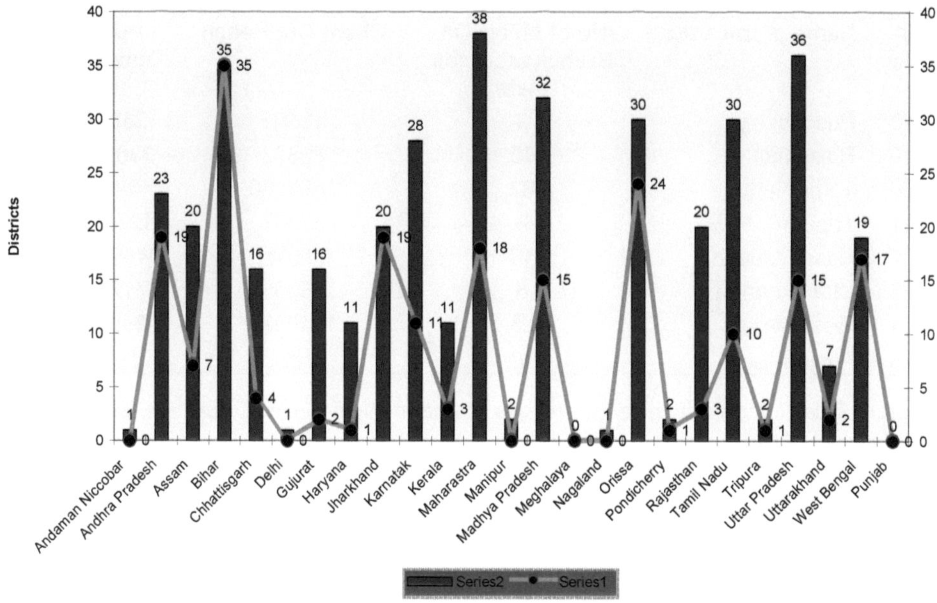

Fig. 2.1. Coverageof Poorand Advanced Districts byMFIs

Table 2.3: Client Outreach and Portfolio Outstanding Scenario of Indian States

Sl. No.	Name of the state	No of MFI (HO& Branches) in the state	Client Out Reach	Portfolio Outstanding
1	Andaman Niccobar	1	239	1786389
2	Andhra Pradesh	38	3653118	19445023878
3	Assam	8	137239	458701844
4	Bihar	26	241651	908233803
5	Chhattisgarh	7	235063	1156423043
6	Delhi	7	27761	439341552
7	Gujarat	14	416375	640302867
8	Haryana	2	1468	11250000
9	Jharkhand	21	115444	519308567
10	Karnataka	25	1963373	11038481244
11	Kerala	11	382001	1277305126
12	Maharashtra	28	1169306	3840548354
13	Manipur	2	651	4560000
14	Madhya Pradesh	17	229300	1045350073
15	Meghalaya	1	N/A	N/A
16	Nagaland	1	12	48000
17	Odisha	35	1577600	5323529239

contd...

Table 2.3: Contd...

Sl. No.	Name of the state	No of MFI (HO& Branches) in the state	Client Out Reach	Portfolio Outstanding
18	Puducherry	3	13161	28785907
19	Rajasthan	13	83873	380384265
20	Tamil Nadu	32	1692560	5754100628
21	Tripura	3	37687	127942065
22	Uttar Pradesh	18	533041	3469441228
23	Uttarakhand	8	16798	71710412
24	West Bengal	28	1540927	5298597926
25	Punjab	N/A	N/A	N/A

Source: Data Compiled from various reports of Bharat Micro Finance, Sa-Dhan, SKS, Access Development Services

Note: N/A stands for data not available

Table 2.4: State-wise per district average Client Outreach and Portfolio Outstanding

Sl. No.	Name of the state	Average Operating District Client outreach	Average Operating District Portfolio Outstanding
1	Andaman Niccobar	239	1786389
2	Andhra Pradesh	158831.2174	845435820.8
3	Assam	6861.95	22935092.2
4	Bihar	6904.314286	25949537.23
5	Chhattisgarh	14691.4375	72276440.19
6	Delhi	27761	439341552
7	Gujarat	26023.4375	40018929.19
8	Haryana	133.4545455	1022727.273
9	Jharkhand	5772.2	25965428.35
10	Karnataka	70120.46429	394231473
11	Kerala	34727.36364	116118647.8
12	Maharashtra	30771.21053	101067061.9
13	Manipur	325.5	2280000
14	Madhya Pradesh	7165.625	32667189.78
15	Meghalaya	0	0
16	Nagaland	12	48000
17	Odisha	52586.66667	177450974.6

Contd...

Table 2.4: Contd...

Sl. No	Name of the state	Average Operating District Client outreach	Average Operating District Portfolio Outstanding
18	Puducherry	6580.5	14392953.5
19	Rajasthan	4193.65	19019213.25
20	Tamil Nadu	56418.66667	191803354.3
21	Tripura	18843.5	63971032.5
22	Uttar Pradesh	14806.69444	96373367.44
23	Uttarakhand	2399.714286	10244344.57
24	West Bengal	81101.42105	278873575.1
25	Punjab	0	0

Source: Data Compiled from the Bharat Microfinance Report-2008.

2.5.4: Trend of Growth and Disbursements of MFIs

The microfinance sector and microfinance institutions in India as of March 2009 is estimated to have outstanding total loans of ₹ 160 to ₹ 175 billion, and ₹ 110 to Rs120 billion respectively. MFIs outstanding loan disbursement have increased to ₹ 114 billion (see fig.2) from just ₹ 60 billion a year ago. The shift in fund disbursement is believed due to the fact of easy acceptance of MFIs in one hand and persistent initiative of GoI. The lunch of NREGS is accepted as one of the drivers for increased demand for fund from MFI and corresponding naïve behaviour of MFIs to broad base the funding operation. The fund disbursement by the Micro Finance Institutions is comparatively more than that of the Self-help group bank linkage programme during 2007-08. An estimation of fund disbursement during the financial year 2008-09 would be nearly ₹ 287 billion (see fig. 2.3). This scenario depicts the easy acceptance of Micro Finance Institution by the public. The growth of microfinance sector in eastern India was driven by capacity enhancement initiatives by the apex MFIs, and tapping of growth opportunities in the eastern market by south-India-based MFIs, and banks.

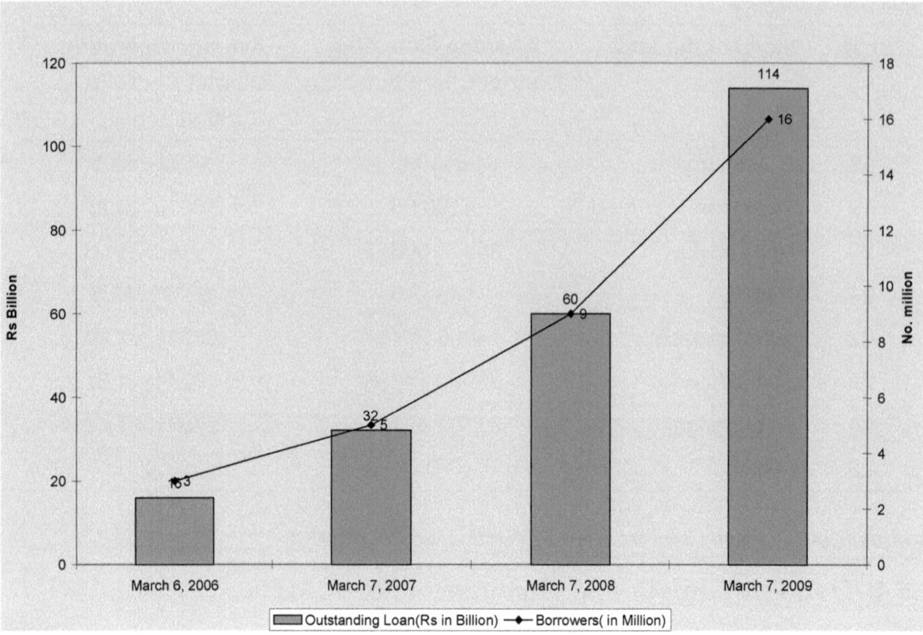

Fig. 2.2: MFI Growth Trend

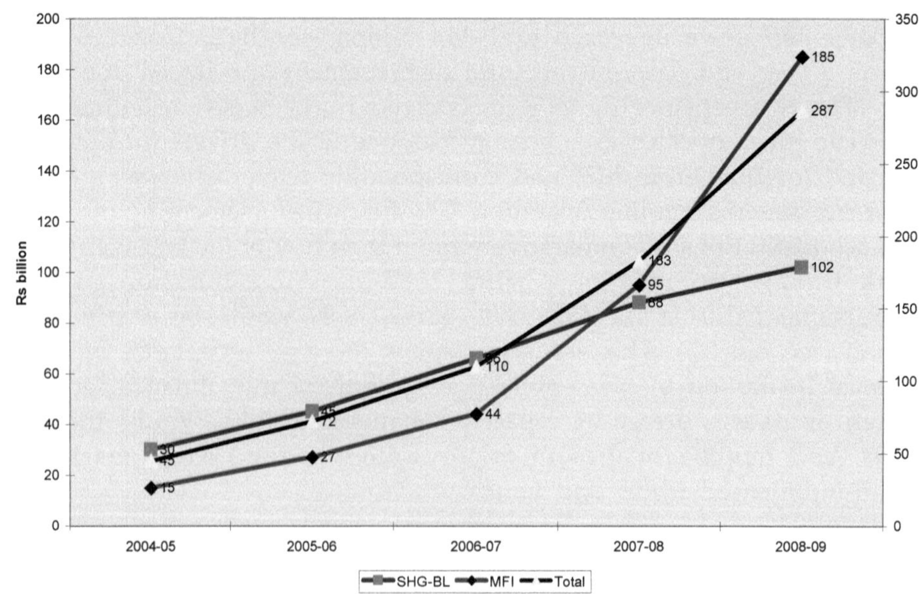

Fig. 2.3: Trend of Loan Disbursement

2.6.0 Origin and Growth of SHGs in India

The origin of SHGs in India dated back to the initiative of NABARD in 1986-87. But the real grounding was made after 1991-92 by linking SHGs with banks. Despite the large rural banking network, vast majority of the rural poor are deprived off access to bank credit. Despite the massive subsidy given under the - credit linked programme under IRDP, poverty levels in India still continue to be high (UNDP 1997) and many people still could not access the service of formal credit institutions. All these factors and forces acted upon to develop a system and mechanism to extend credit support to vast chunk of rural poor administering large number of small loans to accelerate the self-initiated socio-economic development for a better India.

All developing and under-developed countries around the globe have been marching ahead on the path of sustainable socio-economic development and thus have been in search of better and practicable systems to elevate the economy from misery and poverty. The financial system and institutions of the country and Indian banks and financial institutions *per se* are in search of populous methods of intermediation reach out to poor profitably who have until now been 'un-banked'.

Self Help Groups (SHGs) or Rotating Savings and Credit Associations (ROSCAs) appear to be a new distribution channel for financial services which offers a means of achieving the objectives of both parties, and of their clients, in a mutually profitable way, and this form of intermediation is popular with NGOs and banks in India and elsewhere now-a-days. SHGs are now patronized by NABARD for implementation of financial activities for socio-economic and institutional development by leveraging fund from nationalized and co-operative banking sectors in India. Self-help group is a small voluntary association of poor people preferably from the same socio-economic background, either working together or living in the neighborhood, engaged in similar line of activity, working with or without registration for the common good of the member. The minimum number of members to form an SHG is five while the maximum is 20 without registration. They came together for the purpose of solving their common problems through self-help and mutual help. The formation of SHGs is not ultimately a micro-credit project but an empowering process. The concept aims at empowering women and thus uplifting their families above the poverty line. It is a gradual process resulting from interaction with group members through awareness and capacity building.

With regard to group process, all members have to meet regularly once in a week, every fortnight or every month in a specified place at a stipulated time, as decided by the group. The members discuss their socio-economic problems and try to find out solutions of their own. During these meetings small savings have been collected and the savings are used to meet the credit requirements of the members. The individual savings are either kept in a bank or kept with the leader of the self-help groups. In this approach, cash transaction in Self help groups is secondary, and solving the social problems following the participatory mechanism is of primarily importance. The meetings are eye opener to the poor women to understand the socio-economic and institutional issues and the future economic development they could take up for the economic betterment of the larger group members. The meetings also equip the illiterate women to understand their potentiality; support required and derives a plan of action for the activities decided by the group to be taken up. In fact, the thrift and credit associated with the SHG dynamics enables them to think of potential source of resources from their midst and embark upon mobilization of local resources as much as possible.

This SHG mechanism has been widely accepted as an integral part of microfinance credit. Several directives and guidelines from Reserve Bank of India (RBI) and National Bank for Agriculture and Rural Development (NABARD) to commercial banks have clearly mentioned the need to recognize the SHG as a potential tool of micro-credit. The RBI has even launched the programme of linking SHGs with lending institutions like commercial banks so that the credit requirements of the SHGs can be properly attended to by the commercial banks. Considerable improvements have been made over the past ten years.

In the early 1980s, the GOI launched the Integrated Rural Development Programme (IRDP), a large poverty alleviation credit program, which provided government subsidized credit through banks to the poor. It was aimed that the poor would be able to use the inexpensive credit to finance themselves over the poverty line. Also during this time, NABARD conducted a series of research studies independently and in association with MYRADA, a leading non-governmental organization (NGO) from Southern India, which showed that despite having a wide network of rural bank branches servicing the rural poor, a very large number of the poorest of the poor continued to remain outside the fold of the formal banking system. These studies also showed that the existing banking policies, systems and procedures, and deposit and loan products were perhaps not well suited to meet the most immediate needs of the poor. It also appeared that what the poor really needed was better access to these services and products, rather than cheap subsidized credit. Against this background, a need was felt for alternative policies, systems and procedures, savings and loan products, other complementary services, and new delivery mechanisms, which would fulfill the requirements of the poorest, especially of the women members of such households. The emphasis therefore was on improving the access of the poor to microfinance rather than just micro-credit.

Over the past 25 years variety of microfinance programmes has been promoted by the government and NGOs. Some of these programs have failed and the experience learnt has been used to develop more effective ways of providing financial services. In 1999, the GOI merged various credit programs together, refined them and launched a new programme called Swaranjayanti Gram Swarazagar Yojana (SGSY). The mandate of SGSY is to continue to provide subsidized credit to the poor through the banking sector to generate self-employment through a self-help group approach and the program has grown to an enormous size. The emergence of microfinance through self-help route was a pilot project floated by NABARD for promoting 500 SHGs. As the idea gained acceptance from the banking system and the results were promising, the Reserve Bank of India encouraged this positive initiative by issuing instructions to banks in 1996 to cover SHG financing as a main stream activity under the priority sector lending portfolio. A working group set up by the RBI in 1994 came up with wide ranging recommendations on SHG and bank linkage as a potential innovation in the area of banking with the poor. NABARD has played a key role not only in promoting SHGs but also in standing behind the SHG-bank linkage programme. The performance of the SHGs has been extremely encouraging. Repayments by members to SHGs have been exceedingly high and on-time payments have hovered around 98 per cent. Many of the expectations behind the basic philosophy underlying the formation of SHGs seem to have been fulfilled. It has, however, to be noted that there is a concentration of SHGs in southern states the experience so far has shown that the poor can organize themselves and do things to promote the well being. It has also had a tremendous social impact making women more self-confident.

2.6.1 Registration of SHGs

SHGs need not be registered as trusts since members of trust can not become staff of the trust or in any way derive income from the trust's activities. They can not be registered as partnership Firms because such firms with more than 20 members attract the provision of section 1(2) of the companies Act. SHGs whose membership exceeds 20 can only be registered as a society under the Societies Registration Act or as a Co-operative society under the co-operative Societies Act of the State Government or as a company under companies Act. Thus, Indian law requires that organizations of more than 20 persons be formally registered.

This has given rise to SHGs having membership limited to 20. Banks do not lend to groups, which have less than 10 members, which has led to the minimum group size to be 10.

1.6.2 Characteristics of SHG

There is no hard and first rule for formation, nurturing and management of Self-help groups. The SHG groups are treated as a group of people organized for the cause of individual as well as common development barring the caste, clan, and religion differences. In deed, a group of same sex people organized, committed to a cause, work in a contiguous geographic location, having similar socio-economic background, abide by the group cohesion norms are formed in to a group to fight against the common socio-economic and institutional obsolesces on there way of development. In fact, the SHGs bear specific characteristics which can broadly be categorized as Group Constitution, Organization Discipline, Organizational and Leadership Systems, and Financial Management and Performance. The characteristic feature of a sound self-help group is thoroughly discussed the following sub-sections.

2.6.2.1 Group Constitution

Forming a group is very simplest task but nurturing and managing the group to accomplish the purpose for which it has formed is a Herculean task and require specific rules, norms, and above all a good leadership. Therefore, Self-help group formation is usually considered the followings.

(i) The members of the group should be homogeneous in terms of sex, neighbourhood, occupation and community.

(ii) The SHG should have a unique name to give it a distinct identity.

(iii) All members of the SHG should be over 18 years of age and are able to contract.

(iv) The members should understand the purposes of forming an SHG and the long-term benefits, and not short-term benefits like bank loan or gas connection.

(v) Every member has a right to participate in every activity of the group.

(vi) Every member has a right to participate in group meetings and decision- making.

(vii) Every member has a right to access loans from the pooled corpus funds of the group.

2.6.2.2 Organization Discipline

Self-help groups are abided by the following discipline. In fact, this is the set of dos' and don'ts stipulated by the group members to follow rigorously for smooth and sustainable functioning.

(i) Meetings should be convened regularly, on the specified frequency (day and date), time and place. These should be decided as per the convenience of the members, and should be held at least once in month.

(ii) There should be a quorum for the meeting. Ideally all members are to attend all meetings.

(iii) All meeting should have a structured agenda and/or issues to be discussed.

(iv) All member transactions should be conducted during the meeting.

(v) Meetings should be participatory and give scope to all members to share, discuss the issues, and express their opinion and debate.

(vi) Savings amounts and periodicity should be decided mutually among the members, and start from the first meeting of the group.

(vii) Savings should be collected at least once in a month in the group meeting.

(viii) All members should save an equal amount which forms the fixed savings of the members.

(ix) Thrift and any other savings should be accepted only from members.

(x) Mutually decided Interest should be paid on the thrift and added to the thrift.

2.6.2.3 Organizational and Leadership Systems

Structuring an organization and running the institution through a system is the backbone of the organization. The sustainability and decay of the institutions primarily depends on the structure of the organization. Placing the right person in the right place of management is another qualification to be looked at. In general, the following organizational and leadership issues are seriously viewed for putting a vibrant SHG system in place.

(i) SHG leaders President, Secretary and/ or Treasurer should be elected by the members with clearly articulated roles and responsibilities.

(ii) They should promote democratic processes and participatory decision making.

(iii) They should ensure that members share responsibility on rotation. They should be responsible for conducting meetings and resolving conflicts.

(iv) They maintain a bank account on behalf of the group and render truthful and correct accounts to members.

(v) They should maintain books of accounts and records which are available to the members and should be updated.

(vi) SHG records should be reconciled with the bank pass book and individual members pass book at least once in six months.

2.6.2.4 Financial Management and Performance

Management of organization with an efficient system and good team ensures vibrant future of the organization. In deed, management of self-help group is not entirely different from the management of any other institution. Hence, management of SHGs and especially the financial management take the following aspects into consideration.

(i) SHG members should understand that money is fungible and implement the understanding in practice. They should not discriminate between the source of funds, and have different norms for loans given from different sources of funds.

(ii) Members sharing responsibility for cash management in each meeting by rotation not only enhances fund management competence among members but also promotes accountability.

(iii) Funds should be used for lending to members and not to invest in assets or deposits.

(iv) Receipts and payments statement should be prepared at the end of each month.

(v) Loans can be for various purposes *viz.* Agriculture and Allied activities, Consumption, Income Generation Activities (IGAs), Health, Education and for various terms-hand loan, short-term, long-term.

(vi) Loans should be given to members only.

(vii) Preference may be given on sanction of loans to members who are regular in attending meetings, and regular in timely payment of thrift.

(viii) Periodic preparation of micro-credit plans (MCPs) would help improve the SHG members' credit absorption capacity and also enhance the fund management.

2.6.3 SHG Patterns

A variety of group-based approaches relying on social collateral and the enabling and cost-reducing effects are the feature of modern microfinance (MF). On the basis of above parameter SHG are defined as follows;

(i) Groups that are primarily geared to deliver financial services provided by microfinance institutions (MFIs) to individual borrowers (such as the Joint liability groups of Grameen and NGO-banks of Bangladesh);

(ii) Groups that manage and lend their accumulated savings and externally leveraged funds to their members.

While the term 'self-help group' (SHG) can be used to describe a wide range of financial and non- financial associations in India, it has come to refer to a form of Accumulating Saving and Credit Association (ASCA) promoted by government agencies, NGOs or banks. Thus, SHGs fall within the latter category of groups described above. Distinction can be made between different types of SHGs according to their origin and sources of funds. Several SHGs have been carved out of larger groups formed under pre-existing NGO programmes for thrift and credit and broad-based activities. Some have been promoted by NGOs within the parameters of the bank linkage scheme but as part of an integrated development programme. Others have been promoted by banks and District rural agencies (DRDAs). Still others have been formed as a component of various physical and social infrastructure projects.

NGO - promoted SHGs are often nested in *sanghas* or village development groups (VDGs) undertaking integrated development activities. As they develop, SHGs or *sanghas* have been grouped into larger clusters and multi village federations for financial and non-financial activities.

2.6.4. Models of Self-help Groups in India

There are different models of microfinance in operations to meet the credit need of the rural people and especially targeting running up of the institutions involving women. The common models of microfinance operations are narrated in the section below.

2.6.4.1 Model-1: Linkage Model

There are different model of self-help group are in operation in India. All models in general do ultimately object to benefit the poor women and mobilize fund for them at an interest rate decided by the formal banking system for economic advancement. The first model is a "Savings laid SHG Operation Model". Saving is mobilized from the primary members and is deposited in the bank without any intermediary uses

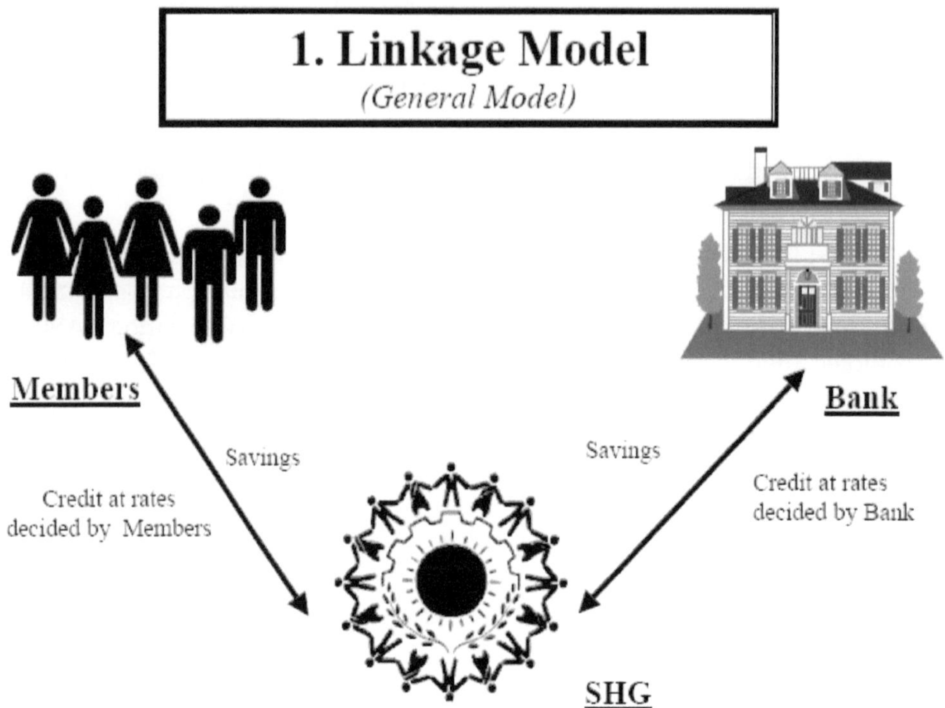

Source: Rakesh Malhotra, DDM, NABARD.

2.6.4.2 Model-2: Modified Linkage Model

The second model is an improved version of the 'savings laid SHG Operation Model'. Savings accumulated and deposited in the bank and bank extend financial support at a cost prescribed by the bank for investment purposes by the self-help group members. The banking institution mearly acts as an institution to collect savings from the people and advance loan at a cost prescribed, without any constructive role in terms of capacity building of SHG members or improvement in group functionality. The fund advanced to the self-help group members are not in proportion to the deposit. The self-help group use to do all activity and functionality of SHG depends on the far-sight-ness and business acumen of the SHG member alone.

2. Modified Linkage Model

(e.g.- OBC Dehradun)

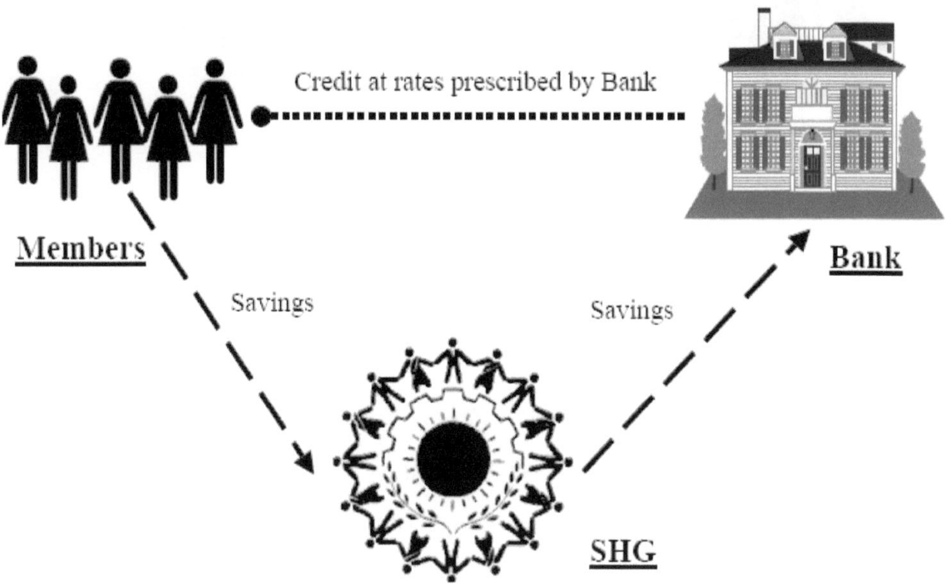

(**Source:** Rakesh Malhotra, DDM, NABARD)

2.6.4.3 Model-3: Non-Government Organization Model

The third "Non-Government Organization Model" is little complicated and there are many players involved in the functioning of the self-help group. SHG member do form the SHG, collect thrift, deposit the thrift in the bank and access financial assistance to carry out IGAs and other developmental activities but receives linkage and support services from the Non-Government Organization (NGOs). NGOs operating in the areas extend support in management of the Self-help Group and economic activities and between the banks and the SHGs. NGOs primarily play an intermediary role between self-help group and banks.

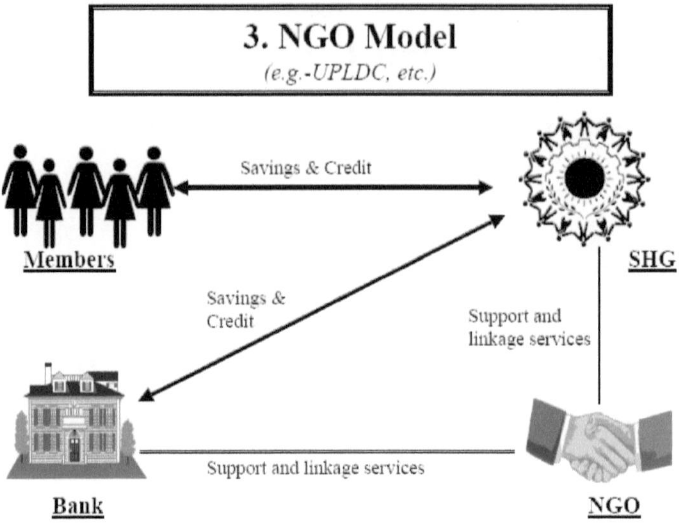

(**Source:** Rakesh Malhotra, DDM, NABARD).

2.6.4.4 Model-4: Modified NGO Models

The "Modified NGO Model" is more complicated in nature and entrust NGOs to perform nearly the role of a bank with social touch. NGOs are viewed as institutions to facilitate collection of deposits, mobilize deposit from the SHG members and SHGs, receives fund from bank and later on advances to SHGs to carry out economic and development activities. Almost all major Micro Finance Institutions are falling the category and this model is a popular model.

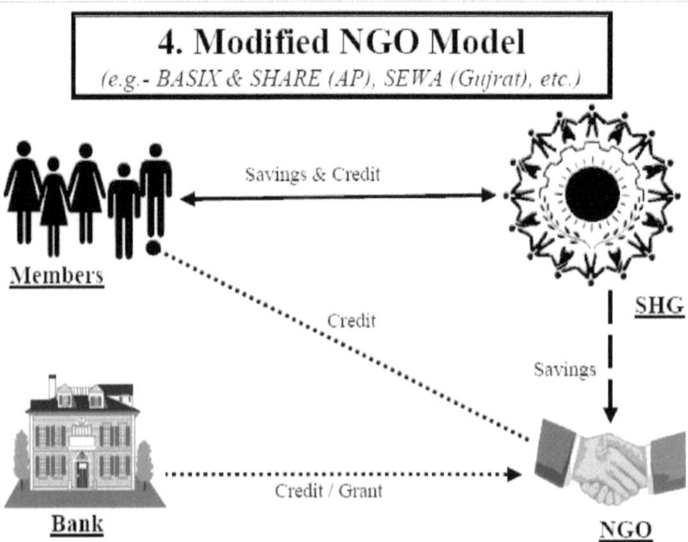

Source: Rakesh Malhotra, DDM, NABARD.

2.6.4.5 Model-5: Direct NGO Model

The "Direct NGO Model" is a different type of NGO involvement in Self-help group activities. NGOs in this model appear as a different agent; independent of SHG management, function, operation, savings mobilization but extend technical and marketing support to carry out income generation programmes and supply the required raw material. Role of Non-Government Organization is viewed as a catalyst agent to imbibe illiterate poor women with technical skill to earn their bread and butter with dignity. The formal financial institutions extend financial support to NGOs for extending training and capacity building programme for the self-help groups. The NGOs upon receiving financial assistance from Banks do in fact extend financial assistance to self-help group members at a competitive cost.

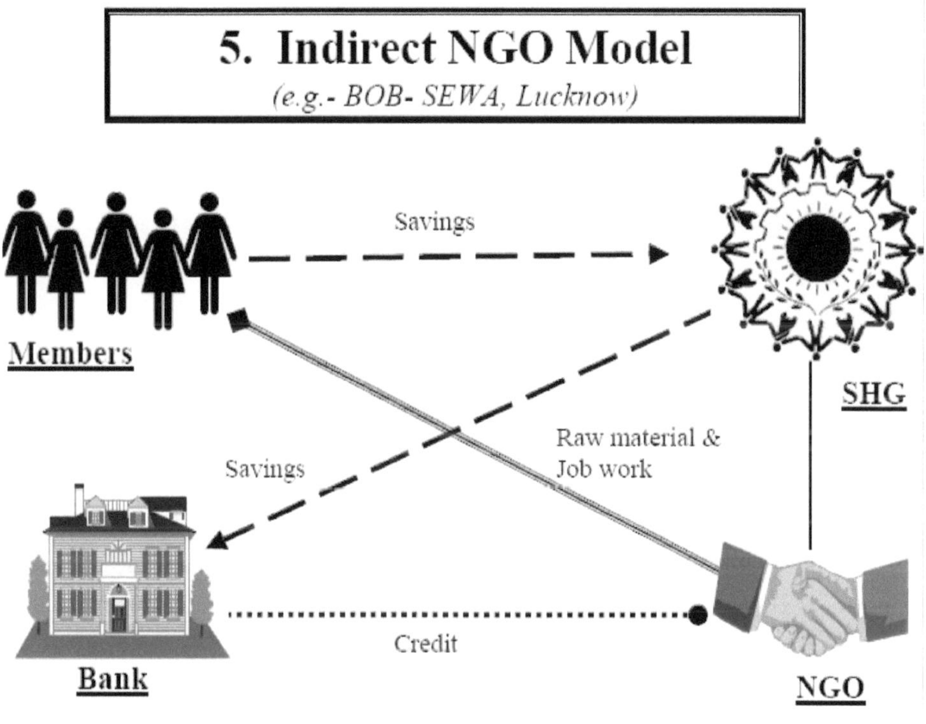

Source: Rakesh Malhotra, DDM, NABARD.

2.6.4.6 Model-6: IFAD Model

The IFAD model is still complicated in nature and added another institution *i.e.* Government line departments, in the functioning and management of self-help group (SHGs). Banks are the centre pin for fund disbursement, while NGOs as well as Line Department extend support services and linkage services. Government as well as NGOs helps women to organize themselves into groups, while banks receive savings and advance credit support to self-help groups. However, NGOs are assigned with a catalyst role to extend linkage service.

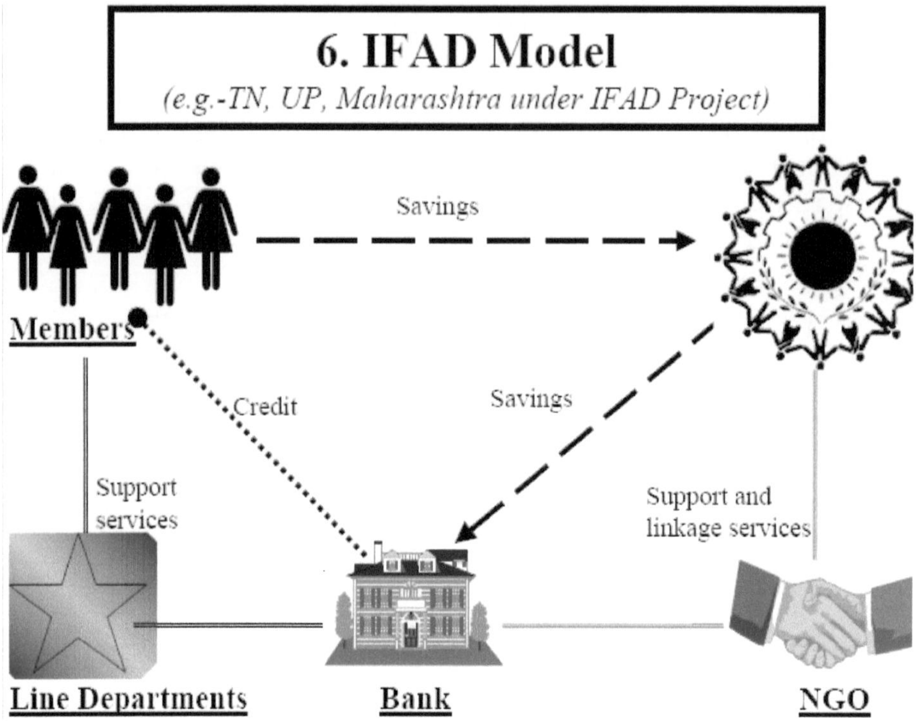

(**Source**: Rakesh Malhotra, DDM, NABARD)

2.6.4.7 Model-7: SGSY Model

This model is basically the extended version of IRDP model of poverty alleviation. The Government and especially the District Rural Development Agency (DRDA) plays a crucial role for balancing the financial performance and the physical activities being grounded. Self-help groups under the model is presumed to be the pace setter for acceleration of economic development through execution of whole lot of development activities in the specific area of their operation. All line departments and the NGOs are mere facilitating agencies and fall in between the DRDA and the SHG performances. As a matter of practice, fund for all developmental activities are routed through banks and banks cross examines the feasibility, viability and eligibility of self-help groups to receive the fund through the SHG rating methodology adopted throughout the country.

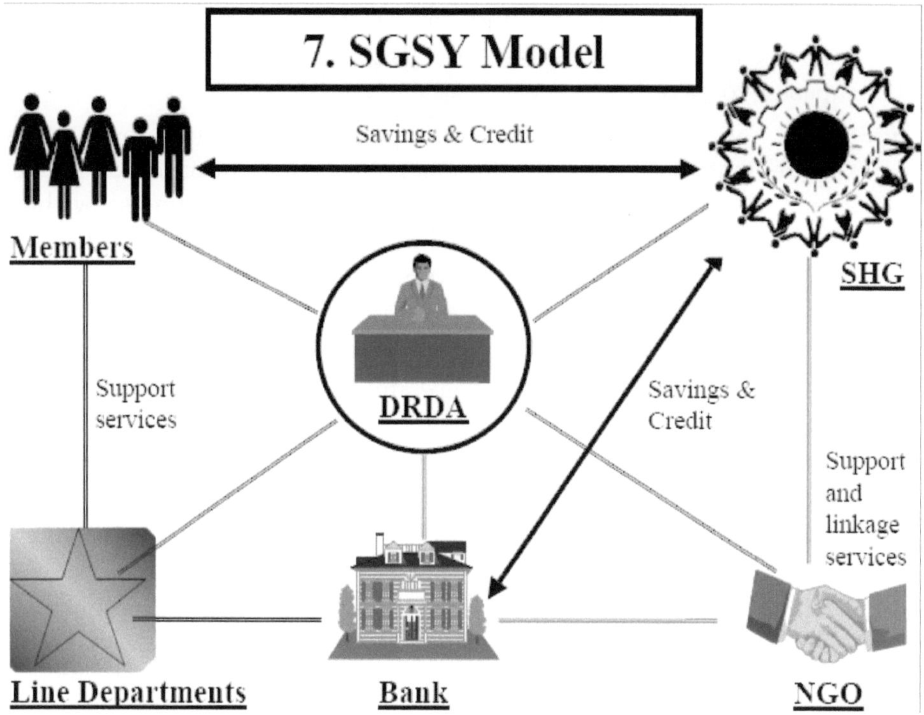

(**Source:** Rakesh Malhotra, DDM, NABARD)

2.6.5 SHG Bank Linkages in India

Given the background of incapacity or disinterest of formal banking sector to serve the needs of low-income clientele, and the increased demand for credit from rural households, the gap between demand and supply of funds in rural economy sought to be bridged by the formal sector which took the initiative to develop a supplementary credit delivery mechanism by encouraging institutional arrangements outside the financial system like SHGs with non-governmental organizations acting as facilitators or intermediaries. This beginning was made with NABARD's pilot project in Karnataka (1986-87) of linking SHGs with formal banks, mediating through NGOs. This project known as the SHG - bank linkage project, was expected to be advantageous to the banking sector from both angles of fulfillment of social goals like reaching out the poor, and achieving operational efficiency by externalizing part of their transaction costs.

The policy interest in the whole concept of SHG linkage with banks was institutionalized with the RBI establishing in 1994 a working group of NGOs and SHGs comprising representatives from NABARD, bankers and development practitioners. The working group has given the following comments on the progress of SHGs in India:

(i) SHGs helped to generate and collect small thrift amounts from a cross-section of people, hitherto considered incapable of having any capacity to save;

(ii) SHGs have facilitated the rural poor in fulfilling their credit requirements, both for meeting emergent consumption needs as well as small production requirements. This can be achieved without requirements of collateral, complex loan procedures and at a low transaction cost without the usurious terms applicable in case of loans from informal channels;

(iii) SHGs have been able to meet successfully the credit requirements of the rural poor as per their choice unlike in the case of borrowing under most of the poverty alleviation programmes implemented through formal credit institutions where the activity to be under taken is not self-chosen and is of unit sizes much greater than their management capabilities.

In the initial period in 1992-93, just 255 SHGs were linked to banks with ₹ 2.89 million as a refinance from NABARD. In a span of couple of years, in 2011-2012, 11.48 lakh SHGs were linked with and refinancing of ₹ 3072 crores. As far as the state-wise outreach was concerned SHG-bank linkage programme covered 572 districts in 31 states and union territories. With over 24.25 million poor households accessing banking services including micro-credit through SHG-bank linkage programme led by NABARD in India it is now the largest and the fastest growing microfinance programme of the world in terms of its outreach. Today, over 2,800 NGOs and 41,082 branches of 560 banks are associated with the programme making it the most cost-effective microfinance initiative as well.

In India, a variety of microfinance schemes and approaches have been practiced by both Government and Non Government organization. Leading national financial institutions like, the National Bank for Agriculture and Rural Development (NABARD), the Small Industries Development Bank of India (SIDBI) and the Rashtriya Mahila Kosh (RMK) have played a significant role in making micro credit a real movement in India. Professional Assistance for Development Action (PRADAN), International Centre for Entrepreneurship and Career Development (ICECD), Mysore Resettlement and Development Agency (MYRADA), Self Employed Women's Association (SEWA); act as resource organizations and have been successful in replicating their experiences in other parts of the country without restricting to one area.

There are three models being followed in the area of Microfinance in India. In model 1 banks themselves take up the work of forming and nurturing the self help groups, opening their accounts and providing them with bank loans. Up to March 2011, about 20 per cent of SHGs financed were from this category. In model 2, where SHGs are formed by formal agencies other than banks, NGOs and others, but directly financed by banks, have occupied the major share with 74 per cent of the total number of SHGs as on 31 March 2011. Model 3 consists of SHGs financed by banks using NGOs and other agencies as financial intermediaries. The share of cumulative number of SHGs linked to this model is relatively small *i.e.* 6 per cent.

2.6.5.1 Objectives of SHG-bank Linkage

The main objectives of linking self-help groups with the banks are to provide:

(i) Provision of supplementary credit to SHGs,

(ii) Reduction in transaction cost for both banks as well as SHGs by reducing paper work.

(iii) Building mutual trust and confidence between banks, NGOs and the rural poor.

(iv) Mobilizing small savings from poor rural women.

(v) Creating healthy relations between SHG members and the linking agencies.

(vi) Constant supervision and monitoring by banks through NGOs.

2.6.5.2 Principles of SHG-Bank Linkage

The broad principles followed under linkage scheme are:

(i) Savings first, No credit with out savings

(ii) Savings as partial collateral

(iii) Bank loans to SHGs for on-lending to members

(iv) Credit decisions by SHGs for on-lending to their members

(v) Interest rates and other terms and conditions for loans to members to be decided by the group

(vi) Joint liability as a substitute for physical and collateral

(vii) Ratio between savings and credit contingent upon credit worthiness and could increase over a period due to good repayment record, if needed

(viii) Small loans to begin with and different credit cycles clearly defined

2.6.5.3 Growth of Self-help groups and Bank Linkage

Experiment of microfinance sector and especially with the SHG led financial devolvement and invest in rural social capital has ever since is following the up-ward increasing trend. Effectiveness of the financial mechanism is experienced due to active involvement of many institutions and injection of several dose of investment through various rural development schemes. The growth and spread of SBLP is thoroughly discussed in chapter six *i.e.* Microfinance Institutions in India and Andhra Pradesh: Spread, Functioning and TFI.

2.6.6 MFIs in India- The Forecasted Fate

Micro finance is the buzz word among the development practitioners, policy makers, researchers and the Government. This sector has been given considerable importance and anticipated to be an effective sure-shot instrument of poverty alleviation and women empowerment. MFIs in India following the SHG rout of fund devolvement has developed and in-built a mechanism to link the rural needy households to access the financial support for carrying out more productive and profitable trajectories of business opportunities. However, the trade-off of more self-help groups' leads to upcoming of more numbers of microfinance institutions and vice-versa is debatable but acclaimed a height and reversible importance among the financial institutions of the country.

The pace of development and growth of Micro Finance Institutions (MFIs) is very fast. The value created in the mind and heart of the rural poor is bearing a lasting impact and more and more numbers of people are slowly atuned with the developmental contribution. Many institutions and agents are getting involved in the process. Over the passage of time the central and state Governments machinery as well as the Reserve Bank of India (RBI), SIDBI and NABARD are becoming extra cautious and more stringent control and regulatory mechanisms are evolved.

Of late, it is recognized that the local and regional level politics is influencing the functioning of the microfinance institutions in its depth and length. Caste and clan based institutions is also regulating the functioning of the MFIs at the local level. With an era of competition, and when large numbers of MFIs are working crossing the state and regional barriers, meeting the business objectives adhering the development objective is cumbersome task. CRISIL believes that the overall asset quality of MFIs will deteriorate marginally over the short to medium term, owing to increase in large-ticket loan disbursements, multiple lending and migration of borrowers. The following sections are presenting the future scenario of microfinance institutions in India.

2.6.6.1. Corrective Measures for Regional Imbalances

Micro finance institutions in India till 2006 have followed an indifferent approach to its operation and expansion, as a result large number of MFIs came up in southern India. The MFIs in Southern India dominated in its absolute number as well as accessing finance. The near saturation of business opportunity in this region encouraged many south based MFIs to scale up operation in Northern and Eastern India; open up a challenging environment for the northern India based MFIs. Available statistics on MFIs reveal that the top 10 MFIs accounted for 74 per cent of the total loan outstanding in the country and vice versa. Regulatory mechanisms should be in place to control and regulate the functioning of the microfinancing operations. Adequate attention need to be given on the appropriate and just competition and stringent control on expansion.

2.6.6.2. Working on Governance

The legal structure and the attendant regulatory requirements of an MFI have a strong bearing on governance practices because they influence management practices and level of transparency. All legal structures, other than the formal company structure, suffer from want of adequate regulation and disclosure standards. This also creates a virtuous/vicious cycle phenomenon: MFIs that have the willingness and minimum capital funds to embrace a corporate structure as an NBFC attract outside investors more easily, which in turn foster better governance and disclosure standards. In contrast, MFIs that are either unable (for lack of adequate sponsor funding) or unwilling to convert to a corporate structure tend to remain 'closed' to transparency and improved Governance standards, and therefore, continue to be unable to attract capital. Moreover, some MFIs, particularly NGO-MFIs, continue to face challenging in striking a balance between their social and business goals, two seemingly conflicting objectives. This often results in poor internal control systems, lack of accountability, and suboptimal performance.

With donor and grant-funds drying up, and related voluntary services dwindling, microfinance has become a key activity for several NGO-MFIs in India. However, their management have not adapted and equipped themselves adequately to manage this evolution, with the result that governance, disclosure, and accountability have suffered in many cases. The boards of these organizations continue to have a strong developmental and social project bias, and limited microfinance or banking experience. They typically comprise chief executive officers of other NGOs, school teachers, academics, parish priests, and other prominent members of the local communities. While the contribution of such members are critical in guiding the MFIs social objectives, to the virtual exclusion of members with banking or accounting experiences, means that MFIs often find it difficult to adhere to commercial and systematic lending activities, such as focusing on a timely recovery of loan.

The present scenario exhibits that unless these NGO-MFIs restructure their boards and management to derive a reasonable commercial orientation into their operating philosophy and mission, the attendant benefits of good internal control systems and transparency are unlikely to materialize. This will ultimately hinder the sustainability of their operation.

2.6.6.3. Pressure on Process and controls on aggressive Growth Plans

It is believed that poor management of operations may pose added credit risks for MFIs, especially as MFIs continue to scale up operations aggressively. MFI's risk management practices have weakened over the past couple of years, on account of a shift in focus towards business growth and network expansion. Some credit sanction and monitoring practices have been diluted. These include lending to clients with multiple loans from different MFIs, reduction in the average waiting period for loans, and doing away with staggered disbursements to joint liability groups (JLGs) and loan utilization check post disbursement. Rapid expansion to new geographies have also put pressure on the internal control mechanisms and audit function, as these have not received adequate focus in the past two years.

Nevertheless, there has been some improvement in MFI's operations: this includes installation of software for monitoring loans, upgrade in cash management services, and availability of banking facilities to MFIs operating in rural and semi-rural areas. Many small and mid-sized MFIs have also benefited from the technical support, such as documentation of internal policies and process mapping, with capacity building support, particularly from apex MFIs. Strong internal audit and control has been the key differentiator between MFIs with high grading and those with lower grading. It is therefore believed that MFIs will need to restructure their internal audit and control processes to minimize operational risks during the growth phase.

2.6.6.4. Heavy Dependence on banks and Financial Institutions (FIs)

MFIs are dependant on borrowings from banks and FIs, and do not raise debt from the capital market. Thus, large NBFC-MFIs face higher costs of borrowing than most large retail finance NBFCs in the country. Banks categorize their lending to MFIs as priority sector advances, which has helped MFIs raise timely resources. Nevertheless, for many MFIs, funding sources are restricted to private banks and apex MFIs; the public sector banks have not been aggressive lenders to MFIs. The large and mid-sized MFIs and NBFC-MFIs primarily borrow from private and foreign banks, while the smaller MFIs borrow mainly from private banks and apex lenders.

The lending model plays a key role in determining MFIs borrowing profile. The Public Sector Banks (PSBs), with their wide spread branch networks , prefer lending directly through the SHG-bank linkage route. Moreover, PSBs prefer to lend to those MFIs that have adopted the SHG Model. Available statistics of MFIs financing as on March 2008 reveals that PSBs accounted for 36 per cent of the total borrowings of societies and trusts as against only 10 per cent for MFIs following the JLG model. Thus NGO-MFIs (societies and trusts) have better access to funds from PSBs than MFIs that are companies. However, the current scenario of PSBs lending predominantly to SHG model based MFIs seems poised for change, with larger NBFCs continuing to aggressively target PSBs to meet their credit demand.

2.6.6.5. Absence of Adequate Regulatory Mechanism

Micro finance activities are undertaken by organizations that are registered under several legal forms. However, currently, only NBFCs are under the regulatory and supervisory purview- the NBFCs are regulated by RBI. The absence of prudential norms and accounting guidelines for non-NBFC MFIs leads to lack of uniformity in accounting practices and highly leveraged balance sheets among MFIs. The financial statements of the microfinance programmes of most non-NBFC-MFIs do not provide the true financial picture. In this context, GoI of India proposed to bring in legislation to regulate the entire microfinance sector in India; the regulator however is yet to materialize. A few NGO-MFIs and non-NGO-MFIs continue to offer in-house insurance facilities by underwriting on their own, although this is a clear violation of insurance regulations.

2.6.7 NABARD-The Microfinance Regulator

Introduction of the Micro Financial Sector (Development and Regulation) Bill, 2007 (henceforth the Bill) invited debates on the regulation of the finance activities and also pointed to the fact of the institutions to regulate the MFI activities. Several private and public institutions participated in the debate and come to a consensus and developed a guideline. The guidelines recommend that as far as possible the regulation should focus on the type of transaction and not the type of institution undertaking it. They also favour a clear distinction between the supervision of depository (deposit accepting) and non-depository MFIs (microfinance institutions). The need for regulation and supervision is greater for those MFIs which are authorized to accept deposits as they pose a greater risk to their members and to the financial system.

The guidelines recommend that for depository MFIs, the supervisory body would be the authority which supervises commercial banks. This leverages existing supervisory skills, and reduces incentives for regulatory arbitrage, namely, structuring operations in such a manner that the organization comes under the jurisdiction of a weaker regulator. The supervisory staff of the regulator, however, must be retrained to suit the different portfolio characteristics of MFIs.

Another alternative suggested for supervision of depository MFIs is "delegated supervision", which refers to an arrangement where the banking regulator delegates direct supervision to an outside body while monitoring and controlling the delegated body's operations. Consistent with the CGAP guidelines, in addition to effectiveness and cost of supervision, the socio-economic and regulatory context of India must be considered in deciding on the regulator for the microfinance sector.

Before considering the choice of regulator in the Indian context, clarity is needed about whether the Bill attempts to regulate the microfinance sector as a whole, or whether it seeks to enable particular institutional forms of MFIs to offer savings services under the supervision of the regulator. If the Bill intends to regulate the sector as a whole, it needs to bring under its ambit the entire range of participants in the sector including banks, non-banking financial companies (NBFCs), section 25 companies (Not for profit companies), non government organizations (NGOs) societies, trusts and cooperatives. The Bill should set ground rules for all participants with regard to their microfinance activity. However, if its objective is to permit MFIs to accept deposits, then it needs to only register and regulate MFIs intending to do so. Any analysis of the Bill is complicated by the fact there is ambiguity with regard to its scope.

The Bill defines a microfinance organization (MFO) as 'including' societies, trusts and cooperatives, leading to varying interpretations. One possible interpretation is that the Bill does not include in its scope legal forms other than those mentioned. This implies that NBFCs and Section 25 companies are excluded from its scope. Another possible interpretation is that the definition of MFO given in the Bill is not an exhaustive one and hence the Bill does, in fact, cover NBFCs and Section 25 companies.

The provisions of the Bill concerning the regulator for the microfinance sector in India are inconsistent with the CGAP guidelines. The Bill designates the National Bank for Agricultural and Rural Development (NABARD) as the single regulator for both depository and non-depository MFOs, while the guidelines suggest separate regulators for each. According to the guidelines, in the Indian context, the Reserve Bank of India (RBI) should regulate depository MFIs, as it already regulates banks and NBFCs. An option worth considering is to create a special category of NBFCs which would be suitable for the deposit taking MFIs.

NABARD is a key participant in the microfinance sector and has been closely associated with one of the two prevailing modes *i.e.* SHG (self help group)-bank linkage mode of delivery of microfinance services in the country. The other is the MFI mode. The developmental role envisaged in the Bill for this institution will further strengthen its role as service provider. It is widely recognized that combining the role of service provider and regulator is not good governance practice as it could lead to serious conflict of interest. NABARD also lacks expertise to regulate and develop MFIs in the urban sector as its role has been confined to rural areas and to agriculture. With India urbanizing at a rapid rate, urban poverty as a public policy issue is becoming more prominent, leading to increased interest in urban microfinance.

The issue of regulatory arrangements for insurance and pension services offered by the MFOs is not addressed in the Bill. Insurance services in India are regulated by the Insurance Regulatory and Development Authority (IRDA), and the pension services are proposed to be regulated by the Pension Fund Regulatory and Development Authority (PFRDA). In the accumulation phase of the micro-pension schemes involving investment of pension savings, asset management companies are likely to be involved. These are currently regulated by SEBI (Securities and Exchange Board of India).

At the minimum, this suggests a strong need for coordination among regulators of different financial and capital market institutions. As the barriers among different financial service providers come down, the role of specialized regulators will require reconsideration in the medium term. The need for efficient regulatory capacity building for the microfinance sector is important for financial inclusion and for providing confidence to customers and investors in the sector. The regulation of the sector requires more extensive consultations and public debate than is reflected in the Bill.

Notes

1. Historical context of microfinance and its evolution, www.globalenvision.org.
2. India top 50 Micro Finance Institutions, CRISIL Ratings October 2009.
3. Srinivasan. N (2008). State of the Sector Report on Microfinance: Models, Policies and Experiences, published by Sage, New Delhi.
4. The number of household facing exclusion has been arrived at by adding rural households facing financial exclusion (93 million) and urban bellow poverty line (BPL) households (18 million). The average credit demand per household has been estimated at ₹ 10,000 per annum. India Top 50 MICROFINANCE INSTITUTIONS, CRISIL RATING, October, 2009.

Chapter 3

Institutional Forms, Policy Framework and Legal Text of Microfinance Institutions in India

3.0 Introduction

Microfinance in India through its major channels served over 86.2 million Indians as on March 2009 with loan portfolio of ₹ 351 billions. Out of this MFI channel client outreach has grown at 60 per cent in 2009 against 41 per cent in 2008. The portfolio of MFIs has grown at 97 per cent during the year as compared to 72 per cent in the previous year. As compared to last year, this year report has made clear distinction between client outreach and active borrowers. Out of 22.6 million clients of MFIs, 17.9 million (79 per cent) are active borrowers. This year's Quick Data is contributed by 233 MFIs as against 223 MFIs in Quick Data 2008.

As on 31st March, 2008, outstanding microcredit portfolio of India Microfinance was about ₹ 22,000 crore, out of which 75 per cent are accounted for by SHG- Bank Linkage Program, 20 per cent by large MFIs and 5 per cent by medium and small MFIs. India's MFIs operate in 209 out of 331 poorest districts of the country; up by 5 per cent over the previous year.

Table 3.1: Types of MFIs and Bank-linkage in 2008

Institution category	MFIs with Loan Portfolio up to 5 crore	MFIs with Loan Portfolio >5 to 50 crore	MFIs with Loan Portfolio over 50 crore	Total	No of MFIs
Society	87.7	543.8	478.3	1109.70	104
Trust	24	149.8	225.4	399.3	31
Cooperative Bank	8.3	-	-	8.3	8
MACS	5.2	11	-	16.2	10
Section 25 Companies	11.6	127.4	543.6	682.6	22
NBFC	10.6	197.3	3312.10	3520.00	25
LAB or any other	13.6	13.7	134.7	162.00	16
Total	161	1043.00	4694.10	5898.20	
No of MFIs	137	57	22	-	216

Source: www.sa-dhan.net

3.1 Need of Controlling Mechanism

There are various forms of financial institutions in operation in India. It is either in the form of a profit making company or a charitable society or trust. But, the fact remains that all these institutions are dealing with money and hence, needs to be regulated at the country level for the larger benefit of the common people. Hence, the following sections are devoted towards this effort to give a fair idea of the institutional arrangements and legal text under which all these institutions are governed.

3.2 Policy Framework and Legal Text for MFIs

It is observed that the extent and depth of MFIs operation in the country is as old as the human civilization but its operation and management is always been adhoc and sporadic in nature. In fact, regulation of MFIs is a big-gray area and there is no such systematic approaches are being taken up, either by central government or the topmost financial institutions of the country to control on the functionality. This section reviews the legal text relating to each of the legal forms being used by microfinance institutions in India, outlines the regulatory framework for those forms and identifies the pros and cons of each form.

3. 2.1 Societies Registration Act, 1860

NGOs are mostly registered under the Societies Registration Act, 1860. Since these entities were established as voluntary, not-for-profit development organizations, their microfinance activities were also established under the same legal umbrella. This sub-section examines the provisions of the Societies Act from the perspective of the practice of microfinance in India.

3. 2.1.1 Purpose

The Societies Registration Act 1860 states that "a society can be formed for the promotion of literature, science or the fine arts or the diffusion of useful knowledge/political education or for charitable purposes." Various state governments have amended these purposes from time to time. However, charitable activities continue to form the core of the stated purpose of the act in each of these cases.

Over a century-and-a-half since the enactment of this law, the Act has been subject to too much legal interpretation. In 1891, a high court judge, Lord McNaughton, defined, the legal sense of the term charity to comprise trust for

(i) relief of poverty

(ii) advancement of education

(iii) Advancement of religion

(iv) purposes beneficial to the community or a section of the community.

Lord Camden, in the same judgment in 1891 , 14 defined charitable purposes as "a gift to general public houses which extends to poor as well as rich, equally. It may be laid down as universal rule that the law recognizes no purpose as charitable unless it is of a public character, that is to say a purpose must, in order to be charitable, be directed to the benefit of the community or a section of the community."

Considering the public nature of microfinance activities and the widely accepted notion of its being help for the relief of poverty and for benefit to the community, microfinance activities are interpreted by most of civil society as being charitable for the purpose of this act. Under this law, as a pre-requisite, the registered society would, however, need to mention microfinance clearly as an activity it would be taking up as part of its "charity" work - as one of the objectives in its Memorandum of Association.

3.2.1.2 Registration of Society

Once the name of the proposed society and also the draft memorandum and rules and regulations of the society have been decided, the procedure for the registration of the society can be taken up. All the subscribers (minimum seven) should sign each page of the memorandum and the signatures should be witnessed by an oath commissioner, notary public (₹ 3 notarial stamp duly affixed), gazetted officer, advocate, chartered accountant or magistrate 1st class with their rubber/official stamp and complete address.

The persons desirous of forming a society should also become members of the first governing body. An outsider cannot become a member of the governing body in the first instance. The following papers should be filed with the Registrar of Societies for registration under the Principal Act or corresponding Acts enacted by various state governments:

(*a*) Covering letter requesting for registration stating in the body of letter various documents annexed to it. It should be signed by all the subscribers to the memorandum or by a person authorized by all of them to sign on their behalf.

(*b*) Memorandum of Association in duplicate along with a certified copy. It should be neatly typed and pages serially numbered.

(*c*) Rules and regulations/bye-laws in duplicate duly signed.

(*d*) Where there is a reference to any particular existing places of worship like Temple, Gurdwara, Church, Masjid or Budhvihar, etc. sufficient documentary proof establishing legal competence and control of applicant society over such places should be filed.

(*e*) Affidavit on non-judicial stamp paper of appropriate value sworn by the president or secretary of the society stating the relationship between the subscribers. The affidavit should be attested by an oath commissioner, notary public or magistrate 1st class.

(*f*) Documentary proof such as house tax receipt, rent receipt in respect of premises shown as registered office of the society or no-objection certificate from the owner of the premises.

Normally a fee of ₹ 50 is payable as registration fee of a society and it should accompany the request for registration payable in cash or by demand draft. Formalities for registration and requirement of documents may differ slightly from state to state. The applicant may, therefore, contact in advance the Registrar of Societies having jurisdiction.

On receiving the documents mentioned above the Registrar shall satisfy himself about the compliance of the provisions of the Act and correctness of the documents and then certify in his hand that the society is registered under the Principal Act 1860 or other corresponding Acts. Presumption that the society was duly registered under the Act arises not on the Certificate of Registration granted by the Registrar but on the copies of the Rules and Regulations and Memorandum certified under Section 19 which constitutes them prima facie evidence of the matters therein contained.

3. 2.1.3 Duties and Responsibilities of the Governing Body

Members of the governing body of a society are the trustees of the property of the organization. Therefore, the management of the assets and liabilities of the organization is the responsibility of the governing body members. Many provisions in the Principal Act provide the manner in which the society is required to function and for all purposes it is the governing body that is responsible for the conduct of the society.

3.2.1.3.1 Investment of Funds of a Society by the Governing Body

Given the financial nature of work of MFIs, such institutions are involved in the transaction of funds of high volume. It is the prime duty and responsibility of the governing body to invest and apply the funds and property of the Society. The governing body also needs to take care that the funds are not misappropriated or misused. No specific provision or guidelines are available in the Principal Act regarding the application of the funds of the Society but it is established law that the governing body of the Society are trustees of the funds of the Society and thus they should apply the funds in the most judicious manner as well as in a manner a prudent man would apply his own funds.

3. 2.1.4 Accounts and Audit

The Principal Societies Act does not provide for the maintenance of accounts or their audit in any manner. State governments have made amendments to the Act, which are applicable to the societies registered in that particular state. Provisions for accounts and audit have also been made in various independent laws enacted by various states. These include provisions for every society to Keep at its registered office proper books of accounts containing entries in respect of all sums of money received and the sources thereof and all sums of money spent by the Society and the objects or purposes for which the sum is spent.

- Maintain the records of all sales and purchases of goods by the Society and also the records of assets and liabilities giving a true and fair view of the state of affairs of the society.
- For this purpose, societies should maintain a cashbook showing daily receipts and expenditure and balance at the end of the day, receipt books containing forms in duplicate, one of each set, to be issued with details of money received by the Society and the other to serve

as counterfoil, voucher files containing all vouchers for contingent and other expenditure incurred by the society, numbered serially and filed chronologically.

To keep an independent check on the accounts maintained by the societies, various provisions have been made by state governments for auditing of their accounts. The auditing of accounts by independent persons with specified qualifications has been specified in these provisions and the auditors are required to prepare and sign a report of their observations after auditing the account books. Every society needs to have its accounts audited once a year by duly qualified auditors and obtain a balance sheet prepared by that auditor. A duly qualified auditor means a chartered accountant within the meaning of the Chartered Accountants Act, 1949 or a person approved by the Registrar of Societies in this behalf. The auditor should submit a report showing the exact state of financial affairs of the society. If the society fails to get its accounts audited, the Registrar of Societies may cause the accounts of the society to be audited and recover the cost from the society. If the society neglects or refuses to furnish records available for audit, the Registrar may order investigation into the affairs of the society.

3. 2.1.5 Powers and Duties of the Registrar of Societies

The Registrar of Societies in each state enforces the provisions of the Principal Act or the corresponding Acts enacted by the other states. The Principal Act does not define the term 'Registrar'. It also does not express the powers which the Registrar may exercise, except for indirect reference to the powers of the Registrar under Sections 1, 2, 3,4,17, 18 and 19. The powers and duties of the Registrar are to

(*a*) certify and inspect documents;

(*b*) call for documents;

(*c*) enquire into and settle disputes;

(*d*) investigate the affairs of a Society (in case the Registrar himself, or upon the instruction of the state governments, is satisfied that the circumstances suggest that the business of the Society is being conducted with intent to defraud its creditors, members or any other person or there is apprehension that the affairs of the Society are being so conducted as to defeat the objects of the society or that the society is guilty of mismanaging its affairs or engaging in any fraudulent or unlawful act.)

(*e*) cancel registration;

(*f*) refuse registration;
(*g*) amend memorandum of association;
(*h*) seize records;
(*i*) enforce attendance of governing body members;
(*j*) order audit;
(*k*) compound offences;
(*l*) settle disputes regarding the election of office bearers;
(*m*) restore property;
(*n*) condone delays; and
(*o*) appoint liquidators.

It is clear from this list of powers that the Registrar does not have overbearing powers in terms of handling of the affairs of a Society. With respect to the functioning of the microfinance operations of a society, the Registrar has no responsibility for any form prudential regulation or determination of its financial performance or solvency. The Registrar can intervene only if a major dispute arises in the management of the Society or there is apprehension of an intention to defraud the creditors of the society or to engage in other unlawful or unauthorized acts.

3. 2.1.6 Taxation

Section 11 of the Income Tax Act, exempts the income of charitable societies from the charge of tax on the fulfillment of certain conditions. Apart from this, sections 12, 12A, 12AA and 13 and certain clauses of Section 10 of Income Tax Act also govern the issue of taxation of such organizations. Since the object of microfinance may be considered a charitable function, local assessing officers of the Income Tax Department often exempt the income from microfinance activities from income tax. However MFIs need to apply to the income tax authorities to get this exemption. The wealth of the societies however, is subjected to wealth tax at 1 per cent of the net value of the wealth with a basic exemption of ₹ 15 lakh. For availing exemption under Section 11, the society is required to fulfill the following conditions:

(a) **Registration:** For registration under Section 12AA with the Commissioner of Income Tax, the Society or Trust or institutions should apply within one year from the date of creation of Society or establishment of institution, in Form No. 10A (in duplicate) along with the memorandum of association or bye-laws of the society in original or the document evidencing creation of the Trust, together with a copy thereof and two copies of the accounts of the society relating to three previous years (or for the year during which the Society or Trust was in existence, in case of a new Society). The Commissioner shall call for documents or information and hold enquiries regarding the genuineness of the Society/ institution. After being satisfied about the charitable or religious nature of its objects and genuineness of its activities, he will pass an order granting registration, and if he is not satisfied, he will pass an order refusing registration, subject to the condition that an opportunity of being heard shall be provided to the applicant before an order of refusal to grant registration is passed and the reasons for refusal of registration shall be mentioned in such order. The order granting or refusing registration has to be passed within six months from the end of the month in which the application for registration is received and a copy of such order shall be sent to the applicant society/institution. If the Commissioner of Income Tax is satisfied that the activities of any institution are not genuine or are not being carried out in accordance with the objects of the institutions, he shall, after giving reasonable opportunity of being heard to the concerned institution, pass an order in writing cancelling the registration granted under Section 12AA.

(b) **Maintenance of Accounts:** The Society should maintain regular books of account, supported by receipts and vouchers. The accounts shall be made on a cash basis. The Society should prepare an 'Income and Expense Account'. Any voluntary contribution received by the society shall be deemed to be income derived from the property held under trust. Where contribution have been made with a specific direction that they shall form part of the corpus it should be so specified on the receipt issued as the same shall be exempt under section 11 (1)(d).

(c) **Compulsory Audit:** Where the total income of the society /institution exceeds ₹ 50,000 in any previous year, the accounts of such society are required to be audited and the audit report which shall be in Form No. 10B is required to be furnished along with the return.

(d) **Income not to be spent for the benefit of certain persons:** No part of the income or property of a charitable society claiming exemption under Section 11 should be used or applied for benefit of any person specified under Section 13/3, subject to certain exceptions.

In the event of refusal to tax exemption, societies need to pay tax on their income, as per the terms laid down in the Income Tax Act. Societies involved in for-profit activities also pay tax as per the terms in the Income Tax act. These tax paying societies can also make investments in the for-profit ventures or bodies, based on their risk appetite.

3. 2.1.7 Acceptance of Foreign Contributions

Acceptance of foreign contribution by MFIs registered under the Societies Registration Act 1860 is regulated by the Foreign Contribution (Regulation) Act 1976 (FCRA). This Act defines foreign contribution as donations or delivery of transfer made by any foreign source

(*a*) of any article, not being an article given to any person as a gift for his personal use, if the market value in India of such article, on the date of such gift, does not exceed one thousand rupees;

(*b*) of any currency, whether Indian or foreign; and

(*c*) of any foreign security.

Besides, 'foreign contribution' includes contribution (referred to above) received from one or more organizations within India whose original source is foreign. In order to access foreign contributions, MFIs need to make applications in form FC-8 of FCRA to the Secretary, Ministry of Home Affairs, New Delhi. Once registered with the Ministry of Home Affairs, these organizations have to agree to receive such foreign contribution only through one of the branches of a bank as it may specify in its application for registration. All such registered organizations need to submit, within a period of four months from the closure of financial year, intimation in Form FC 3 (as per the format prescribed in the FCRA), in duplicate, to the central government as to the following

(*a*) Amount of each foreign contribution received by it;

(*b*) The source from which the contribution was received; and

(*c*) The manner in which it was utilized.

Provided that where such association obtains any foreign contribution through any branch other than the branch of the bank through which it has agreed to receive foreign contribution or fails to give such intimation within the prescribed time or in the prescribed manner, or gives any intimation which is false, the central government may, by notification in the official gazette, direct that such association shall not, after the date of issue of such notification, accept any foreign contribution without the prior permission of the government.

All associations receiving foreign contribution are required to maintain separate accounts for such foreign contributions. The accounts must be maintained on a financial year basis from the first day of April each year and every such yearly account must be furnished in duplicate, duly certified by a chartered accountant to Secretary, Government of India, Ministry of Home Affairs. All these organizations must also get their accounts audited by a chartered accountant. The auditor must certify and issue the following reports:

(*a*) Intimation of foreign contribution in form no FC-3;

(*b*) Balance Sheet as on March 31 of that year; and

(*c*) Statement of receipts and payments account for the year ending March 31.

The auditor has also to review the records including Form No. FC-6 (the articles of accounts) as well as Form No. FC-7 (the securities account) as well as certify the receipt and utilization of foreign contributions in kind such as vehicles, medicines, equipment and so on.

3.2.1.8 Dissolution

Dissolution of an MFI registered under the Societies Registration Act 1856, leads to the cessation of the society's activities. From dissolution arises a need to settle the liabilities of the dissolved societies as also to suitably dispose of its surplus assets. Upon dissolution, steps should be taken by the society for the disposal of the property of the society and settlement of its claims and liabilities in terms of the existing rules of the society. If rules do not indicate the manner of disposal of its properties, the Registrar may decide about the disposal of properties in suitable manner itself with appropriate majority vote or in a manner directed by the general body or a court of law.

The Principal Act provides that if any property remains surplus after the satisfaction of all the debts and liabilities of the society, it cannot be paid to or distributed among the members of the Society or any of them. The members by three-fifth majority may decide to give the surplus properties to some other Society. In the event of any dispute among the members regarding the disposal of surplus property the matter may be referred to the Principal Court of civil jurisdiction for disposal. These provisions are mandatory and if after adjustment of claims and liabilities some property is left, it must go to some other Society with kindred object even if the rules of the society contain a provision for the division of the property of the society upon dissolution among its members. Most of the state Acts have similar provisions for the dissolution of societies. However they differ on the degree of majority votes of the general body of the society. While states like Bihar, Gujarat, Jammu and Kashmir and Maharashtra allow a simple majority vote to take a decision on the modality of disposal of surplus assets, other states like Karnataka, Kerala, West Bengal, Madhya Pradesh, Andhra Pradesh and Puducherry require a three-fifth majority (with Rajasthan requiring a two-third majority).

3. 2.1.9 Compatibility with Insurance Regulations

As per the regulation of the Insurance Regulatory and Development Authority (IRDA), societies are not allowed to take up any type of insurance business on their own. However, they are allowed to function as an agency (but not as a corporate agency) for an insurance company registered with IRDA. Any Society can have an agency for one life insurance and/or one non-life insurance company. The Society needs to fulfill the conditions laid down by IRDA. The proposed regulation on micro-insurance also recognizes NGOs registered as societies as an important vehicle to deliver micro-insurance to a wide cross-section of poor clients.

3. 2.1.10 Advantages and Disadvantages for MFIs of Registration under the Societies Act

Advantages

(a) Simple process of registration;

(b) Simple record keeping and even simpler regulations;

(c) Low possibility of interference by the regulator;

(d) (Largely) Exemption from tax due to the overtly charitable nature of operations; and

(e) Appropriate for taking up micro-insurance (as an agency) on behalf of insurance companies.

Disadvantages

(a) As a charitable institutional form, in essence inappropriate to the for-profit, financially sustainable strategic goal of microfinance operations;

(b) Tax-exemption for surpluses can be (and is) challenged on the question of whether provision of microfinance services is a charitable activity;

(c) No system of equity investment or ownership, thereby, making it less attractive for commercial investors interested in microfinance;

(d) Commercial investors generally regard the investments in such entities risky primarily on account of their lack of professionalism and managerial practices and are, therefore, reluctant to commit large volumes of funds to such MFIs;

(e) In accordance with Section 45S of the RBI Act, 1934, no unincorporated bodies are allowed to accept deposits from the public. Organizations registered under the Societies Registration Act and the Trust Act is considered unincorporated bodies . Therefore, according to the law, they are not even allowed to collect savings from their clients; and

(f) Also vulnerable to the use of 'usurious interest prevention acts' of various state governments as the status of microfinance is debatable as a charitable activity. A discussion on the Moneylenders (Prevention of usurious interest rates) Act is annexed with the report in Annex I.

3.2.2 Indian Trusts Act, 1882

Some MFIs are registered under the Indian Trust Act, 1882 either as public charitable trusts or as private, determinable trusts with specified beneficiaries/members.

3.2.2.1 Formation of a Trust

As per Section 6 of the Indian Trust Act 1882, the essential constituents of a trust are

(a) Three parties - the author, trustees and beneficiary;

(b) Declaration of a trust;

(c) Certainty of the subject matter of a trust; and

(d) Certainty of objects of the trust.

According to Section 3 of the Act, the person who reposes or declares confidence in another person, in some property for the benefits of the beneficiary, is called the 'author' or 'settlor' of the Trust. The author is the creator of the Trust; he gives birth to the Trust. Under the Section 7 of the Indian Trust Act 1882, a Trust may be created by any person/ institution competent to contract. In the context of MFIs, the author of the Trust is an individual with the noble intention of providing financial services to the poor. He arranges for funds from various sources and then deploys the funds to the Trust for the sake of the beneficiaries.

The same section of the Act defines a trustee as a person who, while holding legal ownership and/or possession of, or dominion over, the subject matter of the trust, is bound to allow the beneficial enjoyment of the property to be reaped by another, called the beneficiary. In common parlance however, the word "trustee" is used in a very wide sense so as to include any person who stands in a fiduciary relationship with another, such as the manager, though strictly speaking they are not the trustees. A trustee is liable to indemnify the loss caused to the Trust property or sustained by the beneficiary due to breach of trust committed by him, except where the beneficiary has by fraud induced the trustee to commit the breach or where the beneficiary is an adult and of sound mind and has concurred in the breach or has agreed therein with full knowledge of all the facts. In the context of an MFI, the typical trustees consist of the board or the governing body of the institution that takes policy decisions on behalf of the board and, thereby, holds responsibility for the functioning of the trust.

Also there should be a third person for whose benefit the confidence is reposed by the author or accepted by the trustee. The author of the trust may himself be the beneficiary or one of the several beneficiaries under the Trust. As per Section 9 of the Indian Trust Act, every person capable of holding property may be a beneficiary under the Trust. The capacity to hold property is the only requisite qualification.

Box 3. 1: A note on Private and Public Trusts

A Trust is said to be public when it is constituted wholly or mainly for the benefit of the public at large. The public trusts are essentially charitable or religious trusts. They are of permanent and indefinite character. On the other hand, in a private Trust the beneficial interest is vested absolutely in one or more individuals who are, or within a given time may be, definitely ascertained. In the words of the Supreme Court, the distinction between a private and a public Trust is that whereas in the former the beneficiaries are specific individuals, in the latter they are general public or a class thereof. While in the former the beneficiaries are persons who are ascertained or are capable of being ascertained, in the latter they constitute a body which is incapable of being ascertainment. While the private trusts are governed by the Indian Trust Act, 1882, public trusts are governed by the general law.

A clear declaration of the intention to create a Trust is necessary to form a valid Trust. The evidence of a declaration must be clear and distinct; the acts and writings of the settlor would be sufficient evidence of the declaration of a Trust. In the context of an MFI, the Trust may be declared by a non-testamentary instrument called a Trust deed. There is no prescribed form for a Trust deed. Any words, which indicate the intention of the author to form the Trust, would be sufficient. A Trust declared by a non-testamentary instrument comes into effect from the date specified in the instrument. Once this Trust deed is registered it becomes a valid official document. The following types of registration are associated with a private Trust:

(*a*) Registration under the Indian Trusts Act (for the Trust deed only);

(*b*) Registration under the Societies Registration Act (when the Trust is constituted as a Society), as discussed in the Section 3. 2.1.2.

(*c*) Registration under the Income Tax Act (though private Trusts are generally not required to register under the Income Tax Act as they are not exempted from Income Tax under Section 11, they may seek exemption for that part of their work which is charitable after obtaining registration under the Income Tax Act). Also the Mutually-aided Benefit Trusts (MBTs), a form of private trusts does not need to be registered under this act.

(d) Registration under the Foreign Contribution (Regulation) Act, 1976 (for becoming eligible to received foreign grants for carrying out their activities); as discussed under the Section 3.2.1.7.

3.2.2.2 Management of a Trust

Trusts are managed by their trustees. Though certain rules, liabilities, rights, powers and disabilities of the trustees have been laid out under the Indian Trusts Act, no specific rules/ provisions have been made regarding the management of trusts. Thus, specific provisions may be made in the 'trust deed' or the trustees should themselves agree to a code of conduct for the smooth and efficient functioning of the Trust. A Trust must make rules with regard to

(a) holding meetings;

(b) quorum for meetings;

(c) chairing meetings;

(d) adopting resolutions;

(e) nature of the functions /businesses to be carried out by the trust;

(f) delegation of functions; and

(g) fixing accountability.

3.2.2.3 Accounts and Audit

A Trust must maintain proper and regular books of accounts, to record its financial transactions. The generally accepted accounting norms should be adopted to make the records transparent. The organization must have management policies with regard to:

(a) Maintenance of the bank accounts of the Trust;

(b) Investment of its assets;

(c) Maintenance of the books/records of account (including petty cash book, main cash book, general ledger, journal, investment registers, fixed asset register, stock register) and compliance with the accounting standards;

(d) Maintenance of the records of the trust;

(e) Preparation of financial statements;

(*f*) Compliance of accounting standards; accounting standards relate to the codification of generally accepted accounting principles. The main purpose of accounting standards is to provide information to the users as to the basis on which the accounts have been prepared. The objective of setting standards is to bring about uniformity in financial reporting and to ensure consistency and comparability in the financial statements; and

(*g*) Periodic audit of financial records: An audit is a systematic review of the financial transaction of a Trust. It involves an examination of the validity of supporting documents, certification of the financial statements and submission of an audit report. Every Trust needs to get its books of accounts audit by a qualified chartered accountant, who would be appointed by the governing body of the Trust by passing a resolution.

3.2.2.4 Investment of Funds

There are various apprehensions regarding investment of the funds of a Trust. A Trust deed normally lays down the manner in which Trust funds shall be invested and trustees must invest the Trust funds primarily in the manner thus prescribed. However, a public charitable Trust claiming exemption under certain sections of the Income Tax Act needs to invest its funds in certain specified assets. A charitable Trust is required to invest or deposit its funds in the modes or forms specified under Section 11(5) of the Income Tax Act. Besides, it must not hold any shares in a company (other than a government company or corporation). If any part of its funds is not so invested, the exemption under Section 11 would not be allowed.

Private trusts are required to invest their funds in the manner directed in the Trust deed, otherwise the trusts are free to invest their funds in any manner whatsoever. Such trusts have wide scope for improving their resources by making prudent and creative investments. However, the organization must have a policy on the investment of its funds. This holds significance for the functioning of MFIs that sometimes have a surplus of resources that need to be invested (at least temporarily) in order to optimize returns.

3.2.2.5 Acceptance of Foreign Contributions

The provisions regulating the acceptance of foreign contribution by the societies for charitable purposes would be applicable to the trusts using the funds for similar purposes. The trusts would also need to undergo all the required processes and procedures that a society needs to undergo in order to avail the foreign contribution. The details of these processes and procedures are provided in the earlier part of this report. (Refer to Section 3. 2.1.6).

3.2.2.6 Taxation

Trusts registered under Indian Trust Act, 1882 are also subject to same rules and regulations as discussed under Section 3. 2.1.6 of the report.

3.2.2.7 Determination of Trusts

The dedication of property to a Trust for charitable or religious purpose is permanent. Thus once a Trust is created for religious or charitable purpose it is permanent and cannot be revoked or reversed or terminated. MFIs registered as trusts and though having charitable purpose, they are registered as private trusts. A private Trust will terminate in the following circumstances:

(a) when the purpose for the trust is completely fulfilled;

(b) when its purpose becomes unlawful;

(c) when the fulfillment of its purpose becomes impossible by destruction of Trust property or by otherwise; and

(d) when the Trust being revocable, is expressly revoked.

Thus the Trust would be determined when all the beneficiaries transfer their interests. In such an event, the Trust property shall not be available for fulfillment of its purpose and the Trust will stand dissolved. Similarly, a Trust declared for a specified period of time, will be determined on the efflux of the time period. MFIs registered as trusts would need to have their terms of determination clearly spelt out as part of the Trust deed to deal with any such eventuality.

3.2.2.8 Compatibility with the insurance regulations

In its current shape, IRDA regulations do not recognize the trusts as an agent to deliver the services related to insurance. While the NGOs registered as societies are clearly allowed to carry on the insurance function as an agent, there is no mention of NGOs registered as trusts for this purpose. However given the development role of many of the NGOs registered as trusts, it is quite likely that the proposed legislation on micro-insurance would include the trusts also as the agent for the micro-insurance service delivery.

3.2.2.9 Advantages and Disadvantages of registration under the Indian Trusts Act

The advantages and disadvantages related to MFIs registered under the Societies Act are equally applicable to trusts. However while IRDA recognizes NGOs registered as societies as a distribution channel for the micro-insurance, there is no such clarity with regard to trusts. Apart from this, private trusts may be even more unsuitable as the tax exemption extended to societies may apply to such trusts only to the extent the Income Tax department accepts their activities as being charitable.

Advantages

(a) Simple process of registration;

(b) Simple record-keeping and even simpler regulations;

(c) Low possibility of interference by the regulator; and

(d) (Largely) Exemption from tax due to the overtly charitable nature of operations.

Disadvantages

(a) As a charitable institutional form, in essence inappropriate for the for-profit, financially sustainable strategic goal of microfinance operations;

(b) Tax-exemption for surpluses can be (and is) challenged on the question of whether provision of microfinance services is a charitable activity;

(c) No system of equity investment or ownership, thereby, making it less attractive for commercial investors interested in microfinance;

(d) Commercial investors generally regard the investments in such entities risky primarily on account of their lack of professionalism and managerial practices and are, therefore, reluctant to commit large volumes of funds to such MFIs;

(e) In accordance with Section 45S of the RBI Act, 1934, no unincorporated bodies are allowed to accept deposits from the public. Organizations registered under the Societies Registration Act and the Trust Acts are considered unincorporated bodies. Therefore, according to the law, they are not even allowed to collect savings from their clients; and

(f) Also vulnerable to the implication under the money lenders (prevention of usurious interest rates) acts of various state governments as the status of microfinance is debatable as a charitable activity. A discussion on the Moneylenders' (Prevention of usurious interest rates) Act is annexed with the report in Annex I.

3.2.3 Not-For-Profit Companies Registered Under Section 25 of Companies Act, 1956

An organization given a license under Section 25 of the Companies Act 1956, is allowed to be registered as a company with limited liability without the addition of the words 'Limited' or 'Private Limited' to its name. It is also eligible for exemption from some of the provisions of the Companies Act, 1956. For companies that are already registered under the Companies Act, 1956, if the central government is satisfied that the objects of that company are restricted to the promotion of commerce, science, art, religion, charity or any other useful purpose; and the constitution of such company provides for the application of funds or other income in promoting these objects and prohibits payment of any dividend to its members, then it may allow such a company to register under Section 25 of the Companies Act.

3.2.3.1 Registration

As stated earlier, both a fresh entity and an association as well as an existing company are entitled to be registered as a Section 25 company provided they follow certain procedures and fulfill requirements as provided in the guidelines. The procedure for the registration as Section 25 for both the above types of organizations is discussed below.

3.2.3.1.1 Registration of New Associations

Such associations need to apply to the Regional Director, Company Law Board of the region with a covering letter along with the following documents

(a) Draft Memorandum and Articles of Association of the proposed company in three copies. The memorandum and articles should be typewritten and then printed after having been approved by the Regional Director. No stamp duty is payable on the memorandum and Articles of Association of a company to be registered under these provisions.

(b) List of names, addresses, description and occupation of the promoters in three copies.

(c) List of companies, associations and other institutions in which promoters of the applicant company are directors or hold responsible positions, with description of positions held.

(d) List of members of the proposed Board of Directors.

(e) Declaration in the prescribed format signed by

- An advocate of Supreme Court or of the High Court, or

- An attorney or pleader entitled to appear before the High Court, or
- A Chartered Accountant with a whole-time practice in India, or
- A Company Secretary with a whole time practice in India, on a non-judicial stamp paper of appropriate value.

(f) Copies of accounts, balance sheet and reports on the working of the association for the last two financial years (for one year only if the association has functioned for less than two years), in triplicate.

(g) Statements of assets and liabilities.

(h) Sources of income of the proposed company and estimate of annual income and expenditure.

*(i)*A note on the work already being done by the association and work proposed to be done after being registered under Section 25.

(j) Grounds in brief for making application under Section 25.

(k) Declaration (in the prescribed format) signed by each of the applicants.

Within a week of making application to the Regional Director, a notice declaring the intention of registering a not-for-profit limited liability company with the name of the proposed company, is required to be published in the prescribed format. It should be published one in English, and one vernacular newspaper of the area where the registered office is situated. A certified copy of the notice should be filed with the Regional Director.

The application should also be accompanied by a demand draft of ₹ 500 drawn in favour of "Pay and Accounts Officer, Department of Company Affairs". The fee may also be paid through a treasury challan at specified branches of the Punjab National Bank.

The license is granted by the Regional Director after scrutiny of the applications and considering the recommendation of the Registrar made on the application. The Regional Director may direct the company to incorporate in its memorandum or articles or both, such conditions of the license as may be specified by the Regional Director in this regard. A copy of the application with all enclosures and accompanying papers should be sent to the Registrar of Companies of the State where company proposes to situate its head office/registered office.

After the draft memorandum and articles have been approved, the association should apply to the Registrar of Companies concerned, for its registration as a company along with printed copies of the memorandum and articles, required forms (Form nos. 1, 18 and 32) and also produce the license granted by the Regional Director along with the necessary registration fee of ₹ 50 and filing fee of ₹ 10 per document. As per the opinion of the department, since the quantum of fee payable has been fixed irrespective of the amount of capital, the fee payable at the time of increase of capital will also be ₹ 50 only. The Registrar, after scrutinizing the documents and finding them in order, shall issue the Certificate of Incorporation in Form No. 1 as per Regulation 16(1) f the Companies Regulations, 1956.

3.2.3.1.2 Conversion of Companies already Registered

Such companies should apply to the Regional Director by a covering letter along with

(a) Memorandum and Articles of Association.

(b) List of names and addresses, description and occupation of directors, managers or secretary.

(c) Name of companies, association and of other institutions, in which the directors of the applicant company are directors or hold any responsible position (with description).

(d) Three copies each of profit and loss account, balance sheet, annual report of Board of Directors and audit report of the company for each of the last two financial years (if company has been in existence for less than two years, documents relating to one year may also be submitted).

(e) Assets and liabilities of the company.

(f) Estimate of expenses and sources of income.

(g) A note on work already done by the association and works proposed to be done after being registered under Section 25.

(h) Grounds, in brief, for making application under Section 25.

(i) Declaration (in prescribed format) signed by each of the applicants.

(j) Certified copy of notice published in newspaper. This notice should be published one in English, and one vernacular newspaper of the area where the registered office is situated within one week after or before making the application.

The application needs to be accompanied with a demand draft worth ₹ 500 as discussed in the preceding section. The license is granted by the Regional Director after the scrutiny of the applications and considering the recommendation of the Registrar made on the application. The Regional Director may direct the company to insert in its memorandum or its articles or partly in one or partly in the other, such conditions of the license as may be specified by the Regional Director in this regard.

After receipt of the necessary approval under Section 25 the approved memorandum and articles should be printed without stamping as stamp duty is exempted. Thereafter they should be filed with the Registrar along with required forms (Form no. 1, 18 and 32) for incorporation along with a registration fee of ₹ 50 and filing fee of ₹ 10 per document. The Registrar then issues a certificate of incorporation to the association or club, as per Regulation 16(1) of the Companies Regulations, 1956.

3.2.3.2 Privileges and Exemptions

According to Section 25 (2), registration having been granted to an association, it enjoys all the privileges of a limited company and is subject to all the obligations except where the provisions of Section 25 (or through general or special order of the central government), certain exemptions are granted. These exemptions primarily relate to provisions for the appointment of directors, publication of name and address of the company, holding of annual general meetings, quorum of the board meeting, disclosure of interests by directors. The most important of the exemptions relates to the disclosure of annual returns for income tax and appointment of the company secretary. Section 25 companies are exempted from these two provisions that apply to other types of companies.

The details of these exemptions are thoroughly discussed (see annex 3.1). Further, in January 2000, the RBI issued a notification that Sections 45 IA and IC of the RBI Act will not apply to companies registered under Section 25 of the Companies Act. This exemption effectively means that Section 25 companies can take up the activity of lending without the permission of or registration with the RBI, provided these are within the definition of microfinance loans to business enterprises up to ₹ 50,000 and loans for dwelling units up to ₹ 1,25,000. Deposit taking is, however, not permitted without such registration with the RBI.

3.2.3.3 Taxation

Being an entity engaged in not-for-profit activities and not sharing its surplus amongst shareholders, Section 25 companies are exempt from paying tax under Section 12(1) of the Income Tax Act. This is a major benefit for Section 25 companies. In order to avail this exemption, organizations need to apply to the Commissioner of Income Tax. The process of availing this benefit is discussed in detail in the Section 2.1.7 of this report.

3.2.3.4 Compatibility with Insurance Related Regulations

IRDA allows any company registered under the Companies Act 1956 to become a corporate agent of the insurance company registered with IRDA. Section 25 companies under this provision are allowed to become corporate agents of one life and/or one non-life insurance company. They need to follow all the other relevant regulations related to IRDA.

3.2.3.5 Advantages and Disadvantages of Registering MFIs as Section 25 Companies

Section 25 companies are subject to some of the problems similar to those faced by societies and trusts. However they have certain advantages also. These issues are discussed as under:

Advantages

(*a*) Exemption from Income Tax: Section 25-1A of the Companies Act, 1956, mentions as objectives of charitable Section 25 companies as "the promotion of commerce, art, science, religion, charity or any other useful object". This works as a great advantage for MFIs.

(*b*) Leveraging Capacity: Having a Section 25 company adds to the legitimacy of the institution. This also adds to the credibility of the commercial FIs who are willing to lend to the sector.

(*c*) Easier processes and low regulator interference: As discussed in the preceding sub-sections, Section 25 companies have relatively easier processes to register and they need to comply with very few regulatory requirements. They also need not register with RBI unlike NBFCs. This makes the functioning of the organization very easy.

Disadvantages

(*a*) Deposit mobilization: The same restrictions as for societies/trusts apply to deposit mobilization for Section 25 companies.

(*b*) Not an attractive option for equity mobilization: Being a not-for-profit entity prohibits the Section 25 companies from distributing dividend. Therefore, these entities find it difficult to attract equity investors, even some social investors, to invest equity funds with them. Also the exit route for investors is not smooth since another investor willing to forgo the possibility of a dividend return on the investment must be found.

(*c*) Lack of clarity on ceiling: Although microfinance defined by RBI provides for financial services of very small amounts, the RBI instructions exempting Section 25 companies from registration and reserve requirements for NBFCs stipulate that loans should not exceed ₹ 50,000 for non-housing purposes and ₹125,000 for housing purposes.

3.2.4 Non-Banking Finance Companies

NBFCs have played an important role in the Indian financial sector for a long time, fulfilling the gap in the demand and supply of financial services, particularly from small clients. What has promoted the growth of NBFCs is their localized presence, a higher level of customer orientation than banks and lower documentation requirements albeit at higher interest rates for borrowers. Conversely, depositors earn substantially higher rates of interest as compared to the traditional savings instruments of banks. In this way, NBFCs are able to attract a large volume of deposits from the general public. Lower entry barriers and high returns (in relation to banks) from the NBFC business attract lots of entrepreneurs. Many such companies have been unable to service their debt obligations due to ineffective asset-liability matching and spectacular collapses, like CRB Caps in Ahmedabad and Century Consultants in Lucknow, which has resulted in considerable loss of public confidence in NBFCs' services.

Recognizing the importance of NBFCs in the Indian financial sector and with the objective of integrating these with the financial mainstream, RBI started to regulate them from 1996. These measures include mandatory registration of companies offering financial services with the RBI, compulsory credit rating of deposit taking NBFCs and imposition of prudential norms.

3.2.4.1 Definition

As per the Section 45-I of the RBI Act, an NBFC is a company which carries on any of the following activities as its business or part of its business:

(a) Lending;

(b) Receiving of deposits (as the principal business);

(c) Acquisition of shares, stocks or other securities;

(d) Hire-purchase or leasing;

(e) Insurance;

(f) Chit funds; and

(g) Lotteries

As per this definition, if a company undertakes microfinance, it automatically becomes an NBFC and all related regulations apply.

Pre-requisites for carrying on the business of an NBFC: Under Sub-section 1 of Section 45-IA of the RBI Act, no NBFC can commence or carry on the business of a non-banking company without complying with the following two essential pre-requisites:

(a) Obtaining a certificate of registration from RBI; and

(b) Having Net Owned Funds (NOF) - shareholder equity + internally-generated reserves of ₹ 200 lakh. This limit was earlier ₹ 25 lakh only; it has now been increased to ₹ 200 lakh by RBI. Vide notification no.

DNBS 132/CGM (VSNM)-99, dated 20-4-1999, the RBI specifies the "net-owned fund" (NOF) to be ₹ 200 lakh, for an NBFC which commences the business of an NBFC after April 21, 1999 . Thus, this specification of higher "(NOF)" is not applicable to NBFCs that commenced business before April 21, 1999.

These requirements are cumulative and not alternative. As such, if a for-profit company wants to carry out microfinance operations, it must comply with the above requirements. Though RBI prescribes a set of compliance norms (prudential and non-prudential) for all NBFCs, these are more rigorous (prudential) for NBFCs accepting public deposits. However prudential norms are not applicable to the companies which are not accepting public deposits. It is, therefore, important first to understand the meaning of public deposits.

The definition of public deposits given in the RBI Act, 1934 is exclusive rather than inclusive. Section 45-I (bb) of the Act defines 'deposits' in general terms to include any receipt of money by way of deposit or loan or in any other form'. The Act then excludes the following items from the definition:

(a) Amounts raised by way of share capital;

(b) Amount contributed as capital by partners of a firm;

(c) Amounts received from scheduled banks, cooperative banks or other notified financial institutions;

(d) Amounts received in the ordinary course of business, by way of, security deposits, dealership deposit, earnest money or advance against orders for goods, properties or services;

(e) Amounts received from registered moneylenders, provided that such moneylender is not a company; and

(f) Amounts received by way of subscription in respect of a chit.

Paragraph 2(1) (xii) of the directions also excludes certain deposits from the definition of public deposits. These are amounts received

(a) From the central or any state government or under its guarantee;

(b) From a local authority;

(c) From a foreign government or any foreign citizen, authority or person;

(d) From institutions like LIC, IDBI, ICICI, IFCI, ADB, IFC or any other institutions specified by RBI;

(e) From one company to another company;

(f) As subscription to any shares, stocks, bonds or debentures;

(g) From a director/shareholder of a private company, or a private company which has become a public company under Section 43A of the Companies Act, subject to the director/shareholder furnishing a declaration to the effect that the amount has not been given out of borrowed funds or from the amounts accepted from others;

(h) From mutual funds governed by the Securities and Exchange Board of India (Mutual Funds) Regulations, 1996;

(i) Received as hybrid debt or subordinate debt the minimum maturity period of which is less than 60 months;

(j) From a relative of a director of an NBFC; and

(k) Received by issuance of commercial paper, in accordance with the guidelines issued by the Bank vie circular no IECD 3/08.15.01/2000-2001 dated October 10, 2000.

Barring these exclusions, all receipts of an NBFC come under the category of public deposits.

3.2.4.2 Registration

As mentioned in the preceding sub-section, all NBFCs need to get registration with RBI for carrying on their business. In order to get such registration as NBFCs, MFIs need to fulfill the following two basic criteria:

(a) Should be registered as a company under the Companies Act, 1956; and

(b) Should have a minimum "net-owned fund" of ₹ 200 lakh (if not already registered on 20 April 1999).

If the MFI meets the above two criteria, it can file an application for registration with the RBI's Department of Non-Banking Services. The procedure for obtaining registration is described in following section (see table 3.2).

3.2.4.2.1 Acquisition of an NBFC

Many societies/trusts working in the area of microfinance are buying existing NBFCs. They also transfer their relevant assets and liabilities to these entities in order to transform into a more evolved legal entity. Since the provisions laid down in the Companies Act, 1956, govern the transfer of ownership of NBFCs, it is increasingly being viewed as an easier option for the MFIs. It would be worth mentioning here that in case the existing NBFC was registered before April 20, 1999, the changed provisions on net-owned funds would not apply to them.

Table 3.2: Procedure for registering an NBFC with RBI

Process	Details
Submission of application	Application should be made in the form prescribed by RBI, to be filled in accordance with the instruction contained in the form. Documents required to be enclosed should also be prepared in accordance with the instructions and should be attached to the application.
Where to submit the application	To the regional RBI office under whose supervision the company's registered office is situated.
Processing by RBI	RBI will ensure that the conditions spelt out in the sub-section 45-IA are fulfilled. These conditions broadly cover the following areas: • Capacity of the NBFC to meet the creditor's claims in full, when the claims accrue. • Conduct of the affairs of the NBFC in a manner not detrimental to the interests of depositors. • General character of the management of the NBFC is not prejudicial to the public interest or the interest of its depositors. • NBFC has adequate capital structure and earning prospects. • The grant of certificate serves the public interest. • The grant of certificate will not be prejudicial to the operation and consolidation of the financial sector consistent with monetary stability, economic growth or other relevant factors. • Any other condition which, in the opinion of RBI, is necessary to ensure that the business of the NBFC is not prejudicial to the public interest or the interest of depositors.
Issue of Certificate	RBI will thereafter grant a certificate of registration, with or without specified conditions.
Cancellation of registration	RBI can cancel the registration in the following situations [Section 45-IA (6)]: • When the NBFC is, at any time, found to have failed to fulfill any of the conditions mentioned in *'Processing by RBI'* above. (**b**) • When the NBFC ceases to carry on business. • When the NBFC fails to comply with any condition specified by RBI while granting certificate of registration.(**a**) • When the NBFC is, at any time, found to have failed to fulfill any of the conditions mentioned in *'Processing by RBI'* above. (**b**) • When the NBFC fails to comply with any directions issued by RBI. • When the NBFC fails to maintain accounts in accordance with any law, or with any direction or order issued by RBI. • When the NBFC fails to submit or offer for inspection its books of accounts and other relevant documents when demanded by RBI. • When the NBFC has been prohibited from accepting deposit by an order of RBI, and that order has been in force for not less than three months.

Contd...

Table 3.2: Contd...

Process	Details
Procedure for cancellation of registration	The NBFC should be given a reasonable opportunity of being heard. Where action for cancellation is proposed on the grounds **a** and **b** as stated above, the NBFC must generally be given an opportunity by RBI to take the necessary steps to comply with the conditions, except in cases where RBI is of the opinion that the delay in cancelling the certificate of registration shall be prejudicial to the public interest or the interest of depositors or the NBFC.
Appellate remedy	If the application for registration is rejected by RBI, or if the certificate of registration is cancelled by RBI, the company can prefer an appeal to the central government within 30 days from the date on which the order was communicated to it. The decision of the central government on the appeal, if any, filed by the NBFC shall be final. Where no appeal has been preferred by the NBFC within the stipulated time, the decision of RBI shall be final.
Utilization of funds during the interregnum	Till such time as the certificate of registration is granted by RBI, the NBFC can keep its capital funds invested in any type of deposits with a bank. Investment in any other type of securities will invite penal action.

Source: Taxmann's Statutory Guide for Non-Banking Financial Companies, An authorized publication of Reserve Bank of India (2004).

3.2.4.3 General Compliance Requirements

All NBFCs, being companies registered under the Companies Act, have to fulfill compliance relating to the Board of Directors, Share Capital, Management Structure, Audits, Meetings, maintenance as well as publication of books of accounts and general conduct as per the requirements of the Companies Act 1956. In addition, they have to fulfill the specific requirements of RBI. Some of the general compliance requirements are set out below.

3.2.4.3.1 Submission of Returns

All NBFCs need to report to RBI on a regular basis. Table 3.3 below summarizes the returns to be submitted to RBI along with their due dates.

3.2.4.3.2 Minimum Investment Norms for Liquidity

Not less than 15 per cent of their deposits should be invested by NBFCs in specified securities and approved securities (government securities and bonds guaranteed by the government).

3.2.4.3.3 Opening and closing of branches

NBFCs are required to notify RBI of their intention to open branches. On receipt of such advice the RBI may, after recording relevant reasons, reject such a proposal and communicate this to the NBFC. If no notice of rejection is communicated to the NBFC within 30 days of receipt of such a proposal, the NBFC may proceed to open the branch.

No NBFC shall close its branch/office without publishing such intention in any one national newspaper and in one vernacular newspaper in circulation in the relevant place and without advising RBI, before 90 days of the proposed closure.

3.2.4.3.4 Prudential Norms

Section 45JA of the RBI Act empowers RBI to determine policy and give directions to all NBFC having a net-owned fund of ₹ 25 lakh and above and accepting/holding public deposit, relating to

(a) Income recognition and accounting standards;

(b) Provision for bad and doubtful debts;

(c) Capital adequacy based on the risk weight of assets;

(d) Credit conversion factors for off-balance sheet items; and

(e) Deployment of funds

Important prudential norms for NBFCs are summarized in the following table (see table 3.4)

Table 3.3: Schedule for submission of returns by NBFCs

Nature of return to be submitted	Due Date
Audited balance sheet as on the last date of the financial year and profit and loss account in respect of that year, as passed by the company in its general meeting, together with a copy of the report of the Board of Directors	Within 15 days of the general Meeting
Auditor's report to the Board of Directors, together with a certificate from the auditor to the effect that full amount of liabilities to the depositors, including interest payable, is properly reflected in the balance sheet, and the company is in a position to meet the amount of such liabilities to the depositors	Along with the audited balance Sheet
Information on financial position, as specified in the First Schedule (Return NBS 10) to the Directions	Due date specified in the First Schedule (Return NBS 1)
Half-yearly returns in the format specified (Form No NBS 2)	Within three months of the expiry of the relevant half-year ending in September or March

contd...

Table 3.3: Contd...

Nature of return to be submitted	Due Date
Changes in - Postal address/ tel no/ fax no of the registered office of the company - Names and residential addresses of the directors of the company - Names and designations of the principal officers of the company - Specimen signatures of the officers authorized to sign on behalf of the company	Within one month from the occurrence of the change

Source: Taxmann's Statutory Guide for Non-Banking Financial Companies, An authorized publication of Reserve Bank of India (2004).

Table 3.4: Prudential Norms for NBFCs

Parameters	Provisions
Capital adequacy	Every NBFC is required to maintain a capital adequacy ratio of 15% (in case they are also receiving public deposits), consisting of Tier I and Tier II Capital[1]. The total of Tier II Capital should not be more than 100% of Tier I capital. However there is no minimum bar on capital adequacy if the NBFC is not receiving public deposit.
Prohibition on loans	An NBFC is prohibited from lending against its own shares. Also if an NBFC has defaulted on its deposit obligations, it is prohibited from lending.
Restrictions on investments	An NBFC which is accepting public deposits can invest a maximum of 10% of its NOF in land and building (except for its own use), and 20% of its NOF in unquoted shares of companies other than subsidiaries and companies in the same group.
Ceiling on concentration of credit/investments	Lending to a single borrower – 15% of NOF Lending to a single group of borrowers – 25% of NOF Investment in shares of another company – 15% of NOF Investment in shares of a single group of companies – 25% of NOF
Accounting requirements	The NBFCs have to comply mainly with the following accounting requirements 1. Income recognition 2. Income from investments 3. Accounting standards 4. Asset classification 5. Provisioning requirements 6. Disclosure in the balance sheet
Audit committees	All NBFC having assets of ₹ 50 crore and above as per the last audited balance sheet shall constitute an Audit Committee, consisting of not less than three members of its Board of Directors
Exposure to capital market	Every NBFC holding public deposits of Rs 50 crore and above shall submit a quarterly return within one month of expiry of the relative quarter in prescribed format (Format NBS-6) to the Regional Office of the Department of Non-Banking Supervision of the RBI.

Source: Taxmann's Statutory Guide for Non-Banking Financial Companies, An authorized publication of Reserve Bank of India (2004).

All these prudential requirements are however not applicable to all NBFCs. If the NBFC is not accepting public deposits then it is exempted from the prudential requirement on capital adequacy and ceiling on concentration of credit/investments.

3.2.4.4 Specific compliance requirements relating to NBFCs accepting public deposits

Since the major objective of NBFC regulations is to maintain the confidence of the general public through the protection of their deposits in NBFCs, RBI has stipulated specific compliance requirements for NBFCs accepting/maintaining public deposits.

3.2.4.4.1 Minimum Credit Rating

An NBFC must obtain minimum investment grade rating for fixed deposits from any of the approved credit rating agencies at least once a year, after being in operation for not less than two years. The approved credit rating agencies and minimum credit ratings are as indicated in Table 3.5. Rating agencies use different sets of standards for ascertaining the creditworthiness of FIs. Some of the most common areas universally looked at by these agencies include the capital adequacy, asset quality, liquidity, management strength, profitability, efficiency and corporate governance. If an FI has performed well along all the above parameters; it may attain an investment rating grade from the agency. However a number of such parameters are subjective also; therefore a lot may depend on the methodology and the quality of the risk assessment team. A short note on the rating principle and framework is provided in Annex 3.2.

Table 3.5: Investment Grade Ratings of Approved Rating Agencies for NBFCs

Name of the Agency	Minimum investment grade rating
The Credit Rating Investment Services of India Ltd. (CRISIL)	FA – (FA Minus)
ICRA Ltd.	MA – (MA Minus)
Credit Analysis and Research Ltd. (CARE)	CARE BBB (FD)
FITCH Ratings India Private Ltd.	tA – (ind) (FD)

Source: Taxmann's Statutory Guide for Non-Banking Financial Companies, An authorized publication of Reserve Bank of India (2004).

Any upgrade or downgrade of the credit rating must be intimated to RBI within 15 working days of such change.

3.2.4.4.2 Ceiling on quantum of deposits

The ceiling on public deposits for Loan Companies is thoroughly discussed in this chapter and summarized in the following table (see table 3.6).

Table 3.6: Ceiling on deposit for Loan Companies

Category of Loan Company	Ceiling Limit
Having NOF of more than ₹ 200 lakh[1] and above a. Not complying with all the prudential norms, and not having minimum investment grade rating	Cannot accept any public deposit
b. Complying with all the Prudential Norms and having minimum investment grade rating	
(i) with capital adequacy ratio of at least 15%	150% of NOF
(ii) With capital adequacy ratio of 15% but having credit rating, AAA	Amount outstanding at the close of business on December 18, 1998 or 150% of NOF whichever is more.
(iii) With capital adequacy ratio less than 15% but credit rating, AA	100% of NOF
(iv) With capital adequacy less than 15% but with credit rating, A	50% of NOF
(v) With capital adequacy ratio less than 15% but credit rating below A	Cannot accept any public deposit

Table 3.7: Restrictions and Directions on Deposit Mobilization

Particulars	Nature of restrictions/directions
Demand deposit	Acceptance or renewal of any public deposit which is repayable on demand is prohibited.
Period of deposit	The period of public deposit should be for a minimum of 12 months and for maximum of 60 months from the date of acceptance or renewal.
Ceiling on rate of interest on public deposit	Rate of interest not to exceed 11% per annum. Interest should be paid or compounded at rests which shall not be shorter than monthly rests.
Brokerage	Brokerage, commission, incentive or any other incentive for public deposit paid to a broker should not be more than 2% of the amount of deposit so collected

Source: Taxmann's Statutory Guide for Non-Banking Financial Companies, An authorized publication of Reserve Bank of India (2004).

When the credit rating goes below the minimum specified investment grade, the company should:

(a) Stop accepting public deposits with immediate effect;

(b) Report the position to RBI within 15 working days; and

(c) Within three years from the date of downgrade of credit rating, reduce the excess deposit to nil, either by repayment of matured deposits or otherwise. (See table 3.7 for general restrictions on deposits of NBFC).

3.2.4.5 Audit and Taxation

Being a company registered under Companies Act, NBFCs need to follow the norms laid down by the Registrar of Companies with regard to auditing. They need to get themselves audited once every quarter by an external auditor. They also need to follow the norms on appointment of auditors as laid down by the Registrar.

NBFCs also need to follow the norms on the taxation as laid down in the Income-Tax Act. They are subject to taxation on their surpluses as per the regulation. In order to obtain the grant funds from foreign sources, they also need to have registration under FCRA.

3.2.4.6 Compatibility with Insurance Regulations

As per the guidelines laid by IRDA and RBI, NBFCs are allowed to participate in the insurance business either by setting up their own business or as agents of an insurance company registered with IRDA. Considering the high capital requirement for setting up an insurance company (₹ 500 crore), the first option is not viable for an MFI.

Further, RBI has laid down very strict regulations for NBFCs to carry out the business of insurance as corporate agents of insurance companies. The conditions for the NBFCs registered with RBI who are also doing the insurance business as corporate agent of insurance companies are:

(a) The NBFCs should obtain requisite permission from IRDA and comply with the IRDA regulations for acting as 'composite corporate agent' with insurance companies.

(b) The NBFCs should not adopt any restrictive practice of forcing its customers to go only for a particular insurance company in respect of assets financed by the NBFC. The customers should be allowed to exercise their own choice.

(c) As the participation by an NBFC's customer in insurance products is purely on a voluntary basis, it should be stated in all publicity material distributed by the NBFC in a prominent way. There should be no 'linkage' either direct or indirect between the provisions of financial services offered by the NBFC to its customers and use of the insurance products.

(d) The premium should be paid by the insured directly to the insurance company without routing through the NBFC.

(e) The risks, if any, involved in the insurance agency should not get transferred to the business of the NBFC.

3.2.4.7 Advantages and Disadvantages of Registration as an NBFC

NBFCs are the only type of MFI falling clearly under the purview of the central bank and subject to prudential regulations. As for-profit companies, they need to pay income tax on their lending operations. In the context of the average MFI, some issues that arise in the matter of registering and operating as NBFCs are:

(i) **Minimum Capital Requirement:** The minimum capital requirement (₹ 2 crore) for obtaining a license for an NBFC is high. Given the nature of microfinance in India, and the relatively small size of most MFIs, the mobilization of funds of this volume becomes a major challenge. This is further complicated as the promoters of such NBFCs are largely societies that are not allowed to make equity investments, lest they lose their tax exemption status.

(ii) **Deposit Mobilization:** NBFCs are not allowed to mobilize deposits unless they complete two years of operations and then obtain an investment grade rating. Such a rating is doubly difficult to get for MFIs - even those with strong portfolios - since the conventional credit rating agencies designated for this purpose by the RBI regard lending to the poor and lending in rural areas as 'inherently risky'. Yet, given the needs of the poor, deposit services are an important element of microfinance and if the low-income clients are denied this service, the very purpose of microfinance is defeated.

(iii) **Foreign Direct Investment and Venture Capital restrictions:** The entire financial sector is denied access to venture capital investment and was denied access to external commercial borrowings (ECB) until very recently. Since there are venture capitalists - foreign or domestic - willing to put their money into microfinance, any restriction of this sort places an additional handicap for MFIs. Yet, there are an increasing number of socially responsible international investment funds for microfinance and their investments are effectively barred from India. Foreign direct investment (FDI) in equity is allowed but must be for a minimum of $0.5 million (₹ 2.15 crore today) and cannot exceed 50 per cent of the equity of an NBFC. This means that initially the MFI

must raise more than ₹ 2.2 crore from domestic sources before it can become eligible for foreign equity investment.

However registration with RBI as an NBFC offers advantages also. Some of these advantages are listed below:

(i) **Investor Confidence:** MFIs registered as NBFC because of being recognized and regulated entities for carrying out microfinance operations provide a higher level of confidence to institutional lenders and banks.

(ii) **Limited Regulation, if not Accepting Deposits:** MFIs registered as NBFCs are subject to nominal regulations if they are not accepting public deposits. Though the general poor clients are deprived of a very important financial service in case deposit acceptance is not allowed, it also brings down the level of regulation significantly if the MFI is not accepting deposits.

(iii) **Attractive Option for Social Investors:** Despite having a very high ceiling for equity investment in NBFCs, they are the most preferred options for social investors interested in the sector. With a regulated structure, NBFCs have a relatively disciplined and professional management. With the equity structure being clearly laid out, the exit route for such investors is also relatively safe compared to other types of MFIs. This factor also helps them in placing their donated equity with the NBFC. Considering the disclosure norms that such NBFCs need to follow on various matters, the willingness of such investors to put their funds increases significantly. (However such institutions would need to have necessary clearance/approval under FCRA, details of which have been discussed under Section 2.1.7).

(iv) **Up-scaling of Operations:** With a regulated set-up, disciplined and professional management and a high level of investor confidence, it becomes far easier for the MFIs to upscale their operations. The examples of the MFIs who have converted during the last few years into NBFCs clearly point to this.

3.2.5 Nidhi Companies

Section 620 of the Companies Act allows for the formation of a special type of company called 'Nidhi' company also known as 'Mutual Benefit Society'. Certain provisions of the Companies Act have been modified under this section. Accordingly, provisions on the service of documents, issue of additional capital, annual returns, dividends, loans and remuneration to directors and winding up processes have been modified for companies recognized as Nidhi or Mutual Benefit Society.

3.2.5.1 Acceptance of Deposits

Companies registered under this section will have all clients as members of the company. They are allowed to accept deposits so long as these are from their members only and to grant loans of up to 50 per cent of the market value of property with registered mortgages in favour of the company or up to 100 per cent of the value of jewellery, fixed deposits or government securities physically deposited with the company. The restrictions and relaxation to the Nidhi companies in terms of acceptance of deposits is provided as under:

Restrictions/ Procedures	• Should not accept or renew any public deposit except from its shareholders. • Deposits from shareholders should not be in the nature of current account (savings bank accounts are not prohibited). • Interest rate on deposits should not exceed 14 per cent per annum, and interest can be paid or compounded at rates which shall not be shorter than monthly rests. • Receipts in the prescribed manner must be furnished to depositors. • Must maintain the register of deposits in the manner laid down in the directions. • Should not pay any brokerage/commission/incentive/any other benefit, to any person for public deposits collected by that person. • Should not issue any advertisement in any form and in any media, for inviting or causing to invite deposits from its shareholders. • Security deposits received from employees should be deposited in a scheduled bank or post office in the joint names of the employee and company. The amount should not be withdrawn without the consent of the employee. The amount shall be repayable to the employee along with the interest, unless such amount or any part thereof, is liable to be appropriated by the company for any failure on the part of the employee for due performance of duties. • Submission of returns/documents to RBI as required in the directions.
Relaxations relative to NBFC rules[1]	The requirements laid down in the direction on the following aspect (and also discussed in the preceding section of the report) would not apply: • Minimum credit rating • Prohibition on accepting demand deposits • Ceiling on quantum of deposits • Accepting deposits only for a period of at least 12 months but not exceeding 60 months • Regularization of excess deposits • Renewal of public deposits • Payment of interest on overdue deposits • Joint deposits • Specification of prescribed particulars in application for soliciting deposits • Issue of advertisement and statement in lieu of advertisement • General provision regarding payment of public deposits • Information to be included in Board's report • Safe custody of approved securities

3.2.5.2 Pros and Cons of Registration as Nidhi Companies

Nidhi companies provide very liberal provisions in terms of compliance with several regulatory norms of the Companies Act and, in the sense that they are allowed to accept deposits from members, represent for MFIs a way around the problem of NBFCs needing to be registered with the RBI and then waiting two years or more to get an investment grade rating. However, the requirement is that such companies should lend only against substantial collateral makes them unsuitable as a legal form for MFIs that, by the very nature of their business, must advance loans without collateral. Further, their inability to hold any share capital brings them on par with societies and trusts in other ways.

Therefore, they are also subject to the some of the disadvantages of societies and trusts in the matter of fund mobilization and management. Further, they have the advantage of being governed by the Companies Act, 1956 and that allows them to be a corporate agent of a life insurance and a non-life insurance company. In case they are functioning as MFIs, they can avail of the FDI facility recently announced by RBI. Apart from this, they can also obtain donated equity from abroad; provided they have necessary clearance under FCRA.

3.2.6 Banking Regulation Act as Applicable to Cooperatives

Under the Banking Regulation Act only the State Cooperative Banks, District Central Cooperative Banks and Primary Cooperative Banks (Urban Cooperative Banks - UCBs) are recognized as cooperative banks. The State Cooperative Banks and Central Cooperative Banks form part of the three-tier rural cooperative credit structure promoted by the state governments and NABARD. The relevant categories of cooperative banks suitable for MFIs are Primary Cooperative Banks (UCBs).

Definition of banking: The Banking Regulation Act defines banking as "accepting, for the purpose of lending or investments of deposits of money from the public repayable on demand or otherwise, and withdrawal by cheque, draft, order or otherwise.

Definition of Primary Cooperative Bank: The Banking Regulation Act defines primary cooperative banks as "a primary credit society other than a primary agricultural credit society

(a) The primary objective of or principal business of which is the transaction of banking business;

(b) The paid up capital and reserves of which are not less than one lakh rupees; and

(*c*) The bye-laws of which do not permit admission of any other cooperative society as a member.

The Act also requires that no cooperative society, other than a cooperative bank, can use the words "bank", "banker" or "banking" in its name and no cooperative society can carry on the business of banking in India unless it uses as part of its name, at least one such word.

Urban cooperative banks are registered under the Cooperative Societies Act of the respective governments. UCBs having a multi-state presence are registered under the Multi-State Cooperative Societies Act (MSCA) 2002. The RBI is the regulatory and supervisory authority of UCBs for their banking operations, while their managerial and administrative aspects remain under the respective state governments (for banks registered under the State Cooperative Societies Act), or the central government (for the banks registered under the MSCA 2002).

3.2.6.1 Licensing requirement

To be eligible for a license, a UCB needs to be registered either under any of the state cooperative societies' laws or the MSCA 2002 and must have at least 3,000 members and a capital of ₹1 lakh.

3.2.6.2 Regulation and Supervision of UCBs

Duality of control by the Registrar of Cooperative Societies and the RBI has been pointed out in several studies as one of the primary reasons for most of the problems of the urban Cooperative Banking sector. The administration aspects of such banks are supervised and regulated by state governments while the banking operations are supervised and regulated by the central bank. This has resulted in overlapping jurisdiction of the state governments and the central bank. Moreover, a clear-cut demarcation of the financial and administrative areas for regulation is almost impossible and acts as an impediment to effective supervision. While the central bank of the country has the wherewithal under the Banking Regulation Act for dealing with crucial aspects of the functioning of commercial banks, in the case of cooperative banks it requires the intervention of the Registrar of Cooperative Societies.

Till the early 1990s, apart from basic licensing requirements, RBI did not have much of a role to play in the regulation of UCBs. Following problems in some of these UCBs, some of the prudential norms were made applicable to the UCBs also. In 1993, RBI introduced Income Recognition and Asset Classification Norms for UCBs. In 1995, the prudential exposure norms for single/group borrowers were also made applicable to them.

Subsequently, in a phased time frame, the capital adequacy norms (capital to risk weighted asset ratio) were also applied to UCBs.

The RBI has, from the quarter ended June 2004, extended its off-site surveillance system (OSS) to all non-scheduled UCBs with a deposit size of ₹ 100 crore and above. The OSS reporting system comprises eight reports. While the periodicity of one return is annual, the remaining seven are required to be submitted at quarterly intervals. Prudential concerns monitored through these returns include aspects relating to solvency, liquidity, capital adequacy, asset quality/portfolio risk profile, connected or related lending and concentration of exposures of the supervised institutions.

Box 3.2: Off-site Surveillance (OSS) System for UCBs

A revised set of off-site surveillance returns were prescribed for scheduled UCBs effective from the quarter-ended March 2004 to obtain relevant information on areas of supervisory concern, strengthen management information systems (MIS) within the scheduled UCBs and to sensitize their managements about the prudential concerns of the supervisory authority and thereby help in self regulation. The content and structure of OSS returns have been modified to reduce the volume of data submission, while enlarging the breadth and depth of information being obtained from UCBs. The UCBs now have to submit eight returns including one annual return from the quarter ended April 2004.

The scope of OSS has been extended to cover non-scheduled UCBs with a deposit base of ₹ 100 crore and above from the quarter-ended June 2004. The remaining non-scheduled UCBs would be brought under OSS in a phased manner. A revised application software developed by RBI has been installed in all the UCBs covered under OSS enabling them to submit all the regulatory and supervisory returns in electronic format. The accuracy of data submission by UCBs is ensured through validation checks built in the application package. The application package also enables regional offices to capture the data received from UCBs electronically and the data gets replicated in the server installed in the central office over the INFINET.

Source: Report on Trend and Progress of Banking in India, 2003-04 (Reserve Bank of India)

3.2.6.3 Taxation

Cooperative Banks need to pay income tax as per the provisions of the Income Tax Act. Their income is subject to taxation and any surplus over the expenses, before the payment of dividend and provision to reserves is taxed as per the provisions laid down in the IT Act.

3.2.6.4 Accounting Related Issues

The accounting principles of cooperative banks are governed by the guidelines and notifications issued by RBI from time to time. Some of the important accounting policies where the RBI has provided its guidelines are: income recognition, asset classification, dividend declaration, provision for bad debts, provision on general reserves, provisions for cash and other statutory reserves (*i.e.* Cash Reserve Ratio and Statutory Liquidity Ratio), loans to directors of the company and other stake holders, investment into capital market and securities, etc.

3.2.6.5 Receipt of Foreign Contribution

The issue of receipt of foreign contribution by cooperative banks is governed by the provisions laid down in the FCRA. A cooperative bank can avail foreign contribution upon obtaining necessary clearance from the home ministry and getting itself registered under the FCRA. The details of registration under this Act are discussed under Section 2.1.7 of this report.

3.2.6.6 Compatibility with Insurance-related Regulation

RBI allows the scheduled and licensed cooperative banks to enter into insurance business but without any risk participation. It has laid down the following conditions for them to follow in order to be part of this:

(*a*) The bank should have a minimum positive networth [real or exchangeable value of paid-up capital and reserves as defined in Section 11 of the Banking Regulation Act, 1949(AACS)] of ₹ 50 crore as per the latest RBI guidelines;

(*b*) The bank should have earned net profit for the last three years and should not have any accumulated losses;

(*c*) The gross NPA of the bank should not be more than 10 per cent;

(*d*) The bank should not have violated prudential norms including individual and group exposure norms fixed by RBI/NABARD;

(*e*) The bank should have complied with the instructions issued by RBI/ NABARD on loans and advances to directors/relatives, firms etc;

(*f*) No premium collection accounts will be allowed to be opened with the bank and hence, premium collected should be directly paid to the insurance companies;

(*g*) The bank should comply with regulations of IRDA for acting as a corporate agent;

(h) The bank should submit an undertaking to the effect that banking business will not in any way get contaminated/affected on account of acting as an agent of insurance companies;

(i) The bank should not adopt any restrictive practice of forcing its customers to go in only for a particular insurance company in respect of assets financed by the bank and the customers should be allowed to exercise their own choice;

(j) The bank should obtain prior permission from the concerned regional office of RBI before taking up the insurance agency business. The application should be routed through the concerned regional office of NABARD with their recommendation.

3.2.6.7 Advantages and Disadvantages of Registration as a Cooperative Bank

Advantages

(i) Ability to accept demand and time deposits;

(ii) Low-capital requirement; and

(iii) Access to clearing-house functions

Disadvantages

(i) Duality of control. Problems are greater in case the relevant state does not have a MACS Act;

(ii) Poor public image of cooperative banks following recent governance and management problems resulting in the failure of a number of such banks; and

(iii) On account of these failures it is now also extremely difficult to obtain new licenses from the RBI.

3.2.7 State Acts on Mutually Aided Cooperative Societies

Many states have brought amendments in their cooperative societies acts thereby making the functioning of cooperatives more autonomous and free from government interference. This act was first passed in the Andhra Pradesh and has functioned as a model for many states which have brought fresh legislations to effect changes in their state cooperative society acts. Some of the states have gone a step further and brought in reforms that have far reaching implications for the functioning of MFIs. Noted among these states are Orissa, Bihar, Madhya Pradesh and Karnataka. The Orissa act allows the SHGs to become members of cooperative societies while the other state acts do not have this provision.

AP Mutually Aided Cooperative Societies' Act (APMACS Act) 1995

AP MACS Act of 1995 has emerged as the popular choice for an organizational form for many MFIs in Andhra Pradesh. This law has also worked as a model act for many other states that have either amended or changed their cooperative societies laws. The acts of other states have more or less similar provisions as the APMACS Act.

3.2.7.1 Purpose

The purpose of enactment of the APMACS Act, 1995 was to promote institutions which are based on cooperative principles, are self-reliant, accountable and are formed for the fulfillment of the perceived needs of ordinary people. This was in contrast to the AP Cooperative Societies Act, 1964 which encourages the promotion of largely government controlled and dependent cooperative institutions.

3. 2.7.2 Bye-laws

As the objective of the APMACS Act is to promote independent organizations, it prescribes a minimum set of regulations and restrictions. As such, most of the actions of a cooperative society registered under this Act are defined by its bye-laws. The Act requires the bye-laws to be prepared by the members of the society and is generally non-prescriptive about the content of these bye-laws. As per the Act, the bye-laws need to adhere to cooperative principles and contain names, objectives, eligibility as well as termination criteria for membership, division of profit and other details that govern the relationship of the members amongst themselves.

3.2.7.3 General Body

Subject to the provisions of the Act, the ultimate authority of a cooperative society vests in its general body, which consists of all its members . The cooperative may decide to have a smaller representative general body for more effective decision-making, in case of a large and dispersed membership. The constitution of the representative general body needs to be specified in the bye-laws of the society. The general body of a cooperative society has the powers to amend bye-laws, election and removal of directors, appointment and removal of auditors, review as well as approval of annual accounts and budgets and dissolution of the cooperative society.

3.2.7.4 Board of Directors

The executive authority for running the business of a cooperative society is vested in a Board of Directors. The appointment of the Board of Directors of a cooperative society must be in accordance with its by-laws, though the Act prescribes certain qualifications and disqualifications for eligibility. Some of these are:

Qualifications

(a) Must be a voting member of the society for at least two years (only for societies in existence for more than two years); and

(b) Must have attended two general body meetings of the cooperative society held immediately preceding the elections (only for societies in existence for more than two years).

Disqualifications

(a) Members who have, at any time, lost their voting right at any time;

(b) A director who absents himself from more than three consecutive Board meetings, without leave of absence; and

(c) A director who absents himself from the general body meeting without leave of absence.

3.2.7.5 Mobilization of funds

As per the Act a cooperative society may mobilize funds in the shape of share capital, deposits, debentures, loans and other contributions from its members to such extent and under such conditions as may be permissible under the by-laws of the cooperative society A cooperative society may also mobilize funds in the shape of deposits, debentures, loans and other contributions from other individuals and institutions, to such extent and under such conditions as may be permissible under the by-laws. A cooperative society shall not accept share capital from the government but may accept other funds or guarantee from the government on such terms and conditions as are mutually agreed upon through a memorandum of understanding.

3.2.7.6 Investment of funds

A cooperative society may invest or deposit its funds in any non-speculative manner outside its business.

3.2.7.7 Taxation

Like any other cooperative bodies, MACS are also subject to taxation on their annual surpluses as per the guidelines of the Income Tax Act.

3.2.7.8 External Commercial Borrowing

Following the latest notification of RBI (vide RBI/2004-05/434 A.P.(DIR Series) Circular No. 40), cooperative societies involved in microfinance are eligible to access ECB, provided they have a successful credit history for a period of not less than three years with a scheduled commercial bank; they have a certificate of due diligence on fit-and proper status of the board/committee of management of the borrowing entity from the designated authorized dealer (AD); and they satisfy the definition of NGO, to be explained by RBI in future. As per the above notification, RBI allows NGOs involved in microfinance activities to access ECB.

3.2.7.9 Compatibility with Insurance-related Regulations

Mutually-aided cooperative societies can provide insurance coverage to their clients by having a master policy covering all their clients. However they cannot obtain corporate agency of a life or non-life insurance company as they are not incorporated as a company under the Companies Act, 1956.

3.2.7.10 General administrative guidelines

The Act prescribes guidelines regarding some of the administrative issues also. Some of these are:

a. **Election:** The society will ensure that the Board of Directors is elected before the expiry of the term of the existing directors. These elections need to be in accordance with the by-laws.

b. **Meetings:** The general body needs to meet at least once a year and the board of directors at least once in a year. The by-laws of the society may prescribe a higher frequency of meetings.

c. **Accounts and records:** Some of the important books of account that the Act requires of all cooperative societies are:

    1. Cash book;
    2. Accounts of assets and liabilities;
    3. Accounts of all purchases and sales of goods;
    4. Updated register of members;
    5. Copies of audit reports and special audits; and

d. **Audit:** A cooperative society may get its accounts audited by a qualified chartered accountant or by any auditor from the office of the Registrar.

3.2.7.11 Submission of returns

A cooperative society needs to submit the following returns within 15 days of conduct of the annual general meeting:

(*a*) Annual report of activities;
(*b*) Annual audited statements of accounts with auditor's report;
(*c*) List of members at the close of the year under reporting with services provided to each member;
(*d*) Statement on the disposal of surplus or on the allocation of deficit;
(*e*) List of names of directors, their addresses and their terms of office; and
(*f*) Compliance reports relating to audit, special audit and inquiry, if any.

3.2.7.12 Functions and Powers of the Registrar

The Registrar does not have much power in the case of cooperative societies registered under the APMACS Act. The Registrar cannot refuse registration of a cooperative and its by-laws if they are in consonance with the Act. He also registers all amendments to the by-laws if they are in accordance with the law.

(*i*) *Power to conduct general meetings:* Registrar has the powers to convene general body meeting where a board fails to do so in the stipulated time.

(*ii*) *Power to order inquiry:* Registrar can order an inquiry into the affairs of the cooperative on an application of a federation to which the cooperative society concerned is affiliated, or a creditor to whom the cooperative society is indebted.

(*iii*) *Recommend dissolution to a tribunal:* Registrar can recommend dissolution of a cooperative society, to a tribunal set up for the purpose by the state government, when he has reasonable grounds to believe that the society has

- obtained its registration by fraud or mistake;
- is serving illegal purposes;
- has willfully, after notice by the Registrar, violated any of the provisions of this Act or its by-laws;
- is no longer operating in accordance with principles of cooperation and the provisions of this Act;
- has not commenced business within two years of the date of registration, or
- has not carried on business for the past two consecutive years.

3.2.7.13 Advantages and Disadvantages

Advantages

(i) Ease of formation and management;

(ii) Ability to accept deposits;

(iii) Minimal interference from regulators - unlike cooperative societies formed under the traditional cooperative laws; and

(iv) Client centric and therefore mobilize equity also from the clients.

Disadvantages

(i) Virtually no regulation. May also encourage risk-taking behaviour by the promoters/managers resulting in the collapse of the institution;

(ii) May affect scaling up, as cooperatives are not perceived as a trustworthy organizational form by the formal financial sector; and

(iii) Unlike other entities like societies or trusts, they do not have the advantage of tax- exemption.

3.2.8 Producer Companies

Amendments in the Companies Act 1956 have provided for a new type of company to be owned by the primary producers.

3.2.8.1 Formation of a Producer Company and its Registration

The Act defines the primary producer as any person engaged in any activity connected with or relatable to any primary produce. As per the proposed amendment, primary produce means

(a) Produce of farmers, arising from agriculture (including animal husbandry, horticulture, floriculture, pisciculture, viticulture, forestry, forest products, revegetation, bee-raising, and farming plantation products), or from any other primary activity or service which promotes the interest of the farmers or consumers; or

(b) Produce of persons engaged in handloom, handicraft and other cottage industries;

(c) Any product resulting from any of the above activities, including by-products of such products;

(d) Any product resulting from an ancillary activity that would assist or promote any of the aforesaid activities or 'anything ancillary' thereto; and

(e) Any activity which is intended to increase the production of anything referred to in sub-clauses 'a' to 'd' or improves the quality thereof.

The Act states the objects of the producer company as

(a) The objects of the producer company shall relate to all or any of the matters as may be prescribed;

(b) Every producer company shall deal primarily with the produce of its active members for carrying out any of its objects specified in the rules (as prescribed above in (a).

Any 10 or more individuals, each of them being a producer or any two or more producer institutions, or a combination of 10 or more individuals and producer institutions desirous of forming a producer company having its object specified in the rules prescribed under Section 209 and otherwise complying with the requirements of this part and the provisions of this part and the provisions of this Act in respect of registration, may form a company as a producer company under the Companies Act.

Application for registration shall be presented to the Registrar of Companies of the state in which the registered office of the producer company is, as stated by the memorandum of association, to be situated

(a) The memorandum of the producer company; and

(b) Its articles, duly signed by the subscribers to the memorandum.

If the Registrar is satisfied that all the requirements of this Act have been complied with in respect of registration and matters precedent and incidental thereto, he shall, within 30 days of the receipt of the documents required for registration, register the memorandum, the articles and other documents, if any, and issue certificate of incorporation under this Act. A producer company so formed shall have the liability of its members limited by the memorandum to the amount, if any, unpaid on the shares respectively held by them and be termed a company limited by shares. On registration, the producer company shall be incorporated as a private limited company provided the provisions of the private limited companies with reference to the limit to the number of members shall not apply to a company registered under this part and that under any circumstances whatsoever, the producer company shall not become a public limited company.

A multi-state cooperative society with objects not confined to one state can also make an application to the Registrar for registration as a producer company. The procedure for the registration of multi-state cooperative society would be same as any other entity. Once registered as a producer company, the provisions of the new Act shall govern the new entity. Also, the Registrar of Companies who registers the company shall forthwith intimate the Registrar with whom the erstwhile multi-state cooperative society was earlier registered for appropriate deletion of the society from its register. All properties and assets, movable and immovable, of, or belonging to, the multi-state cooperative society as on the transformation date, shall vest in the producer company. The same would happen to the rights, debts, liabilities, interests, equity, privileges and obligations of the producer company.

3.2.8.2 Share Capital, Special User Rights, Transferability of such Rights and Benefits to Members

The share capital of a producer company shall consist of equity shares and the shares held by a member, shall as far as possible, be in proportion to the patronage of that company. The producers who are active members may, subject to the provision in the articles, have special rights and the producer company may issue appropriate instruments to them in respect of such special rights. The equity shares of the producer companies would be transferable to an active member of the company subject to the prior approval of the board. Any producer company may, subject to such conditions as may be prescribed, issue bonus shares by capitalization of reserves in proportion to the shares held by the members on the date of such issue. The new Act also provides for the disbursement of dividend and patronage bonus as per the provisions in the article.

3.2.8.3 Membership, voting rights and general meeting

Every producer company shall in each year, hold, a general meeting, as its annual general meeting (AGM) and not more than 15 months shall elapse between the date of one AGM and that of the next. The various procedures of the AGM of the producer company would be governed by the provisions under the Company Act. However the voting rights of the members would be different from those of the Company Act. In a case where the membership consists solely of individual members, the voting rights shall be based on single vote for every member, irrespective of his shareholding or patronage of the producer company. In a case where the membership consists of producer institutions only, the voting rights of producer institutions shall be based on the basis of the shareholding by such producer institutions during the first year of registration and on the basis of the participation in the previous years' business during the subsequent years. In a case where the membership consists of individuals and producer institutions, the voting rights shall be computed on the basis of a single vote for every member.

3. 2.8.4 Finance, Accounts and Audit

All the provisions of the Companies Act would apply to the producer companies with regard to the financial management, accounts and book keeping and audit. Every producer company shall have internal audit of its accounts by a chartered accountant at such intervals and in such manner as may be prescribed. Every producer company would need to maintain a reserve, in addition to any reserve maintained by it, as may be specified in the articles. In case the producer company does not have a surplus in any financial year for transfer to the reserves, the contribution to the reserve shall be shared amongst the members in proportion to their patronage in the business of that company in that year.

3. 2.8.5 Donations, Loans and Investments

A producer company may by special resolution make donation or subscription to any institution for the purpose of promoting social and economic welfare of producer members or producer or general public but not for any political purpose whether directly or indirectly, subject to such conditions as may be prescribed. It can also provide financial assistance to the members provided that any loan or advance to any director or his relative shall be granted only after the approval of the members in the general meeting.

It shall apply its reserves for investment in the approved securities, as may be prescribed, or by a special resolution or for promotion of its objectives by acquiring the shares of another producer company whether by formation of a subsidiary company or by joint venture or in any other manner. Any producer company, either by itself or together with its subsidiaries may invest, by way of subscription purchase or otherwise, shares in any other company, other than a producer company, for an amount not exceeding such proportion of the aggregate of its paid-up capital and free reserves as may be prescribed.

3.2.8.6 Amalgamation, Merger or Division to Form New Producer Companies

A producer company may, by a resolution passed at its general meeting, decide to transfer its assets and liabilities, in whole or in part, to any other producer company, which agrees to such transfer by a resolution passed at its general or special meeting, for any of its objects. It may also decide to divide itself into two or more new producer companies. Similarly two or more producer companies may, by a resolution passed at any general meeting of its members, decide to amalgamate and form a new producer company or merge one producer company with another. Every such resolution needs to have the approval of a majority of its members with at least two-third of its voting rights attendance. In the case of transfer of whole assets of one producer company to another, the registration of the first producer company shall stand cancelled and that company shall be deemed to have been dissolved and shall cease to exist forthwith as a corporate body. When the two producer companies amalgamate into a new producer company and the new producer company is duly registered with the Registrar, the older producer companies shall be deemed to have been dissolved and shall cease to exist. The amalgamation, merger or division of companies under the this Act, shall not in any manner whatsoever affect the pre-existing rights or obligations and any legal proceedings that might have been continued or commenced by, or against, the concerned resulting company, or merged company, as the case may be.

3. 2.8.7 Taxation

At this point in time, the Income Tax Act does not have any special relief for the proposed producer companies. Upon notification from Parliament, it is quite possible that the producer companies would have similar treatment under the Income Tax Act as cooperatives or companies.

3. 2.8.8 External Borrowing

With the recent notification of RBI regarding ECB by the NGOs (vide notification no. RBI/2004- 05/434 A.P.(DIR Series) Circular No. 40 dated 25 April 2005), it is quite likely that MFIs registered as producer companies would also have access to cheaper external funds for borrowing provided they meet the conditions laid down in the above notification.

3. 2.8.9 Compatibility with Insurance Regulations

With its member-focused approach, producer companies are likely to provide a good opportunity for insurance companies to be tapped as channels for product distribution. It is quite likely that they will have similar status as cooperative societies and act as channels for the distribution of insurance products.

3.2.8.10 Advantages and Disadvantages for MFI Registration as Producer Companies Advantages

(i) Simple process of registration;

(ii) Simple record-keeping and even simpler regulations;

(iii) Low possibility of interference by the regulator; and

(iv) Member-focused approach.

Disadvantages

(i) As a member-owned organization, bound to face the problems related to professional management and governance; also the profit-centric approach necessary for the long term sustainability of the microfinance programme may take a back seat;

(ii) Tax exemption for surpluses not likely; and

(iii) Commercial investors could shy away - as equity investment may not be possible in such organizations. Also debt investment would have its limitations; as such organizations by their design would have problems in mobilizing equity.

3.2.9 Institutional Issues with MFIS in India

The above discussion clearly underlines the regulatory limitations of the legal framework for the microfinance sector in India. Though many innovative initiatives have been undertaken by Indian MFIs over the past five to seven years, their efficacy has been limited and their operations hamstrung by the absence of a supportive regulatory environment.

MFIs would like to provide financial services to low-income clients with the objectives of providing financial services to large numbers of low-income clients, and ensuring long-term sustainability, while operating within the boundaries of the legal framework.

As the discussion above suggests, the problems of these MFIs are manifold. While societies and trusts provide an initial breakthrough to start microfinance operations; these structures are not appropriate for supporting the scaling-up of operations. MACS and producer companies may offer better options, but their lack of professional governance structures raises serious doubts in the minds of investors. Mobilizing equity support for such institutions is also a problem. These factors hamper the scaling-up of operations as these entities would find it hard to mobilize the substantial volumes of funds required for expansion. Section 25 companies also suffer from such lacunae - though their problem is related mainly to equity mobilization and thereby to the leverage capacity of such institutions. This brings the discussion to the two remaining options - NBFCs and cooperative banks. Following the problems with the cooperative banks over the last few years, RBI has been extremely reluctant to provide fresh licenses for such entities. NBFCs have a similar problem; RBI has been very reluctant to offer registration to new NBFCs for one reason or another, compounding the problem of raising the substantial capital requirement of ₹ 2 crore.

The RBI has recently shown some willingness to provide registration to NBFCs established by MFIs. NBFCs established by Spandana in Guntur and SKS in Hyderabad have been registered during 2004-05. However, the problem of ownership structure is an issue for NGO-MFIs even if RBI starts being generous in providing registration to such entities. Societies are not allowed to promote companies since the Societies Act bars them from investing in equity. Some of the leading MFIs have found a 'noble' solution in the formation of Mutual Benefit Trusts (MBTs). Societies promote these MBTs through their clients and the MBTs, in turn, invest in the equity of NBFCs. The parent NGO-Societies could provide grant support to MBTs and this money may be used for equity investment in the NBFC. This appears to be a very safe solution despite MBTs not having the benefit of tax saving. However, some of these MBTs are using clients' savings also for investing in NBFCs as equity.

The ICICI Bank has tried to bring some innovation in order to find a solution to this problem. According to their arrangement, they enter into a partnership with an existing NGO-MFI or Section 25 company. While ICICI Bank provides the funds for client mobilization and lending, these institutions, using their community mobilization skills, form the groups and function as an agency for the bank. These loans to the clients or groups remain on the accounts of ICICI Bank and the MFIs provide the front-end services to the clients in return for a commission. This seems to be meeting both the ends - while the sanctity of the legal system prevails, clients also benefit from receiving credit services. However, since the ICICI Bank is currently the only bank applying this "partnership" model, this arrangement - which by-passes the capital adequacy norms that normally inhibit investors - heightens the dependence of the institution on a single entity and limits the ability of the institution to engage in product innovation since aspects like repayment conditions and interest rate to retail clients are specified by the agreement with the bank. A summary of the issues affecting the choice of institutional form by MFIs is presented below (see Table 3.8).

Table 3.8: Features of Legal Forms for Microfinance Institutions in India

Legal Entities	Registration	Start-up Capital	Regulatory Requirements	Investor Confidence		Deposit Mobilization	Governance	Scaling up of Operations
				Equity	Debt			
Societies	Very easy and can be obtained within weeks	No minimum requirement	Cursory	Not possible however grants could be provided	Low	Not allowed as per the guidelines of RBI though the central bank turns a blind eye if there are no problems[1].	Mostly informal	May be problematic; as the funds for up-scaling would be hard to raise and ineffective management could hamper growth

contd...

Table 3.8: Contd...

Legal Entities	Registr-ation	Start-up Capital	Regulatory Requir-ements	Investor Confidence		Deposit Mobiliz-ation	Gover-nance	Scaling up of Oper-ations
				Equity	Debt			
Trusts	Same as above	Same as Above	Cursory	Same as above	Low	Not allowed in public charitable trusts - but can be mobilized in the case of private and mutually-aided trusts	Mostly Informal	Same as above
Section 25 Companies	Easy, but not as easy as those of trusts and societies; especially for an existing company to convert into a Section 25 company	Not required	Exempted from most of the regulatory requirements of NBFCs	Not facilitative as Sec 25 companies are not allowed to declare dividends	High	Not allowed	Relatively better and more formal	Possible to scale up due to professional governance and high lender confidence. Problems may still come up due to lack of equity from commercial investors as the leveraging capacity of the organization would be affected.
NBFC	Difficult - may take up to a year as RBI has been very cautious in providing licenses	₹ 2 crore	Substantial – especially if the entity is allowed to mobilize deposits	High as the entry and exit (perceived) routes are easier	High	Allowed if it meets certain criteria - especially the need for an investment grade rating from a recognized agency[2]	Likely to be professional	It provides very sound background for scaling up of the operations as both the investor confidence and the management strengths are normally good with these entities.

Contd...

Table 3.8: Contd...

Legal Entities	Registration	Start-up Capital	Regulatory Requirements	Investor Confidence		Deposit Mobilization	Governance	Scaling up of Operations
				Equity	Debt			
Cooperative Societies (MACS)	Easy	Nominal	Nominal	Low; due to lack of a clear exit route	Moderate	Allowed from members	Doubtful - as it comes from the board members who are mostly beneficiaries. Raises issues of corporate governance as well.	Relatively easier. However, here again leverage can be a problem as there is no commercial equity investment
Cooperative Banks	Extremely difficult - RBI has not been entertaining any request for registration	Moderate – ₹1 lakh	Very high. Also subject to multiple regulations – by RBI and the state Registrar of Cooperatives	Low; due to lack of a clear exit route	High	Allowed	Same as above	Relatively easier. But better then MACS as the entry capital requirement itself is very high.
Producer Companies	Relatively easier	No minimum ceiling	Nominal	No external investors	High; as these entities would have more freedom in operations than cooperatives	Allowed from the members.	Same as above; as the basic character of such companies would be similar to that of cooperatives	Same as MACS.

3.3.0 Recent Initiatives in the Field of Microfinance Regulation

A number of recent initiatives have dealt with the issue of microfinance regulation. This can be taken as a sign of growing interest in the field. Below, some of the most important initiatives are briefly summarized. Furthermore, some of the regulatory changes triggered by this debate are listed.

3. 3.1 NABARD Task Force

In the late 1990s, it was felt that a suitable national policy framework was essential for the orderly development of the microfinance sector in the country. Various critical issues such as ways to scale up microfinance programmes in India, the division of roles between existing mainstream financial institutions, development banks and newly emerging microfinance institutions (MFIs) and the need for an appropriate policy and regulatory framework for their operations were engaging the attention of microfinance practitioners, the Government and the RBI. It was at this juncture that in

November 1998 a high-powered Task Force on Supportive Policy and Regulatory Framework for Microfinance (henceforth referred to as the Task Force) was set up by NABARD at the instance of RBI. The Task Force comprised senior officials from GoI, RBI, NABARD, banks and chief executives of prominent NGOs implementing various microfinance models in the country. The objectives of the Task Force were, among others, to come up with suggestions for a regulatory framework that brings the operations of the MFIs into the mainstream, to assess the possible role of self-regulatory organizations and to explore the need for a separate legal framework for microfinance. To date, this has been the most comprehensive study about legal issues for microfinance.

The Task Force (1999) made various recommendations, based on which RBI issued a number of important instructions to microfinance practitioners. For the first time, a definition of microfinance was agreed upon. The definition is broad enough to allow for different types of institutions offering various types of financial services .

With regard to the hitherto unregulated microfinance NGO sector, setting up a system of self-regulatory organizations (SROs) features prominently among the recommendations. The Task Force recommends a classification of MFIs according to the question of whether deposit facilities are offered or not and if so, what kind of facilities these are. Both the amount of savings mobilized (below or above a pre-determined cut-off limit) and the origin of savings (from clients or from the general public) determine the extent of external control by a supervisory authority. The strong reliance on self-regulation even for NGOs mobilizing deposits from the public does not go in line with more recent recommendations from other international experts, who stress the enforcement problems of a self-regulatory framework and the potential conflicts of interest of the SROs . Basing the extent of regulation on the origin of savings is similar to van Greuning's (*et al.*) approach of looking at the liability side of the balance sheet to determine the need for external regulation (van Greuning, Gallardo, and Randhawa, 1999).

3. 3.2 Other Working Groups on Regulatory Issues

Following a meeting at the Prime Minister's Office in October 2001, seven working groups comprising government, NGOs and banking sector representatives were set up, one of them with the mandate of exploring the legal and regulatory challenges currently faced by the microfinance sector in India. In December 2002, the reports of six of the seven working groups were finalized. The Indian microfinance association Sa-Dhan - as one of the members of the group looking into regulatory issues - prepared a report summarizing the group's results (Sa Dhan 2002a). This report gives clear recommendations regarding how to make the current legal framework more microfinance-friendly. Some of these need immediate attention; some are for the longer term. In addition, the document indicates a way forward for the debate with a strong emphasis on gathering information about the sector as a precondition for better regulation.

RBI has continuously watched the progress of microfinance. In October 2002, the RBI constituted four informal groups for addressing various aspects of microfinance including regulatory arrangements and funding issues. Senior bankers, NGOs and prominent microfinance practitioners and all stakeholders are represented in these informal groups. The recommendations of these informal groups have recently been made public. They are of great importance as some of the changes favoured by a number of stakeholders require the attention and involvement of the RBI.

Major recommendations of these informal working groups are to create a separate category of non-deposit taking NBFCs with a lower capital requirement of ₹ 2.5 million (US$ 56,000), and to require NGOs and federations of SHGs either to wind up their savings business and transform into Section 25 companies or to transform into NBFCs or MACS.

3.3.3 Recent Regulatory Amendments for Microfinance

Due to the increased attention to regulatory issues for microfinance, several legal and regulatory changes were initiated over the recent years. The most important changes are a number of instructions issued by the RBI, which for the first time specifically deal with microfinance .

3.3.3.1 Mainstreaming of SHG-Bank Linkage Model

A Working Group on NGOs and SHGs set up by RBI in 1995 looked into various aspects of the linkage between informal SHGs and banks. According to the model, formal banks lend either directly or through an intermediary such as an NGO or an SHG federation to groups of poor people, mostly women. Based on the recommendations of the Working Group, RBI in 1996 issued instructions to the banks to treat the SHG-bank linkage model as part of their mainstream lending .According to the circular, lending to SHGs or to NGOs for on-lending to SHGs is taken as part of their directed lending .Furthermore, SHGs are allowed to open savings accounts with banks. The operational guidelines, earlier issued by NABARD in 1992 for lending to SHGs, envisaged complete flexibility for the banks in regard to lending norms like interest margin, amount to be lent, use of collateral security, repayment period etc . These guidelines are well accepted by the banking system for providing financial services to the poor. Over the years, the linkage programme has made very good progress.

In April 1999, RBI set up a 'Micro Credit Special Cell'. This demonstrates RBI's increased interest in the microfinance sector.

3.3.3.2 Exemptions for Non Deposit-Taking Section 25 Companies

As regards Non-Banking Financial Companies, the RBI issued instructions on 13th January 2000 to the effect that if NBFCs are engaged in 'micro-financing activities', registered under Section 25 of the Companies Act, 1956, and not accepting public deposits, they would be exempt from the purview of Section 45-IA (registration and minimum net owned funds), 45-IB (maintenance of liquid assets) and 45-IC (transfer of profits to reserve fund) of the RBI Act, 1934 .These instructions essentially created a new legal form for providing microfinance services without having to comply with the strict regulatory requirements for NBFCs. Section 25 companies are not subject to any capital or liquidity requirements. According to Section 25 of the Companies Act, they must use potential profits for promoting its objects .Although they are not explicitly exempted from supervision by the RBI, in practice they only report to the Registrar of Companies. The price for this is that they are not permitted to go into deposit-taking business.

3.3.3.3 Allowing More Flexibility for Banks Offering Microfinance Services

In a circular letter to the banks dated 18th February 2000, RBI recognized the definition of microfinance as proposed by the Task Force . The central bank advised that for mainstreaming microfinance and to enhance the outreach of microfinance service providers, banks are free to choose their own model for extending microcredit. Further, microcredit extended by banks to individual borrowers directly or through any intermediary would be taken as part of their priority sector lending. Banks are expected to make microcredit an integral element of their corporate credit plans, which are to be monitored by the central bank. To increase microlending, the RBI recommended simplification of loan documentation and decentralization of the loan decision by delegating more power to branch managers. Other than for small loans given directly by banks to small borrowers, the interest rate for loans to MFIs and for MFIs lending to the ultimate borrower were fully liberalized

3.3.3.4 A Self-Regulatory System for MFIs has not Found Sufficient Support

There were some initiatives from the microfinance sector to build a consensus among the industry and to advocate for putting in place a broad-based system for regulation of the microfinance sector and, more particularly, the MFIs in the country. The prospect of instituting a self-regulatory regime as proposed by the Task Force was discussed very actively .Yet as of now, the RBI has not delegated any regulatory power to a self-regulatory organization. Despite their initial enthusiasm and the recommendations of the Task Force, the MFIs have recently lost interest in establishing SROs. In the recommendations of the informal group set up by

RBI, SROs do not figure at all.

Against this background, the following chapters develop different scenarios that have taken shape during the discussions with different stakeholders in the industry. Keeping in view the broad objective of providing financial services to the hitherto un-served rural poor in the shortest possible time, the different scenarios bring out the prospect of achieving this objective in different ways. The focus is on the prevailing status of regulation in each scenario and the need and scope for change there under.

3.3.3.5 Regulatory Regime

The picture of the regulatory regime below relates to the general regulatory powers applicable to different micro credit entities in India. This means that while there are no specific regulations applicable to micro credit, there are regulations applicable to the providers thereof. These regulations may be summed in the table below:

Categories of Providers	Legal Framework governing their activities
(a) Domestic Commercial Banks, Public Sector Banks, Private Sector Banks & Local Area Banks	(i) RBI Act 1934 (ii) BR Act 1949 (iii) SBI Act (iv) SBI Subsidiaries Act (v) Acquisition & Transfer of Undertakings Act 1970 & 1980
(b) Regional Rural Banks	(i) RRB Act 1976 (ii) RBI Act 1934 (iii) BR Act 1949
(c) Co-operative Banks (i) Co-operative Societies Act	(ii) BR Act 1949 (AACS) (iii) RBI Act 1934 (for sch. banks)
(d) Co-operative Societies	(i) State legislation like MACS
(e) Registered NBFCs	(i) RBI Act 1934 (ii) Companies Act 1956
(f) Unregistered NBFCs	(i) NBFCs carrying on the business of a FI prior to the coming into force of RBI Amendment Act 1997 whose application for CoR has not yet been rejected by the Bank (ii) Sec. 25 of Companies Act
(g) Other providers like Societies, Trusts, etc.	(i) Societies Registration Act, 1860 (ii) Indian Trusts Act (iii) Chapter IIIC of RBI Act, 1934 (iv) State Moneylenders Act

3.3.3.6 The Micro Financial Sector (Development and Regulation) Bill, 2007

In March 2008, the Finance Minister tabled the bill in the Lok Sabha, which was then referred to the Lok Sabha Standing Committee on Finance.

3.3.3.6.1 Key features of the Bill

It provides for the regulation and supervision of cooperative societies and non- profit institutions (including societies and trusts) that are providing microfinance. The regulator for all of these institutions would be the National Bank for Agriculture and Rural Development (NABARD).

- Microfinance is defined to include loans, savings, insurance and pension services. Loans cannot exceed more than ₹ 50,000 (₹150, 000 for housing purposes).
- The bill defines an MFO (Micro Finance Organization) as any organization that provides microfinance services and includes societies, trusts and cooperative societies.
- All MFOs that accept deposit from 'eligible clients' need to be registered with
- NABARD. Minimum experience of three years and minimum net owned fund of ₹ five lakhs has been fixed as a condition for registration.
- NABARD has to specify the form and manner of accounting of business operations of microfinance organizations.
- Registered MFOs will be required to submit reports to the regulator.
- MFOs will also be subject to inspection by the regulator in case of complaints of harmful practices.
- It also proposes to set up a corpus fund called the Micro Finance Development and Equity Fund for the development of the sector
- Every MFO that accepts deposits has to create a reserve fund by transferring a minimum of 15% of its net profit every year.
- The central government may establish a Micro Finance Development Council to advise.

NABARD on formulation of policies related to the micro financial sector. The Lok Sabha Standing Committee on Finance raised many objections to this bill and suggested that the Bill needs to be dropped and an appropriate Bill needs to be evolved.

3.3.3.7 The Andhra Pradesh Micro Finance Institutions (Regulation of Money Lending) Rule, 2010

In exercise of the powers conferred under sub-section (1) of section 23 of the Andhra Pradesh Micro Finance Institutions (Regulation of Money Lending) Ordinance 2010 (Andhra Pradesh Ordinance No.9 of 2010) the Governor of Andhra Pradesh hereby makes the following rules:

1. Short title and commencement

(1) These rules may be called the Andhra Pradesh Micro Finance Institutions (Regulation of Money lending) Rules, 2010.

(2) These rules shall come into force from the date of publication in the official Andhra Pradesh Gazette.

2. Definitions:- In these rules unless the context otherwise requires, -

(a) "Effective rate of interest (r)" shall be the effective rate of interest as specified in column (17) of Form - 1.

(b) "Forms" means Forms appended to these rules.

(c) "Ordinance" means the Andhra Pradesh Micro Finance Institutions (Regulation of Money Lending) Ordinance, 2010.

(d) "Pawn" includes Ration Card, any Health Scheme Cards, property title documents, household articles, pension cards, collateral security; or any benefit provided by Government to the borrower.

(e) "Slum Level Federation" means and includes federation of Self Help Groups in a slum or habitation facilitated by the Mission for Elimination of Poverty in Municipal Areas (MEPMA).

(f) "Village Organization" means and includes federation of all Self Help Groups in a village or habitation facilitated by Society for Elimination of Rural Poverty (SERP).

(g) "Words" and "Expressions" used but not defined in the Ordinance or under rules thereof shall have the same meaning assigned to them under any other law relevant with the objectives sought to be achieved under the Ordinance.

3. Application for Registration

(1) Every Micro Finance Institution operating in the State as on the date of the commencement of the Ordinance shall, apply to the respective Registering Authorities in rural and urban areas seeking Registration in Form-I furnishing the details.

Provided that Micro Finance Institution operating in more than one District shall file separate applications for Registration in each District.

4. Verification by the Registering Authority

(1) The Registering Authority shall verify whether the details submitted by the Micro Finance Institution are in accordance with the requirements of Rule 3 or not.

(2) In case, the application is not in accordance with the requirements of Rule 3, it shall be returned for rectification duly stating the reasons for the same.

(3) If the application is in accordance with the requirements of rule 3, the same shall be published in the village/municipal ward at the office of the Village Organization / the Slum Level Federation, for information and to get feed-back, if any.

5. Consequences of furnishing incorrect information

If any of the information furnished in Form - 1 is found to be incorrect, the Registering Authority shall also launch prosecution under the relevant provisions of the Indian Penal Code 1860.

6. Undertaking to be given by Microfinance Institution before according Registration

Every Microfinance Institution shall give an undertaking in Form-2 signed by the Chief Executive Officer or a person of equivalent designation of the Micro Finance Institution.

7. Time limit for Registering

If all the details furnished by the Microfinance Institution are in accordance with the requirements of Rule 3 and Rule 6, Registration in Form - 3 shall be issued within seven working days from the date of submission of Form-1 and Form - 2.

8. Renewal of Registration

(1) Every Microfinance Institution registered as stated above and desirous of continuing operation of money lending after completion of the one year period, shall apply for renewal of Registration as per the provisions of sub-section (4) of Section 3 of the Ordinance in Form - 4 listing therein name-wise details of the borrowers along with the purpose for which loan has been given and the interest rate charged.

(2) Every application for renewal shall be verified by the Registering Authority by publishing the relevant details in the villages/wards through the Village Organization or Slum Level Federation or any other person or agency authorized by the Registering Authority inviting objections if any, for such renewal within such date as shall be specified.

(3) All objections received from the Self Help Groups or the Village Organizations' or the Slum Level Federations, shall be got verified by the Registering Authority to satisfy himself about the compliance of the Micro Finance Institution to the provisions of the Ordinance and these rules.

(4) If any of the details furnished by the Micro Finance Institution in the application filed for renewal, either based on the verification done by the Registering Authority or based on complaints received from public, are found to be incorrect, or if the Micro Finance Institution is found to have violated any of the conditions of the Ordinance, the Registering Authority shall refuse renewal in writing stating the reasons therein.

(5) The Registering Authority shall also launch prosecution under the relevant provisions of the Indian Penal Code 1860 in all cases of furnishing false information.

9. Register of Microfinance Institutions operating in a district

(1) Each RA shall maintain the list of Micro Finance Institutions registered in the district in Form - 5, which shall be available for inspection by general public during the office hours.

(2) The relevant particulars of the Form - 5 concerning the Gram Panchayat or the slum area shall be published at least twice a year on the notice board of the concerned Gram Panchayat or the Ward Office concerned, apart from exhibiting them on the website.

10. Cancellation of the Registration

(1) The Registering Authority shall cancel the Registration accorded to an Micro Finance Institution on violation of any condition of the Ordinance, or any of these Rules or any clauses of the points of the Undertaking submitted in Form - 2, or on any complaint filed and found to be valid under Section 14, or suo moto under Section 5 (1), including the following:

(a) Charging of rates of interest higher than what has been stated in the application for registration.

(b) Attempt to recover any money not permitted under the Ordinance or under these rules on the loans given.

(c) Operating in areas not specifically permitted in the Registration.

(2) Before passing an order for cancellation of the Registration, the Registering Authority shall cause a notice to be served upon the Micro Finance Institution, requiring it to show cause within fifteen days from the date of service of such notice as to why it's Registration should not be cancelled.

(3) The Registering Authority shall, if the Micro Finance Institution files objections before the date fixed in the notice, after considering the same and if the Micro Finance Institution does not file objections before the date fixed, than after the said date, pass an Order in accordance with the Ordinance, which shall be communicated to the Micro Finance Institution.

11. Suspension of Registration

(1) In case of commission of offences under Section 11(7) and Section 16 of the Ordinance, the Registering Authority can issue an order for suspension of the Registration pending enquiry after recording the reasons for such suspension which shall be communicated to the Micro Finance Institution.

(2) Every suspension of Registration shall immediately be followed by a notice under Rule 10 (2) and the enquiry shall be completed within 15 days from the date of receipt of explanation of such Micro Finance Institution.

12. Effect of cancellation or non-renewal of Registration

In case of cancellation of Registration as per Rule 10 or non-renewal of Registration as per Rule 8 (4), such Micro Finance Institution shall not extend any fresh loans to the borrowers, but will be entitled to recover the dues of the subsisting loans subject to following the provisions of Section 16 of the Ordinance.

13. No dual membership

(1) No member of a Self Help Group shall be a member of another Self Help Group or any group formed by any Micro Finance Institution.

(2) For this purpose, each Registering Authority shall maintain a complete and updated database of the all Self Help Group members and the list of borrowers furnished by the Micro Finance Institutions operating in that district.

14. Disclosure of interest rates

(1) All Micro Finance Institutions registered under this Ordinance shall display the effective interest rates being charged on the loans given by them prominently in all their offices; and the same shall be furnished to the borrowers in writing.

(2) Non-display of the effective rate of interest shall be an offence punishable under Section 18 of the Ordinance.

15. All Charges to be subsumed under interest only

(1) No Micro Finance Institution shall levy any charge or impose any cost by whatever name called, on the borrowers except the interest rate as disclosed in the letter granting the loan and as reported in the Form - 1 and in Form - 4.

(2) No Micro Finance Institution can collect any amounts not being a part of the interest rate and any such collection is a violation under Section 8 (2) of the Ordinance punishable under Section 18.

16. Discharge of certain loans

(1) Based on the database submitted by Micro Finance Institutions registered under these Rules, each RA shall, identify such loans where the interest paid has exceeded the principal borrowed, and issue a direction in Form - 6 for discharge of such loans and for return of such interest portion that exceed the principal amount to the borrower.

(2) Any attempt to recover these loans discharged in the above manner shall be an offence under Section 18 of the Ordinance and the Registration of such Micro Finance Institutions is liable for cancellation as per Section 5 of the Ordinance.

17. Procedure to be adopted in case of excess recovery

(1) Every Registering Authority, on any complaint of excess recovery of the loan by the Micro Finance Institution or in cases where the interest exceeded the principal amount, shall cause notice to be served upon the said Micro Finance Institution, fixing a date requiring the Micro Finance Institution to show cause why the Registration should not be suspended or cancelled or renewal thereof be prohibited for the said contravention.

(2) The Registering Authority shall consider the explanation filed if any, received before the last date stated in the notice, and pass order for refund of such excess money to the borrower.

(3) Non-implementation of the Order passed in accordance with Section 9 of the Ordinance shall be an offence punishable under Section 18 of the Ordinance.

18. Procedure in case Micro Finance Institution wants to advance loan to an Self Help Group member

(1) Micro Finance Institution proposing to advance loan to an Self Help Group or its member already having an outstanding loan taken from any Bank through the Self Help Group, shall apply for prior approval in Form-7 giving details of the member of the Self Help Group as laid down in Section 10 (2) of the Ordinance.

(2) The Registering Authority or any person or Agency authorized by him shall verify the details within the time limits laid down in the Ordinance; and conduct such verification with the Self Help Group to satisfy himself that the purpose for which the loan is proposed would generate additional income needed to service the loan.

(3) Every permission granted under section 10 (3) of the ordinance shall be in Form - 8.

19. Granting multiple loans to be an offence

Any Micro Finance Institution granting a loan to any Self Help Group or its members having an outstanding loan except as laid down in Section 10 shall be liable for punishment as laid down in Section 18 of the Ordinance.

20. Manner, form and format of the loan

(1) The loan contracted between a borrower and the Micro Finance Institution shall be repayable in the manner as laid down below:

(a) The repayment of the loan shall be on a periodicity which shall not be less than one month.

(b) The interest shall be calculated on diminishing balances.

(c) The equated installments calculated in the above manner shall be intimated along with all other terms and conditions of the loan and shall be duly acknowledged by the borrower.

(2) The letter sanctioning the loan shall be in Form - 9.

21. Issue of receipts for the repayments

Every payment received from a borrower shall be duly acknowledged by issuing a written receipt and a copy of the same shall be available in the records of the Micro Finance Institution.

22. Furnishing statement of account to the borrower

The Micro Finance Institution shall deliver a copy of the statement of account on demand free of cost to the borrower.

23. Place of collecting repayments in urban area

No Micro Finance Institution shall collect repayment of the installment due in urban areas except in a public place like office of Slum level Federation or ward office.

24. Microfinance Institution not to employ recovery agents

(1) No Microfinance Institution shall employ any recovery agent for recovery of the loans except the employees named in the Form-1 filed at the time of Registration.

(2) For the purpose of this Rule, any person who is not named in Form 1 but effecting recoveries shall be treated as recovery agent; and Micro Finance Institution responsible for such collection shall be liable for action under Section 18 of the Ordinance.

25. Monthly Statement by Microfinance Institution to RA

(1) The monthly Statement as per Section 12 of the Ordinance shall be in Form-10 of these rules and shall contain all the loans granted and recoveries made during the previous month.

(2) Non submission of Form-10 shall be an offence punishable under Section 18 of the Ordinance.

26. Procedure for summoning Witnesses and requiring production of documents

For the purposes of implementing the provisions of the Ordinance, the Registering Authority shall have the same powers with regard to summoning of Witnesses, requiring production of documents, recording evidence and other proceedings as are provided in Chapter X of the Land Revenue Act and as per provisions of Code of Criminal Procedure 1973 as the case may be.

27. Power to search and seizure

(1) The Registering Authority or any other officer authorized by him can conduct the search of such premises and seize such records as required under Section 13 of the Ordinance.

(2) The Registering Authority or any other officer authorized by him can summon the representative of the Micro Finance Institution located in the district by issuing summons as per Form-11.

28. Complaints

(1) Every complaint under Section 14 of the ordinance shall be either in writing before the Registering Authority or by using the Toll Free telephone provided for this purpose.

(2) All complaints registered shall be enquired into by the Registering Authority or an officer authorized by him within seven days from the date of receipt of the complaint.

(3) If, after due enquiry into the complaint, the Registering Authority is satisfied that there is sufficient evidence of a violation of a provision or provisions of the Ordinance, the Registering Authority shall, issue a show cause notice to the Micro Finance Institution in Form - 12 and taken action as laid down in Rule 10.

29. Prosecution of cases

(1) In case of any offence committed under Sections 16 and 17 of the Ordinance the Officer in charge of the Police Station under whose jurisdiction the offence has occurred shall register the case either on complaint or suo moto and proceed in accordance with the provisions of the Code of Criminal Procedure 1973.

(2) The progress of investigation and prosecution in all such cases shall be reported to the Registering Authority on a monthly basis.

Notes

1. SCIT vs Pemsel (1891) 3 TC 53 96 cited in Nabhi, 2004. Formation and Management of a Society. 16th edition.
2. As per the guidelines of the Insurance Regulatory and Development Authority (IRDA), a Corporate Agency would be a company registered under the Companies Act, whose primary business would be other than distribution of insurance products, and insurance product distribution should be a subsidiary business. In exceptional cases, the license for corporate agency is granted exclusively for the insurance distribution. However in such cases also the agent should be a public limited company with a minimum share capital of Rs15 lakh and to be managed by insurance professionals.
3. Section 45 of RBI Act provides regulatory guidelines on the issues related to deposit mobilization by non-banking institutions and financial institutions.
4. Unincorporated bodies are those who are not registered under the Companies Registration Act 1956.

5. Henceforth referred to as Section 25 companies.
6. These sections of RBI Act deal with the requirements that the companies involved in the business of financial services need to have.
7. However this specification of higher "net-owned fund" shall not be applicable to such company whose application for certificate of registration under Section 45-IA of the said act is submitted to Reserve Bank of India on or before April 20, 1999.
8. Source: Taxmann's Statutory Guide for Non-Banking Financial Companies, An authorized publication of Reserve Bank of India (2004).
9. Tier I and Tier II capitals form part of the net worth of the organization. Tier I capital means the owned fund (sum total of paid-up equity capital, preference shares which are compulsorily convertible into equity, free reserves, balance in share premium account, capital reserve representing surplus arising out of sale proceeds of asset, excluding reserves created by revaluation of assets and accumulated surplus or deficit) reduced by investment in shares of other NBFCs and in shares, debentures, bonds, outstanding loans and advances including hire purchase and lease finance made to and deposits with subsidiaries and companies in the same group exceeding, in aggregate, ten per cent of owned fund. Tier II capital includes preference shares other than those which are compulsorily convertible into equity; revaluation reserves at discounted rate of 55%; general provisions and loss reserves to the extent these are not attributable to actual diminution in value or identifiable potential loss in any specific asset and are available to meet unexpected losses, to the extent of one-and-one-fourth per cent of risk weighted assets; hybrid debt-capital instruments and subordinated debt, to the extent the aggregate does not exceed Tier I capital.
10. As per the directives of RBI on the NBFC Acceptance of Public Deposits (Reserve Bank) Directions, 1998, dated 31st January 1998; Loan Company is the one which is a financial institution carrying on as its principal business the providing of finance whether by making loans or advances or otherwise for any activity other than its own, but does not include an equipment leasing company or a hire-purchase finance company.
11. As mentioned in the preceding sub-sections of the report, vide notification no DNBS 132/CGM (VSNM)-99, dated 20-4-1999, the Reserve Bank of India specifies the "net owned fund" to be Rs200 lakh, for an NBFC which commences the business of a non-banking financial institution on or after 21st April 1999. This specification of higher "Net Owned Fund" shall not be applicable to such company whose application for certificate of registration under section 45-IA of the said act is submitted to the Reserve Bank of India on or before 20th April, 1999.
12. The position on the external commercial borrowing has changed following the introduction of the Finance Bill for the year 2005-06. Microfinance: Models, Policies and Experiences institutions registered as NBFCs are likely, in the near future, to be allowed to have access to external commercial borrowings.

The details of this would, however, be known only after the approval of the Finance Bill by Parliament.

13. As per the directives of RBI on the NBFC Acceptance of Public Deposits (Reserve Bank) Directions, 1998, dated January 31, 1998.
14. Provision laid down vide Notification no DNBS 164/CGM (CSM)-2003, dated 8 January, 2003.
15. Brought down from the previous ceiling of ₹ 100 Crore vide UBD.PCB.Cir. 33/09.116.00/2004-05 dated January 5, 2005.
16. As per the Article 3 of the APMACS Act the membership of a cooperative society shall be voluntary and available without restriction of any social, political, racial or religious discrimination, to all persons who can make use of its services and are willing to accept the responsibilities of membership.
17. Relates to the requirements for registration of companies.
18. For details on the topic refer to the discussions in **Section 2.1** of the report.
19. Refer to **Section 2.5** for discussion on the topic.
20. Microfinance: Models, Policies and Experiences is defined as: "Provision of thrift, credit and other financial services and products of very small amounts to the poor in rural, semi-urban or urban areas for enabling them to raise their income levels and improve living standards". Mahajan and Ramola have criticized, among others, that "very small amount" has only been defined for credit (and only unsatisfactorily); cf. Mahajan and Ramola (2003: 4f.)
21. Cf., for example, CGAP's Guiding Principles on Regulation and Supervision of Microfinance: Models, Policies and Experiences, which conclude: "Self-regulation of financial intermediaries in developing countries has been tried many times, and has virtually never been effective in protecting the soundness of the regulated organizations." Christen, Lyman and Rosenberg (2003)
22. It is important to note that the Reserve Bank of India often issues very comprehensive notifications in circular letters giving detailed instructions, explaining its policy or stipulating regulations. To give an example, between 1998 and today the RBI has published 35 notifications regarding NBFCs alone.
23. RBI Circular Letter RPDC. No. PL.BC.120/04.09.22/95-96 dated April 2, 1996 (available at http://www.nabard.org/whats/tf/ann3.htm).
24. 40% of net bank credit must be lent to the priority sector, of which 25% are to be lent to the 'weaker section'.
25. NABARD Circular Letter No. NB.DPD.FS.4631/92-A/91-92 dated February 26, 1992 (available at http://www.nabard.org/whats/tf/ann2.htm).
26. RBI Notification Ref.DNBS.(PD).CC.No. 12 /02.01/99-2000 dated January 13, 2000 (available at http://www.rbi.org.in/sec14/14935.pdf).
27. Sec 25 (1) lists describes such a company as an association "formed as a limited company for promoting commerce, art, science, religion, charity or any other useful object". It is at least questionable whether microfinance falls under any of these activities.

28. RBI Circular RPCD.No.PL.BC. 62/04.09.01/99-2000 dated February 18, 2000. Following the terminology of the Microcredit Summit Campaign, the RBI uses the term microcredit, although it explicitly includes 'thrift' (*i.e.* reduced/deferred expenditure) and other financial services.
29. RBI Circular RPCD.No.PL.BC 94/04.01.01/98-99 dated April 24, 1999. Reportedly, some banks are still stipulating a 3 per cent maximum spread for onlending by NGOs.
30. At this time, Sa-Dhan published a book on microfinance regulation in India, which very clearly advocates for a self-regulatory approach as earlier recommended by the Task Force; cf. Fukaya and Shadagopan (2001).

Chapter 4

Microfinance Institutions in India and AP: Spread, Functioning and TFI

4.0. Introduction

Microfinance is changing. What began many years ago as a philanthropic movement to bring finance to poor people is rapidly becoming a big business and, at the same time, a more risky business (Lascelles, 2012). Today, unsustainable microcredit indebtedness is commonplace across developing countries: in India; in Bangladesh (Banking with Poor, 2009); and in Peru (Kevany, 2010); and also in transition countries, notably in the Balkans (Bateman, 2011) and especially in Bosnia and Herzegovina (Cain, 2010).

Home to the largest population of poor in the world, India has been a natural candidate for experimenting with microfinance as a tool for poverty alleviation. With a nationalized formal banking sector that has emphasized rural and developmental banking for several decades now, India's involvement with small credit targeted primarily at the rural poor is hardly new. However, recent years have generated unprecedented interest in microcredit and microfinance in the form of group-lending without collateral (Chakrabarti, 2004). Microfinance companies are offering small-scale financial services in both the forms - credit and savings, especially to the poor in rural, semi-urban and urban areas. These financial services are meant to help rural poor in undertaking economic activities, mitigating vulnerabilities to income shocks, smoothening consumption, increasing savings and supporting self-empowerment. Microfinance involves extension of small loans and other financial services to low income groups, is a very important economic conduit designed to facilitate financial inclusion and assist the poor to work their way out of poverty. It has the potential to fill the critical gap left by formal financial institutions in providing financial services to low income groups (Sinha, 2012).

The Indian population can broadly be divided into four categories based on household income levels. The Rich who make up 0.4% of the households have an annual household income greater than $20,000. The Middle Class comprises 11 million households, or 5.9% of the total households, and has an annual household income between $4,000 and $20,000. The Aspirers make up nearly 22% of the households and have an annual household income between $1,800 and $4,000. Lastly, the deprived segment, the prime target of the microfinance industry, comprises 135 million or 72 per cent of the households and has an annual household income below $1,800 . There are a number of microfinance companies in India, which play some pivotal roles to the development of India. India's microfinance sector is fragmented with more than 3000 microfinance companies (MFIs), NGOs and NGO-MFIs to focus the deprived section.

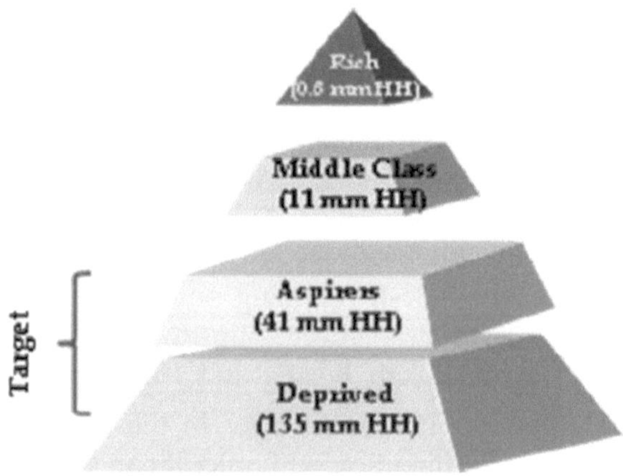

Fig. 4.1: Targeted Households

Source: Lok Capital.

4.1. Microfinance Institutions and Services in India

Microfinance institutions are present in all states of India and serving the community in varied forms. Initiated with the objective of helping poorest of the poor to earn a livelihood with dignity has gradually been moved away from the prime goal. Most of these institutions have become business enterprises with an assured as well as earmarked market segment with the rural poor and deprived section of the community in semi-urban and rural India. The penetration of microfinance institutions is varying across the states and territories of India (see fig. 4.2). States like Tamil Nadu, Karnataka, Andhra Pradesh and Orissa exhibits more than 15 per cent penetration of microfinance institutions. In contrast, states and union territories like J&K, Himachal Pradesh, Punjab, Arunachal Pradesh, Nagaland, Manipur and Mizoram represent very little presence of microfinance institutions. Booming with the objective to raise income to a level where the families in the lower income bracket would be out of the danger-zone of deprived of three square meals a day, to a level to access basic amenities of life. The commercial outlook of MFIs is mostly attributed to the nature of investments and bulk loans that they had received. It is important therefore to critically understand the role of funding stakeholders *viz* investors and lenders in the governance of MFIs and their social orientation.

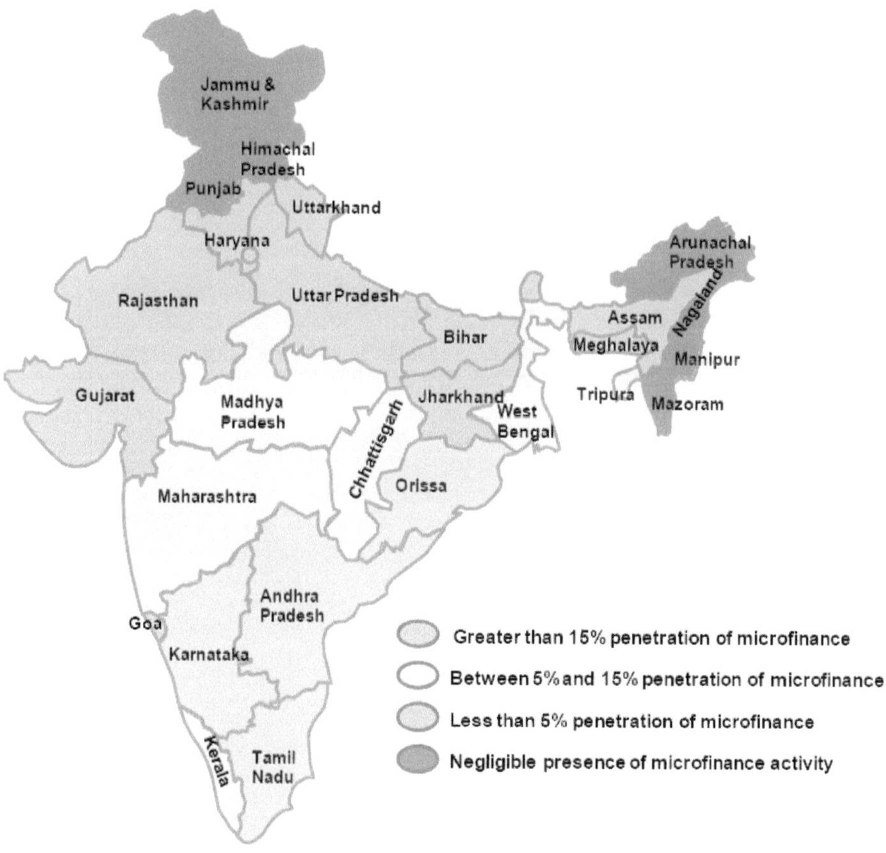

Fig. 4.2: Geographic Spread of Microfinance in India

Source: Lok Capital

There are many institutions created and many programmes as well as schemes floated by the state and central government to increase access of fund to the poor and deprived through institutional mode to help enable a secure life and livelihood (see fig. 4.3). However, the microfinance operation is taking place using two major models for microfinance delivery- the MFI model and the SHG Bank Linkage Programme (Sinha, 2009). Almost 75 per cent microfinance institutions and the volume of operation occur in five south Indian states. The MFI data reveals that for-profit MFIs encompass nearly 90 per cent of the total client outreach and loan portfolio (see fig. 4.4). The entire effort is executed through a customized package to waive out the difficulties present in rural and semi-urban areas of the country.

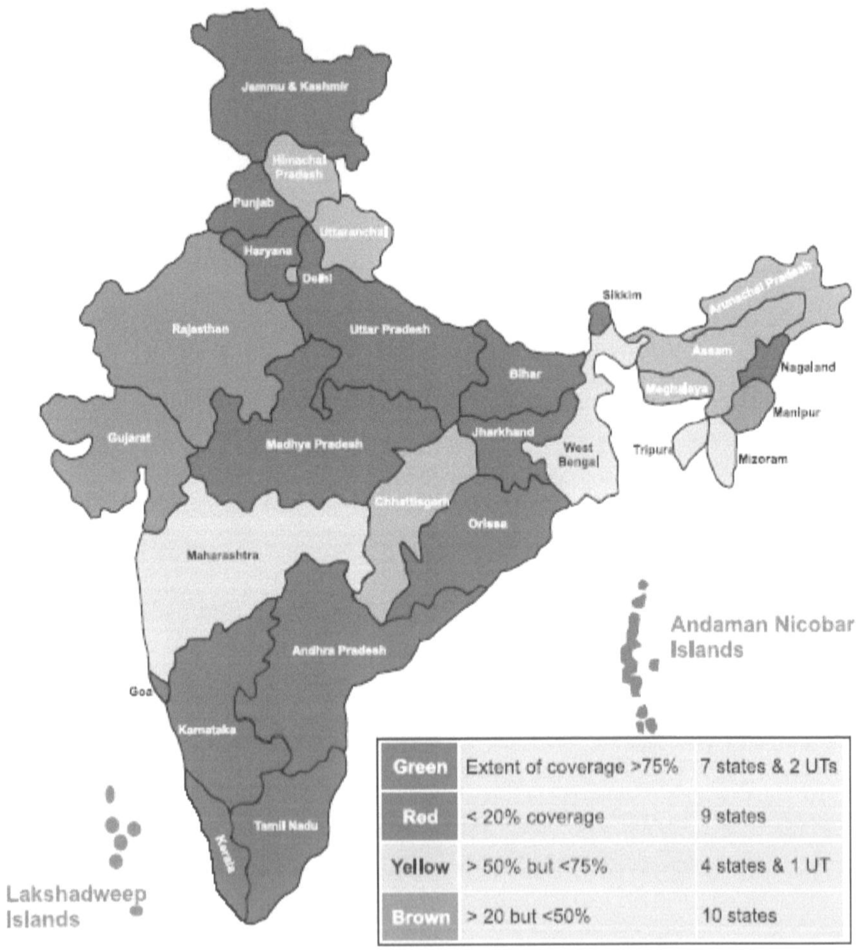

Fig. 4.3: Geographical Spread of SHG-Bank Linkage Programme in India as on 31.03.2011

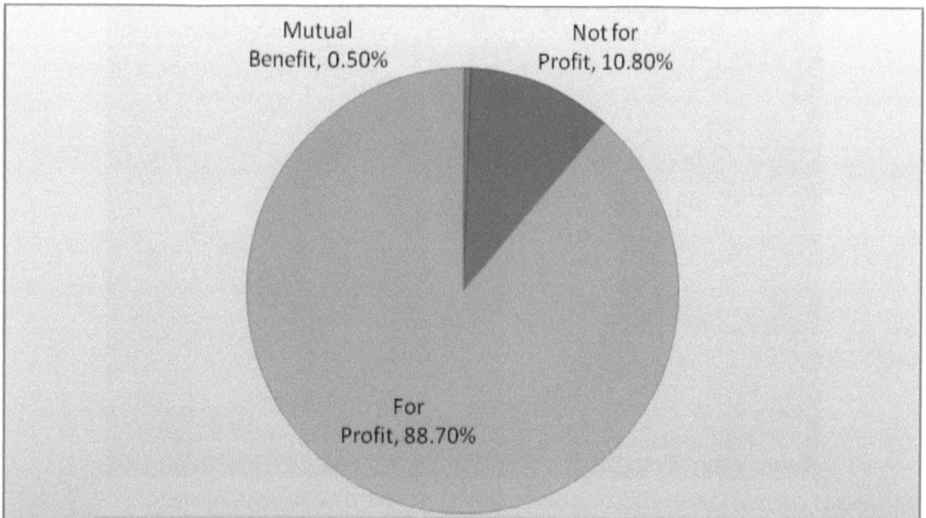

Fig.4.4: Client Outreach by MFIs as Per Legal Form

There are a vast number of microfinance institutions operating in the country. Morethan half (52%) of the active clients of the industry belong to the Southern states, while almost one quarter (23%) belong to the Eastern States . Western States contribute 10 per cent to the total client base, and Central and Northern states contribute 6 per cent each. The Northeastern States seem to be the most underserved region of the country, responsible for a meagre 3 per cent of the total number of clients (See Fig.4.5). A similar trend is also seen in loan portfolio- the southern states contributing 54 per cent, Eastern states 22 per cent, Western states 10 per cent, Central states 7 per cent, Northern states 5 per cent and Northeastern states only 2 per cent (See fig. 4.6).

Fig.4.5: Region-wise MFIs Client Outreach

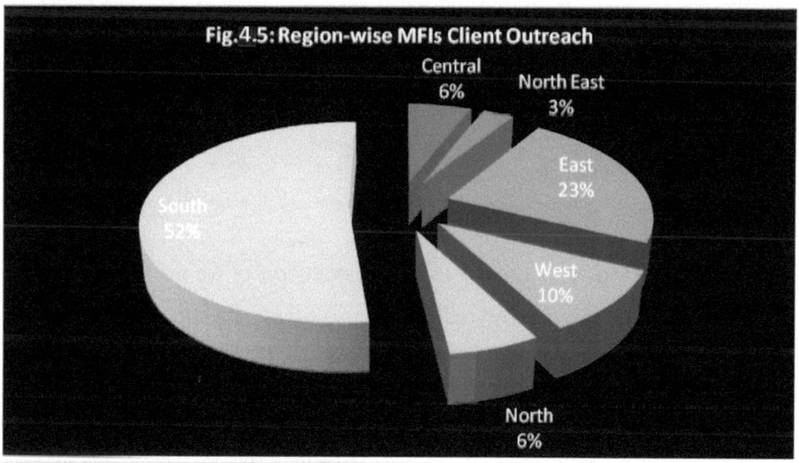

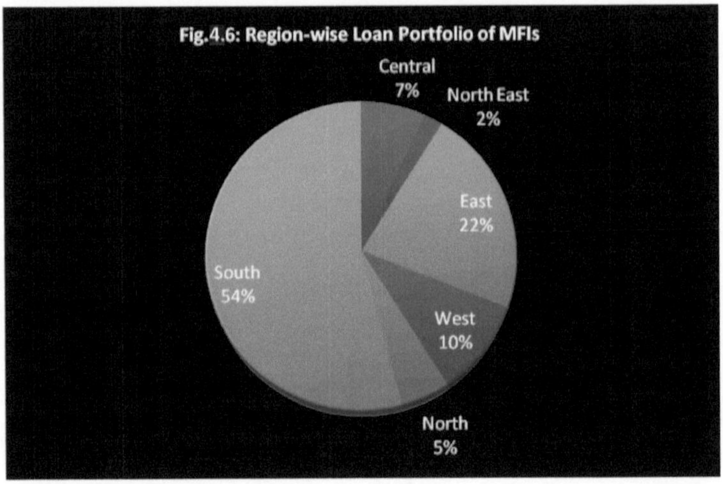

Further, the distribution of sources of funds for microfinance, presented in (see Fig.4.7) shows that the share of debt in MFI finances climbed sharply. The current level of debt, amounting to 69.2 per cent of total funds raised by the leading MFIs represents a reduction from the highest level of around 80 per cent reached in 2008. The extent to which commercial debt continues to dominate the financing of Indian microfinance is apparent.

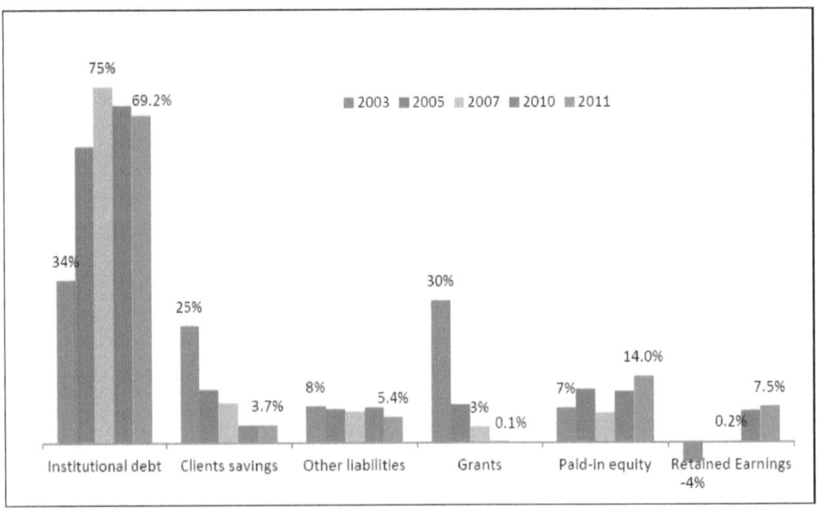

Fig. 4.7: Loan Portfolio distribution of MFIs

Indeed, the domination of Commercial Bank funds in Indian microfinance is under played in this since it excludes off?balance sheet financing via portfolio sales and securitisation of portfolios undertaken by some of the leading MFIs to the Commercial Banks. A separate compilation of the portfolio managed by MFIs for others - securitised portfolios that are not on MFI balance sheets - shows that the amount added some 10.5

per cent of the portfolio to the MFIs' managed portfolio. The share of net worth/equity in MFI balance sheets and the distortionary effect of inadequate provisioning is also discussed in the report (M-CRIL, 2011) .

The un-accounted MFIs are mostly working in difficult locations and against the adverse conditions and without the financial or/and institutional support. The forms of the institutions do not fall into the organisational ambit of the so called microfinance systems and procedures. However, the numbers of microfinance institutions involved in micro-lending to the micro-people are increasing persistently over the years in terms of number as well as business volume (see fig.4.1). The fund support to the formal microfinance institutions are supported by different national and regional banks. The volume of finance flown into various types of MFIs is ever increasing and the major volume of support is extended by the Commercial Banks (see fig.4.2).

Table 4.1: Loan Disbursement to MFIs during 2006-07 to 2010-11

Sl. No.	Financing Agency	Period	No. of MFIs	Amount (in ₹ Cr)
1	All Commercial Banks	2006-07	327	1151.34
		2007-08	497	1968.6
		2008-09	522	3718.93
		2009-10	645	8038.61
		2010-11	469	7605.18
2	Regional Rural Banks	2006-07	7	0.22
		2007-08	8	1.51
		2008-09	59	13.4
		2009-10	46	24.14
		2010-11	9	4.16
3	Cooperative Banks	2006-07	0	0
		2007-08	13	0.04
		2008-09	NA	NA
		2009-10	0	0
		2010-11	NA	NA
4	SIDBI	2006-07	NA	NA
		2007-08	NA	NA
		2008-09	NA	NA
		2009-10	88	2665.75
		2010-11	2	843.78
5	**All Agencies**	**2006-07**	**334**	**1151.56**
		2007-08	**518**	**1970.15**
		2008-09	**581**	**3732.33**
		2009-10	**779**	**10728.5**
		2010-11	**471**	**8448.96**

Source: Compiled from various Status Reports of Microfinance in India, NABARD.

Table 4.2: Outstanding Loans with MFIs during 2006-07 to 2010-11

Sl. No.	Financing Agency	Period	No. of MFIs	Amount (in ₹ Cr)
1	All Commercial Banks	2006-07	541	1584.27
		2007-08	1072	2745.24
		2008-09	1762	4977.89
		2009-10	1407	10095.32
		2010-11	2176	10688.85
2	Regional Rural Banks	2006-07	8	0.2
		2007-08	24	3.58
		2008-09	153	31.2
		2009-10	103	52.22
		2010-11	23	42.01
3	Cooperative Banks	2006-07	1	0.01
		2007-08	13	0.02
		2008-09	NA	NA
		2009-10	3	0.01
		2010-11	NA	NA
4	SIDBI	2006-07	NA	NA
		2007-08	NA	NA
		2008-09	NA	NA
		2009-10	146	3808.2
		2010-11	139	3041.77
5	**All Agencies**	**2006-07**	**550**	**1584.48**
		2007-08	**1109**	**2748.84**
		2008-09	**1915**	**5009.09**
		2009-10	**1656**	**13955.75**
		2010-11	**2315**	**13730.62**

Source: Compiled from various Status Reports of Microfinance in India, NABARD.

The other mode of microfinance operation is through the Self-help Group and Bank Linkage Programme. The programme is gaining momentum over the years and mostly identified as an established means of community reach and widely accepted by all. The number of self help groups formed is phenomenal and mostly centered around women as the kingpin for leveraging the fund. It is taken for granted that women initiated enterprises are more sustainable, run with vision and dedication and hence, more profit fetching.

Under the SBLP, NGOs usually act not as MFIs but as SHG promotion agencies (SHPAs)- a different model of microfinance in which the NGO role is to provide developmental support to the SHGs, that open group deposit accounts in a bank and then can then borrow directly from the bank. This model was initially promoted by NGOs such as MYRADA (in Karnatak) and PRADAN (in Tamil Nadu and in Bihar). Since mid-nineties, when the model began to be scaled up, promotional work has been undertaken by specialized government agencies, such as District Poverty Initiative or the Velugu project in Andhra Pradesh and the Kudumbashree project in Kerala, the Women's Development Corporation in the states of Tamil Nadu and Maharashtra, the Women and Child Development Department and the District Rural Development Agencies (DRDAs) in most other states (Sinha, 2009).

Livelihood promotion and IGPs centering around tiny and/or small activity are primarily focused to supplement and complement the income of the people in the self-help initiative. The groups are designed by default to generate own savings as well as access financial services from the financial institutions to sustain the activities undertaken.

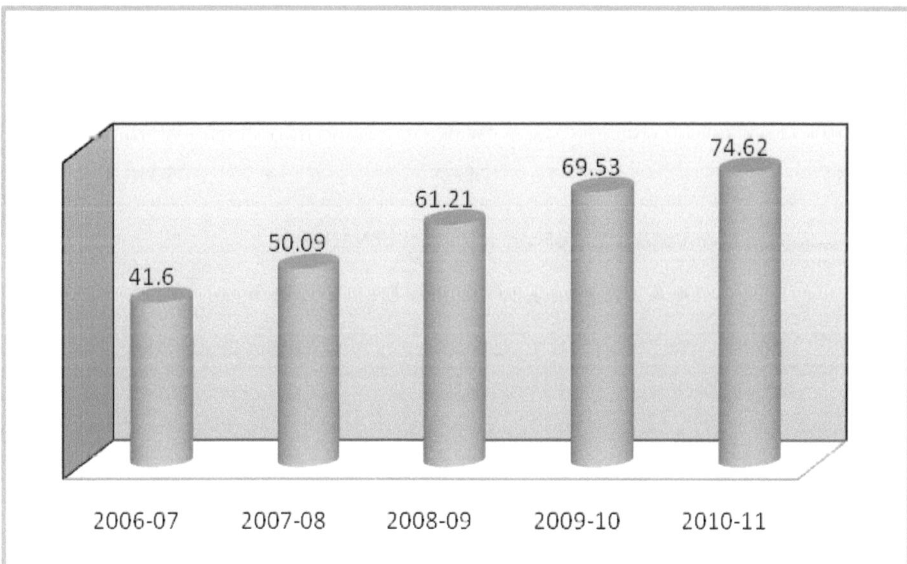

Fig. 4.8: Growth of SHGs in India

The number of self-help groups is increasing at a faster rate (see fig.4.8). The volume of saving has increased on an average of rupees 1000 crores for the last five years (see table 4.3). The increased volume of saving symbolizes the intent and spirit of the people to undertake the IGPs and the commitment. The number of Self-help Groups that had outstanding loans at the end of March 2011 had increased by about 4.9 per cent. The outstanding loans against these groups had increased by about 12.3 per cent. The number of clients added by the SBLP is about 2.9 million. The MFIs in comparison added 6.4 million clients during the year. In terms of incremental loans outstanding the SBLP added ₹ 33.6 billion. Savings by SHGs members continued to increase; 7.54 million groups had saved ₹ 69.25 billion in banks which is about 10 per cent more than the previous year's level (Srinivas, 2011).

The spread of the microfinance through SHG-BL programme is all time high in the southern region. The number of self-help groups formed across the states of the northern region is morethan 52 per cent of the total self-help groups formed in the country so far (see table 4.4). It is also visible that the volume of savings mobilized and deposited in the banks is also increasing over the years in the southern region in comparison to all other regions of the country (see fig 4.9).

Table 4.3: Status of SHGs and Mobilization of Savings

Year	No. of SHGs	Savings
2006-07	41.6	3512.71
2007-08	50.09	3785.39
2008-09	61.21	5545.62
2009-10	69.53	6198.71
2010-11	74.62	7016.3

Source: Compiled from Various Microfinance Reports of NABARD

Fig.4.9: Region-wise Trend of SHGs and Savings

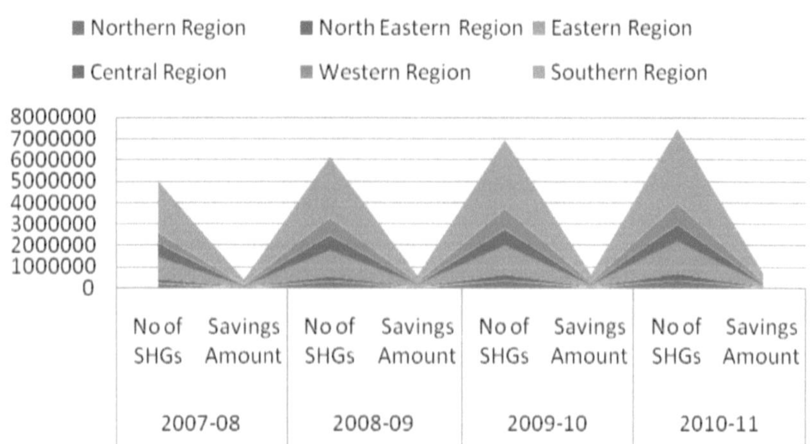

Table 4.4: Year and Region-wise Savings of SHGs with Banks during 2007-08 to 2010-11

	2007-08		2008-09		2009-10		2010-11	
	No. of SHGs	Savings Amount	No. of SHGs	Savings Amount	No. of SHGs	Savings Amount	No. of SHGs	Savings Amount
Northern Region	208166	12767.69	310998	22703.92	351801	34207.31	372772	32857.16
North Eastern Region	203045	8633.18	240093	10210.16	292188	12167.09	324739	13104.6
Eastern Region	1074043	80599.65	1233635	159688.04	1374242	112014.6	1527618	140837.61
Central Region	644896	33531.01	712915	38679.2	765965	51363.57	786436	60338.01
Western Region	472734	33063.23	796262	66428.4	945620	92667.48	960921	82901.13
Southern Region	2406910	209944.2	2827244	256852.1	3223434	317450.84	3489460	371591.77
Total	5009794	378538.96	6121147	554561.82	6953250	619870.89	7461946	701630.28

Source: Status of Microfinance in India, 2010-11, NABARD

4.2. MFIs in Andhra Pradesh-Spread, Coverage and Access

Andhra Pradesh is the fourth largest state by area and fifth largest by population in India. The origins of microfinance in Andhra Pradesh can be traced back to government-led attempts to form "Self Help Groups (SHGs)," or groups of 12 to 20 (mostly) women which collect regular savings from members and make loans internally to members. The government first formed SHGs through the Development of Women and Children in Rural Areas (DWCRA) program. Nationally, 2,73,000 groups (2.73 lakhs) were formed under the program until its absorption into the larger Swarnajayanti Gram Swarojgar Yojana (SGSY) programme in 1999 . However, the formation of SHGs on a large scale did not take of until the creation of the SHG-Bank Linkage Program (SBLP) in 1992. Under the program, India's apex agricultural development bank, the National Bank for Agriculture and Rural Development (NABARD) provided a set of incentives for banks to lend to SHGs that adhered to certain guidelines, such as collecting regular savings from group members.

The state of Andhra Pradesh in several ways can be described as the microfinance capital of India. It accounts for 30 per cent of borrower accounts and the outstanding loan portfolio in case of MFIs. In the case of SHG linkage programme, the state had 25 per cent groups linked and 39 per cent of loans outstanding as at march 2010 (Srinivasan, 2011) and nearly 25 per cent of the ₹ 30,000-crore microfinance trade in the country (Business

Standard, 2011). As many as 24, of the total 44 MFIs recognised by the RBI are operating in the State spread over a branch network of 525 across districts.

Over the last five years, as MFIs chased growth, the nature of microfinance delivery has changed (Joseph, 2007). The job of income generation and social capital building was left for others to tackle. As competition intensified, and markets reached saturation, MFIs took on the additional risk of lending to households with uncertain cash flows. If this trend persists, there is a huge likelihood that the industry will follow some of the flawed practices of the sub-prime lending models, resulting in huge collective losses over a period of time.

About 9.2 million borrowers in the southern state have defaulted in repaying money borrowed from microfinance institutions (MFIs)-the largest number of defaulters in any location in the world. As Andhra Pradesh has a population of 84.6 million (2011 census provisional figure), theoretically one in every 11 Andhraites is a defaulter. This, however, is not the actual case as most such borrowers have more than one account. By industry estimates, around four million people of the state-almost all women-have turned defaulters. As the female population in Andhra Pradesh is 42.1 million, one in every 10 women in this state has borrowed from MFIs but not repaid.

To put the Andhra Pradesh phenomenon in perspective, in 2008, when the Indian government announced a Rs60,000 crore debt-waiver scheme for farmers across the nation, some 40 million farmers didn't repay debt. Indeed, the money involved in Andhra Pradesh is minuscule compared with the national debt-waiver scheme, but the point to mention here is that in this case, the debt has not been waived even though a section of the local politicians has created that impression and encouraged borrowers not to pay back.

The MFI industry's total exposure to Andhra Pradesh was around ₹ 7,200 crore in 2010 and they have been able to collect from the borrowers only 10 per cent of this money. They are carrying on their books ₹ 6,500 crore worth of bad assets, and when they write off this amount in March, at the end of the current fiscal, many of India's big MFIs' net worth will turn negative. Unless the Reserve Bank of India (RBI) relaxes the prudential norms that stipulate a 15 per cent capital adequacy ratio (₹ 15 capital for every ₹ 100 worth of loans) for this set of financial intermediaries, many will have to face the sever consequence of bankruptancy.

The MFIs that will bear the brunt will include India's lone listed microlender SKS Microfinance Ltd., Spandana Sphoorty Financial Ltd., Share Microfin Ltd, Asmitha Microfin Ltd and Vijay Mahajan-promoted Bhartiya

Samruddhi Finance Ltd.-India's oldest MFI. The SKS Microfinance will be the least affected as it had raised money from the public through its capital float in August 2010; the others would be in a bad shape. Some are restructuring the debt taken from banks. But even then, keeping themselves afloat is not an easy task as commercial banks, which provide 90 per cent of resources that MFIs need for business, are not forthcoming in giving money for fear of piling up bad assets.

Even the MFIs which do not have any direct exposure to Andhra Pradesh are being shunned by the banks. For instance, Arohan Financial Services Ltd, an MFI in West Bengal, hasn't got any money from banks ever since the Andhra crisis broke out, even though it doesn't have a single borrower in the southern state. For lack of resources, the MFI is witnessing shrinking assets and hence, slowly closing down offices. Utkarsh Micro Finance Pvt. Ltd. of Varanasi is another victimized in the process to note. It currently lends to the poor people in eastern Uttar Pradesh and Bihar. In 2010 and early 2011, banks withdrew all lines of credit. Subsequently, a few banks have started sanctioning loans but they are not releasing money in time and the required volume.

With no fresh money in the kitty and non-payments of loans by borrowers across the of Andhra Pradesh, MFIs are forced to shrink loan books, close down offices and retrench employees. The outstanding loan book of the industry, which was ₹ 30,000 crore in October 2010, has shrunk to ₹ 15,000 crore in January, 2011 and at least 30 per cent of 1,50,0000 employees have been retrenched. MFIs have adopted several means to down size the business and shot down at many places besides downsizing the staff in phased manner.

A study carried out by the Centre for Microfinance in 2009 stated that while 93 per cent of households had contracted debt from some source, 82 per cent households reported loans from informal sector at a significant level. Thus, the ground level situation in Andhra Pradesh was that of a 'deluge of debt' coming to households fuelled by competitive lending of MFIs and an unprecedented level of lending by banks through the SHGs.

The data shows (see table-4.5) that the gap for serving the un-served areas and population is jointly served by the MFIs and the SHG loans. The overall, coverage has been increased over the years. The volume of loan leveraged through the SHG mode is quite revealing. The MFIs client base has been increased at a steady rate of about 1.3 million in 2009 and 2010. The addition of client base is none the less for SHGs. The SHG programme added an unprecedented 4.3 million members during 2008-09, 1.5 millions members in 2009-10 and 4.5 millions in 2010-11.

Table 4.5: Progress Trends of Microfinance in Andhra Pradesh

Sl. No.	Parameters	2008	2009	2010	2011
1	SHG Loans (₹ in Millions)	53,857	89,021	117,395	128,694
2	MFI Loans (₹ in Millions)	19,445	35,652	52,107	52,045
3	SHG Members	10,506,639	15,819,427	17,310,000	21,891,909
4	MFI Customers	3,635,115	4,949,393	6,244,648	5,751,000
5	Total MF clients	14,141,754	20,768,810	23,554,648	27,642,909
6	Total No. of Poor households	2,520,000	2,520,000	2,520,000	2,520,000
7	Average loan per account	5,183	6,002	7,193	7,982
8	Average loan per poor households	29,088	49,473	67,226	71,721
9	No. of loan accounts per poor household	5.6	8.2	9.3	10.9

Source: Microfinance India, State of the Sector Report 2011

It can also be seen from the table that the volume of outstanding loan by MFIs and SHGs is noteworthy. In fact, the volume of outstanding SHG loan has been steeper compared to MFIs during the period. MFIs loan outstanding has grown by above 16 000 million in absolute terms during the initial two years and become near stagnant for the last two years. The SHG loan outstanding had a quantum jump in 2009 and later on had a sluggish coverage for the rest of the years. SHGs increased clients' base by 108 per cent and loans by 103 per cent during 2008-2011.

4.2.1 Geographic Spread of MFIs across Districts of Andhra Pradesh

The previous section while opening the eye on the value, volume, access and state of health of MFIs and Self Help Groups during 2010-11, the current section is attempting to give district-wise scenario of MFIs in the state of Andhra Pradesh. The volume of financial transactions and the spread of MFIs across the districts had been controlled by the recently launched ordinances. Further, the activities of microfinance institutions are controlled and regulated through opening up of separate departments and/ or divisions in rural and urban areas. MFI activities in urban and municipality areas are being monitored by MEPMA while DRDA regulate the functioning of MFIs in rural and semi-urban areas of the state.

The overall focus of government is to control the ill-means being adopted for sanctioning of financial assistance to innocent rural people and also to help enable people to minimize heavy dependency of people on the MFIs to pursue income generation avenues for survival. The system and procedure is keeping control of unwarranted incidents like one person accessing multiple financial assistances from multiple institutions of formal and non-formal nature and its recovery.

There are 525 microfinance institutions are running business over 23 districts of the state of Andhra Pradesh. All these institutions are registered with either DRDA or MEPMA of the respective operating regions. It is surprise to mention that not- a single district of the state is devoid of presence of MFIs, which ensures the scope of business at one hand and the presence of formal financial institutions to the meet the financial need of the people on the other. Poor presence as well as no presence of formal financial institutions is provisioning the 'niche' as well as acting as the 'zone of comfort' for MFIs to safe play with leaving financial assistance to the un-served and under covered people with financial alternatives in need. There are 280 MFIs registered with DRDA and 245 MFIs registered with MEPMA catering to the financial need of people.

Each district of the state is covered with on an average 12.17 number of MFIs registered with DRDA and 10.17 numbers of MFIs listed with MEPMA. The ratio of coverage of DRDA and MEMPA is almost equal 53: 47 for DRDA MFI's and MEMPA MFI's. Guntur district reveal presence of highest number of registered MFI's across districts of the state and to the dismay just less than half of them are functioning in the urban centers. The coverage for the rural villages and un-touched areas by the MFIs are hardly 55 per cent of the total MFIs listed by the Government machinery.

District Chittoor stand 2nd in terms of presence of MFIs while district Kurnool stood in 3rd position in terms of MFI numbers as on 2011 (see table-4.6). There are a fewer number of MFIs in operations recorded for districts like Adilabad, Nizambad, Khammam, Mahabubnagar etc. However, 12 districts of the state are recorded of operation of more than 20 MFIs to cater the financial need of rural, semi-urban and urban poor.

Hyderabad is the state head quarter for the state of Andhra Pradesh. Almost all formal banks are having offices in the state headquarter. All formal institutions are running as well as regulating their operations from the offices located at the state head quarter. All formal financial institutions are working with the mandate and commitment of addressing issues of poverty and make people self-reliant and help enable people to choose as well as adopt livelihood means suitable to their skill, potentiality, and market and lend all possible support to withstand on their own and be self-sufficient. Despite all arrangements, there are 17 MFIs are in operation in the state head quarter to cater the financial requirement of the people living in the slums and suburb area of the city.

Table 4.6: No. of Registered MFIs in DRDA's and MEPMA's across Districts

Sl. No.	District	DRDA	MEPMA	Total
1	2	3	4	5
1	Adilabad	8	6	14
2	Ananthapur	10	10	20
3	Chittoor	25	15	40
4	East Godavari	12	11	23
5	Guntur	25	20	45
6	Hyderabad	0	17	17
7	Karimnagar	10	8	18
8	Khammam	9	6	15
9	Krishna	14	8	22
10	Kurnool	18	16	34
11	Mahabubnagar	10	5	15
12	Medak	9	10	19
13	Nalagonda	15	11	26
14	Nizamabad	7	7	14
15	Prakasam	16	9	25
16	Ranga Reddy	14	14	28
17	Nellore	12	11	23
18	Srikakulam	11	9	20
19	Visakhapatnam	13	11	24
20	Vizianagaram	9	9	18
21	Warrangal	11	11	22
22	West Godavari	13	11	24
23	Kadapa	9	10	19
	Total	280	245	525

Source: apmfi.cgg.gov.in, MFI information for the year 2011

Astonishingly, 69, 12, 179 beneficiaries across the districts of the state of Andhra Pradesh are accessing financial assistance for some requirement or the other. All these beneficiaries have utilized ₹ 7895.43 crores. The money in circulation is very huge comparing to the formal institutional support extended to people to carry on business and/or activities of economic in nature for sustenance. The huge money in circulation across the beneficiaries is amounting ₹ 13,607 on an average. High doses of money in circulation are correspondingly revealed with high outstanding of ₹ 4933.07 crores to be collected back from the borrowers which in fact, is amounting to ₹ 7,137 per head of borrower. Indeed, nearly 50 per cent of money injected into the economy in different forms is lying idle and would be collected back from the borrowers.

In terms of coverage, there are almost 12 per cent beneficiaries are covered with 5 per cent of total operating financial institutions in East Godavari District and 9 per cent of beneficiaries by equal number of MFIs in West Godavari districts. The loan coverage in each district varies from a minimum of 3 per cent till a maximum of 12 per cent across the districts. While access to loan is universally seen across the districts, per head loan amount is as maximum as ₹ 12,257 for Kurnool district, and as minimum as ₹ 10,789 for Nellore districts of the state (see table 4.7).

Table 4.7: Fund Devolvement Position of Registered MFIs with DRDAs

Sl. No.	District	No. of MFIs	No. of Borrowers	Loan Given	Out Standings
1	2	3	4	5	6
1	Adilabad	14	388930	4342560875	2990226055
2	Ananthapur	20	242582	2810716325	1889843061
3	Chittoor	40	277958	3254858550	1933837003
4	East Godavari	23	823091	9324644200	5278302294
5	Guntur	45	489322	5702921356	3364682534
6	Karimnagar	18	240637	2621898250	1567005496
7	Khammam	15	315442	3735510725	2547552488
8	Krishna	22	246736	2750148900	1698846178
9	Kurnool	34	263856	3234127305	2175011564
10	Mahabubnagar	15	210540	2472412487	1470503223
11	Medak	19	185505	2214067300	1489807390
12	Nalagonda	26	308650	3709608850	2325049534
13	Nizamabad	14	194210	2307604950	1672096544
14	Prakasam	25	238102	2752596375	1600461045
15	Ranga Reddy	28	281782	3308374770	2412912375
16	Nellore	23	373132	4025944200	2457362154
17	Srikakulam	20	210352	2444990100	1572152905
18	Visakhapatnam	24	288685	3257264500	2264154595
19	Vizianagaram	18	225321	2468304700	1445439560
20	Warrangal	22	288712	3114545525	1814781942
21	West Godavari	24	591597	6539587125	3636658210
22	Kadapa	19	227037	2561711050	1724052751
	Total	508	6912179	78954398418	49330738901

Source: www.apmepma.gov.in

Total principal paid by the borrower is amounting to ₹ 2976.03 crores and the interest paid is ₹ 441.14. The principal to interest collection is in the proportion of 87.09 per cent to 12.91 per cent. The overall interest payment constitutes 12.91 per cent to the total loan sanctioned to the borrowers. Per head interest payment made by the borrower as revealed from table 6.8 is amounting to ₹ 638.22 and the principal paid is ₹ 4305.49. This statement is inducing the fact of a similar amount of interest payment for the nearly 50 per cent amount of outstanding to be collected and equal amount of interest burden to be borne by the beneficiaries for the years to come (see table-4.8).

Table 4.8: Loan Recovery Status of Registered MFIs with DRDAs

Sl. No.	District	Loan Given	Principal Repayment	Interest Repayment	Out Standings
1	2	3	4	5	6
1	Adilabad	4342560875	1352334820	244231640	2990226055
2	Ananthapur	2810716325	920873264	128435845	1889843061
3	Chittoor	3254858550	1321021547	198188796	1933837003
4	East Godavari	9324644200	4046341906	604443397	5278302294
5	Guntur	5702921356	2338238822	353888429	3364682534
6	Karimnagar	2621898250	1054892754	156355878	1567005496
7	Khammam	3735510725	1187958237	162660579	2547552488
8	Krishna	2750148900	1187958237	162660579	1698846178
9	Kurnool	3234127305	1059115741	155741651	2175011564
10	Mahabubnagar	2472412487	1001909264	140717520	1470503223
11	Medak	2214067300	724259910	102458425	1489807390
12	Nalagonda	3709608850	1384559316	203571804	2325049534
13	Nizamabad	2307604950	635508406	98679205	1672096544
14	Prakasam	2752596375	1152135330	170063457	1600461045
15	Ranga Reddy	3308374770	895462395	125504017	2412912375
16	Nellore	4025944200	1568582046	227492041	2457362154
17	Srikakulam	2444990100	872837195	131182448	1572152905
18	Visakhapatnam	3257264500	993109905	148131356	2264154595
19	Vizianagaram	2468304700	1022865140	157945853	1445439560
20	Warrangal	3114545525	1299763583	191813530	1814781942
21	West Godavari	6539587125	2902928915	428713560	3636658210
22	Kadapa	2561711050	837658299	118608112	1724052751
	Total	78954398418	29760315032	4411488122	49330738901

Source: www.apmepma.gov.in

Distribution of loan is the simplest activity in microfinance operation and is encountered with high risk. An appropriate client with worth knows as well as realizes the value of money and do things in time and do clear up debts in timely manner is the business mantra of microfinance. Loan disbursement and repayment statistics from table 4.7 reveals that higher loan amount would not necessarily be associated with higher rate of default. 37.51 per cent of circulated money is being effectively utilized. This has been confirmed by the beneficiary repayment worth of Nizamabad district amounting ₹ 8,610, ₹ 8,563 of Ranga Ready district, ₹ 8,076 of Khammam district, ₹ 6,147 of West Godavari District and ₹ 6,283 of Warrangal District against the loan availed.

Recovery of loan is the key challenge for all MFIs operating in the state of Andhra Pradesh. The MFI Ordinance 2010 *i.e.* Andhra Pradesh Micro Finance Institutions (regulation of money lending) and Andhra Pradesh Micro Finance Institutions (regulation of money lending) rules 2010 is further restricting the scope of operation of MFIs and the debt recovery means adopted. Instead, self help group approach with linkage to formal financial institutions is tried as an alternative means to serve the financial need of the uncovered area and population.

The DRDA based self help groups began with the motto to equip and empower rural folk and especially the women towards 'helping themselves', the SHGs are found almost in all villages and across the communities of the state. Though started sporadically in selected pockets, execution of APRPRP has accelerated the pace and space for promotion of self-help initiatives with financial assistance of the World Bank. The year 2004, marked as the year for formation and promotion of self-help groups in the state, is recognized for leveraging as well as congregating maximum women for the common cause of economic empowerment.

Facts also corroborate that Non-Governmental Organizations (NGOs) have played a significant role in grounding and spread of the programme. A catalystic support has been extended by the NGOs towards formation, normation and capacity building of self-help groups. Hence, spread of self-help groups is believed to be directly proportionate to the presence of the number as well as effectiveness of the community based organization.

All districts of the state are home to a minimum of 30,000 self help groups. District West Gadavari has the highest self help groups (80650) while Ranga Reddy is the lowest SHG concentration district with 31513 SHGs. There are 8 districts with concentration of less than 40,000 SHGs, while another 8 districts are concentrated with lessthan 60,000 SHGs, and there are 5 districts representing concentration of lessthan 50,000 SHGs (see table 4.9).

Table 4.9: District-Wise SHG based Credit Plan

District	No. of SHGs	SHGs linked during 2010-11	SHGs linked during 2011-12	% increase
Adilabad	32365	16292.18	18743.64	15
Anantapur	51694	35211.59	30724.52	-13
Chittoor	57088	56277.78	63755.88	13
East Godavari	80650	58298.24	35417.18	-39
Guntur	54667	33313.31	30166.19	-9
Kadapa	31997	24140.60	24039.44	0
Karimnagar	51423	31101.94	24157.53	-22
Khammam	46297	25847.56	21852.43	-15
Krishna	53286	42295.97	37621.54	-11
Kurnool	44164	26622.03	26298.34	-1
Mahabubnagar	46121	27996.72	29335.71	5
Medak	38920	24934.76	23782.44	-5
Nalgonda	54150	25428.74	24547.89	-3
Nellore	34624	21043.07	24364.83	16
Nizamabad	35297	18338.47	19020.19	4
Prakasam	49925	35844.09	38713.75	8
Ranga Reddy	31513	24873.37	32030.94	29
Srikakulam	38559	34014.64	37302.83	10
Visakhapatnam	42322	25697.62	26868.86	5
Vizianagaram	32026	30155.44	31660.05	5
Warangal	52083	29855.82	26208.72	-12
West Godavari	59036	36553.43	31808.89	-13
Total	**1018207**	**684137.37**	**658421.79**	**-4**

Source: SERP, Hyderabad, 2012

Mere presence of SHGs is not the parameters of better economic well-being of the area and population. The effectiveness and potentiality of the group is visible and measurable on the value and volume of linkages initiated during the period. Statics on SHG linkage during 2010-11 and 2011-12 reveals that there are a sizable self-help groups being linked to banks and generated a good volume of business. However, a comparative year-wise performance across the self-help groups over the districts concludes that SHG performance has touched a negative performance of -39 per cent to a maximum performance of +29 per cent.

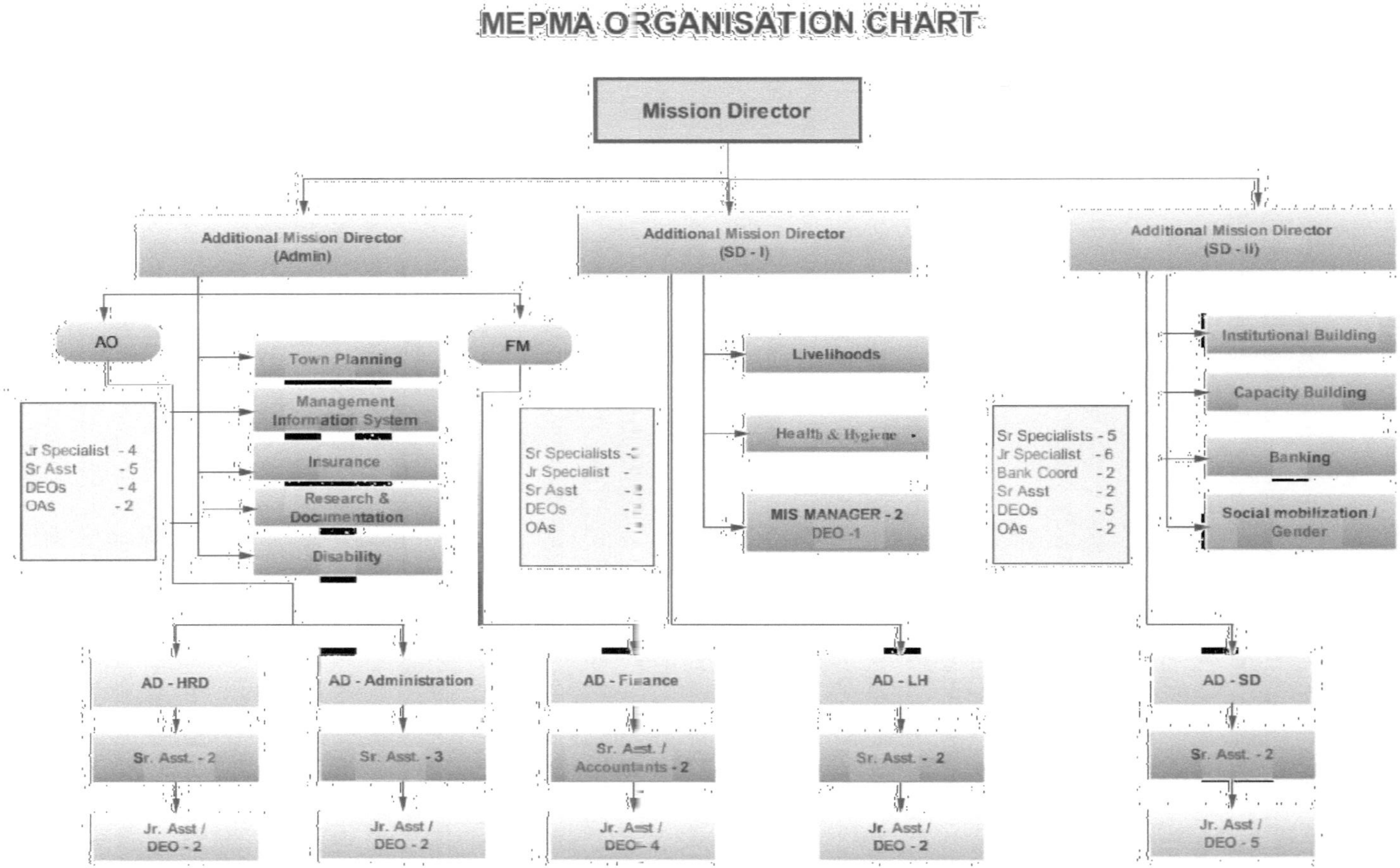

Fig 4.10: Functioning of MEMPA at Municipalities and/or Urban Centers

4.3 Role of Development Sector in Microfinance-Cognizance of NGOs

Non-Governmental Organizations has gained global presence and recognition for the welfarestic activities executed in a holistic way and with adoption of holistic approaches to rural development and empowerment. Over the decade, NGOs are being treated as instrumental in grounding up of livelihood programmes for enhancement of economic wellbeing of the people beyond the lines of caste, creed and religion. The models followed are purely with a holistic intent and better equipped towards 'empowerment of the vulnerable and keeping away vulnerable from debt traps' (see figure-4.10).

A significant proportion of microfinance institutions (MFIs) in developing countries operate either as nongovernmental organizations (NGOs) or as project run by international NGOs. Many of these NGO-MFIs plan to 'transform' into a for-profit company-often, a regulated financial institutions (Lauer, 2008). This gradual and evolutionary growth process has given a great opportunity to the rural poor in India to attain reasonable economic, social and cultural empowerment, leading to better living standard and quality of life for participating households (NABARD, 2009-10).

NABARD has been instrumental in facilitating various activities under microfinance sector, involving all possible partners at the ground level in the field. NABARD has been encouraging voluntary agencies, bankers, socially spirited individuals, other formal and informal entities and also government functionaries to promote and nurture SHGs.

As of 2010, there are 112 institutions of various forms involved in catalyzing the microfinance activities in the state of Andhra Pradesh and are enlisted in NABARD. There common types of institutions involved are Society, MACS, Trusts, Section 25 company, NBFCs and local area banks. Of the total institutions involved in facilitating functioning and/or running of self-help groups, 69.09 per cent are MACS, 14.5 per cent society and rest are NBFCs and Trusts (see table 4.11).

4.4 Government Initiative on Total Financial Inclusion (TFI)

As a part of Financial Inclusion Programme (FIP) of RBI, all formal financial institutions are working to meet the inclusion mandate laid on them. With a mandate of covering under the fold of service to the uncovered villages and population, the drive has gone ahead to meet the allotted figures and numbers. No frill -zero balance account are being opened for beneficiaries from remote villages.

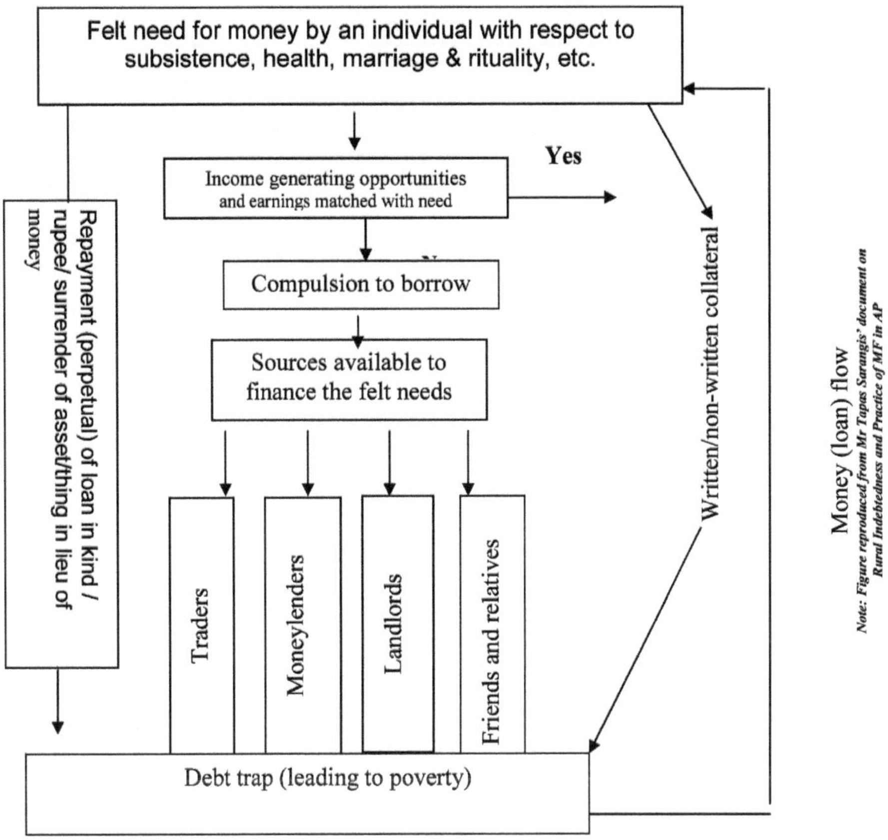

Fig. 4.11: Tribal/Rural Indebtedness and the Debt Trap

The entire initiative is mass drive for making people literate on the functioning of financial institutions as well as bridging the gap of financial service delivery and its access. Promotion of small savings and theft habits are the beginning of a better future. 67.19 per cent of self help groups in the state are having bank accounts and hence, the formal banking services have to reach the unreached with a better 'leap and bound' approach.

4.5 The Road Ahead-Opportunities and Challenges

The death of the MFI industry will push the poor into the grip of moneylenders and deal a blow to the government's financial inclusion drive. The government and RBI must draw an MFI survival strategy before it's too late. We need them at least till such time the banks are ready to reach out to the masses.

Table 4.10: MFIs in Andhra Pradesh as on 2011

Sl. No.	Name	Address	Legal Form	Lending Model
1	Adarsha MACTS	Tangutur (V)& (M), PRAKASAM DT	MACS	SHG
2	Adarsha Mahila MACS	Nadendla Village & Mandal	MACS	Individual
3	Adarsha Mahila MACS	Unguturu, Gandhibomma Centre, KRISHNA DT	MACS	SHG
4	Adarsha Mahila Vikasa MACS	Undrajavaram, WEST GODAVARI	MACS	SHG
5	Adarsha	Managalagiri Mandal, KRISHNA DT	MACS	SHG
6	Sangamam Bahujana Mahila MACTS Ltd	Mangalagiri (V)&(M)	MACS	Individual
7	Adarsha welfare Society	Teachers Colony, MAHABUBNAGAR DT	MACS	SHG
8	AMMACTS	3-13/3, Behind Govt. Hospital, Gandhi Nagar, V.Kota, Chittor, AP - 517424	MACS	NA
9	Anjani MACTS	Addanki (V) &(M), PRAKASAM DT	MACS	SHG
10	Annapurna Mahila MACTS	Yedlapalli Village, Tsunduru mandal	MACS	Individual
11	APMAS	Plot No.20, Road No.2, Rao & Raju Colony (near LV Prasad Eye Institute), Banjara Hills, Hyderabad - 500034	Society	NA
12	ASSIST	G.T.Road, Chilakaluripeta, Guntur District	Society	SHG, MACS
13	Bapatla Mahila MACTS	D.No.1-2-67/1, Railpeta, Baptla	MACS	Individual
14	BASIX	501/502, Nirmal Towers, Dwarakapuri Colony, Punjagutta, Hyderabad - 500082	NBFC	NA
15	Bharatamata,MACTS	Vemulapadu(V), Santhanuthalapadu(M), PRAKASAM DT	MACS	SHG
16	Bharathi MACTS	Doddavaram (V) ,Maddipadu (M), PRAKASAM DT	MACS	SHG
17	Bhartiay Samruddhi Finance Ltd.	Third floor, Surabhi Arcade, Bank Street, Troop Bazaar, Koti, Hyderabad - 500 001	NBFC	
18	BIRDS	26/130 B2, Eva Nest, Gnanapuram, Nandyal - 518502	Society	SHG
19	Centre for Good Governance	Dr. MCR HRD IAP Campus Road No.25, Jubilee Hills, Hyderabad - 500033	Society	NA

Contd...

Table 4.10: Contd...

S. No.	Name	Address	Legal Form	Lending Model
19	Centre for Good Governance	Dr. MCR HRD IAP Campus Road No.25, Jubilee Hills, Hyderabad - 500033	Society	NA
20	Chaitanya Bharathi MAHILA macts Ltd.	Gullapalli village, Chilakaluripeta Mandal	MACS	Individual
21	Chaitanya MACTS	Galijeruggula(V), Bestavaripeta(M), PRAKASAM DT	MACS	SHG
22	Chaitanya MACTS	Yarrathivaripalle, Reddyvaripalle, Nimmanapalle, Chittoor District	MACS	SHG
23	Chaitanya Mahila MACS Ltd	11-137/1-10, By Pass Road, B.Kothakota 517 370, CHITTOOR DT	MACS	Individual
24	Chaitanya Progressive Farmers MACS Ltd	11-137/1-10, By Pass Road,B.Kothakota 517 370, CHITTOOR	MACS	Individual
25	Chaitanya Tadikonda Mandal sangamam Bahujana Mahila MACTS Ltd.	Tadikonda Village & Mandal	MACS	Individual
26	Chetana Mahila MACTS Ltd	GT Road, Chilakaluripeta	MACS	Individual
27	Chittoor Mahila Kalanjia Samakhya	#24/73, Lakshmi Nagar Colony, Chittoor 517101	Trust	SHG
28	Confederation of Voluntary Association	20-4-10, Near Charminar Bus Stand, Hyderabad - 500002	Society	SHG
29	Desabandhavi Duvvurisubbamma WMAC Society	Uppada, EAST GODAVARI DT	MACS	SHG /Individual
30	Dhanalakshmi MACTS	Pullalacheruvu (V) & (M) PRAKASAM DT	MACS	SHG
31	Grama Siri	J.P.Nagar, Nandirajuthota PO 522120, Bapatla, Guntur	Society	SHG
32	Gudupalle Mahila Kalanjia Samakhya	Near Lakshmi Movie Land, North Palace Road, Kuppam 517425	Trust	SHG
33	GUIDE	74-24-15/1, Ground floor, Yarlagadda Apparao Ashok nagar, Vijaywada, Krishna - 520007	Society	Others
34	Indian Grameen Services	406, Nirmal Towers, Dwarakapuri Colony, Punjagutta, Hyderabad - 500082	Section 25 Company	Others
35	Indira Priyadarshini Women Welfare Association	Jadcherla Mandal, MAHABUBNAGAR	MACS	Individual
36	Indur Intideepam MACTS	Nizamabad	MACS	Federation to MACS to SHGs

Contd...

Table 4.10: Contd...

Sl. No.	Name	Address	Legal Form	Lending Model
36	Indur Intideepam MACTS	Nizamabad	MACS	Federation to MACS to SHGs
37	Integrated Rural Development Society	37/2 Kottakotavari Veedhi, Amadalavalasa, SRIKAKULAM	Co-operative	Grameen
38	Jana Chaitanya MACTS	Karavadi(V) & Ongole (M) PRAKASAM DT	MACS	SHG
39	Janani MACTS	Kanigiri (V) &(M), PRAKASAM DT	MACS	SHG
40	Jeevana Jyothi	Thulluru Mandal		
41	Sangamam Bahujana Mahila MACTS Ltd	Thulluru Village and Mandal	MACS	Individual
42	Jeevani MACTS	Karedu (V) Ulavapadu(M), PRAKASAM DT	MACS	SHG
43	Kandukuri Rajyalakshmamma WMAC Society	Kondevaram, EAST GODAVARI DT	MACS	SHG /Individual
44	Kasthurba Gandhi WMAC Society	Komaragiri, EAST GODAVARI DT	MACS	SHG Individual
45	KBSLAB	7-4-58/1, 1st floor, PC Towers, Venkateshwar, Mahabubnagar - 509002	Local Area Bank	Others
46	Kiranmai Pedakurapadu Mandal Sangamam Bahujana Mahila MACTS	Pedakurapadu (V)&(M)	MACS	Individual
47	Kolleru Rural Development Service Organisation	24A-20-17, Ashok Nagar, Near Andhra Bank, Eluru 534 002, WEST GODAVARI DT	Society	SHG
48	KRUSHI	H.No.2-9-119, Mukarampura, Karimnagar - 505001	Society	SHG
49	Kuppam Kalanjia Samakhya	#6/A Saikrupa Nilayam, Rajiv Nagar Colony, Krishnagiri Road, Kuppam 517425, Chittoor District	Society	SHG
50	Manasa MACTS	Lingalasamudram (V) & (M) PRAKASAM DT	MACS	SHG
51	MARI	H.No. 1-8-499, Behind Ekasila ark, Balasa - Mudram, Hannam Konda	Society	NA
52	Matrusri MACTS	Singarayakonda (V) &(M), PRAKASAM DT	MACS	SHG
53	MHIPL	107, Prasasan Nagar, Jubilee Hills, Road No.72, Hyderabad -	Ltd. Company	NA

Contd...

Table 4.10: Contd...

Sl. No.	Name	Address	Legal Form	Lending Model
54	Modern Architects of Rural India	1-8-499, Balasamudram, Hanamkonda 506 001 WARANGAL DT	Society	SHG
55	Mother Teresa MACT	Bestavaripeta (V)&(M), PRAKASAM DT	MACS	SHG
56	Mother Teresa Mahila MACTS	D.No.3-62, Bodduvari Centre, Gorantla Village, Guntur Rural Mandal	MACS	Individual
57	Mother Theresa MACTS	Avanigadda, 9th ward, KRISHNA DT	MACS	SHG
58	Mother Theresa Mahila Bank	Muppala, Chandarlapadu, KRISHNA DT	MACS	SHG & Individual
59	Pitlam Intideepam MACTS	Nizamabad	MACS	NA
60	Pragathi MACTS	Darsi (V) & (M) PRAKASAM DT	MACS	SHG
61	Pragathi MACTS	Pernamitta (V)& S.N.Padu (M) PRAKASAM DT	MACS	SHG
62	Pragathi Mahila MACS Ltd.	Komatinenivaripalem (V), Chilakaluripeta Mandal	MACS	Individual
63	Pragathi Seva Samity	Bethel 2-5-898, Circuit Road House, Hanamkonda, Warangal - 506001	Society	SHG
64	Prasanthi MACTS	Prasant Nagar, Near MRO Office, Kurabalakota, Chittoor District	MACS	SHG
65	Pratibha Mahila MACTS	Yedlapadu Village & Mandal	MACS	Individual
66	Punganuru Mahila Kalanjia Samakhya	#31-95-5, Kothalndlu, Punganuru 517247, Chittoor District	Trust	SHG
67	PWMACTS	Siva Surya Apartments, Ground floor, Main Road, Payakaraopeta, Vishakapatnam - 531126 (Andhra	MACS	Grameen
68	Rashtriya Seva Samithi	Seva Nilayam, Annamaiah Marg, A.I.R Bye-Pass Road, Tiripati - 517501	Society	SHG
69	Ratnamani Mahila MACTS	Vemuru (V) & (M)	MACS	Individual
70	Rural Development Trust (RDT)	Bangalore Highway, Anantapur	Trust	Individual
71	Saadhana MicroFin Society	50/760, A-116, II Floor, Gayatri Estates Kurnool - 518002	Society	Grameen
72	Sahaya MACTS	Ongole (V) & (M), PRAKASAM DT	MACS	SHG
73	Sai MACTS	Ammanabrolu(V), N.G.Padu(M), PRAKASAM DT	MACS	SHG

Contd...

Table 4.10: Contd...

Sl. No.	Name	Address	Legal Form	Lending Model
74	Samatha MACTS	Reddigudem, Near Mandal Office, KRISHNA DT	MACS	SHG
75	Sanghamithra Mahila Mutually Aided Cooperative Society Ltd.	28/16/14, Kota Indlu, Punganur 517247, Chitoor District	MACS	Individual
76	Santoshi Mahila Vikasa MACS	Peravali Mandal, WEST GODAVARI DT	MACS	SHG
77	Sarada Mahila MACTS	Pasummaru (V), Chilakalripeta Mandal 522616	MACS	Individual
78	Seethalamba MACTS	Poluru(V) Yeddanapudi(M) PRAKASAM DT	MACS	SHG
79	SERP	Hermitage Office Complex, III & IV floor, HUDA Complex, Hill fort road, Adarshnagar, Hyderabad - 500004	Society	SHG
80	SEVA	H.No.11-24-36, M.V.Colony, Road No.2, Deshaipet Road, Warangal - 506002, Andhra Pradesh	Sec.25 Company	SHG
81	Shakti Mahila MACS Ltd.	4-67, Opp Agrl College Staff Quarters, Tirupati, CHITTOOR DT	MACS	SHG
82	SHARE MICROFIN	1-224/58, Rajeev Nagar, Nacharam, Hyderabad - 500076	NBFC	Others
83	SKS	Flat No.301, III floor, Babukhan Estate, Basheer-bagh, Hyderabad-500001	Ltd. Company	Grameen
84	Sneha MACTS	Peddapavani(V) Lingasamudram(M), PRAKASAM DT	MACS	SHG
85	Sneha Tadepalli Mandal Sangamam Bahujana Mahila MACTS	D.No.14-7-02, Near Srinivas Cinema Hall, Krishna Nagar, PO, Tadepalli (V)&(M)	MACS	Individual
86	Spandana Sphhoorty Innovative, Financial Services Ltd., #	7-1-19/5, Vamsiram's Jyothi Bhupal Apartments, Opp.: Country Club Road, Begumpet, Hyderabad – 500016	Grameen	
87	Spandana MACTS	Mopadu(V) Kandukur(M), PRAKASAM DT	MACS	SHG
88	Spoorthi Sangaman Bahujana Mahila MACTS Ltd.	Amaravathi Village & Mandal, Ph. 08645-254041	MACS	Individual
89	Sri Anjaneya MACTS	Jarugumalli (V) & (M) PRAKASAM DT	MACS	SHG
90	Sri Chaitanya MACTS	Giddalur (V) & (M) PRAKASAM DT	MACS	SHG

Contd...

Table 4.10: Contd...

Sl. No.	Name	Address	Legal Form	Lending Model
91	Sri Chaitanya MACTS	Santhanuthalapadu (V)&(M), PRAKASAM DT	MACS	SHG
92	Sri Lakshmi MACTS	Karamchedu (V) & (M) PRAKASAM DT	MACS	SHG
93	Sri Nidhi MACTS	Vetapalem (V)&(M), PRAKASAM DT	MACS	SHG
94	Sri Padmavathy Mahilabhyudaya Sangam (SPMS)	Society	SHG	
95	Sri Pragathi MACTS	Marturu (V) & (M) , PRAKASAM DT	MACS	SHG
96	Sri Vani MACTS	Marripudi (V) & (M) PRAKASAM DT	MACS	SHG
97	Sri Vyshnavi MACTS	Parchur (V) & (M) PRAKASAM DT	MACS	SHG
98	Star Youth Association	17-178-17/14, Beside Telephone Exchange, Velgode (K) -518533	Society	Grameen
99	Stree Sakthi MACTS	Uppugunduru(V) N.G.Padu(M), PRAKASAM DT	MACS	SHG
100	Swasakthi	MACTS , Chimakurthy (V)&(M),PRAKASAM DT	MACS	
101	SWAWS	6-1-132/103/c2, Navratna Apartments, Skandagiri, Secunderabad - 500061	Society	Grameen
102	Tajeswari Mahila MACTS	Kakumanu Village and Mandal 522112	MACS	Individual
103	Tejaswi Mahila MACTS Ltd.	Old Panchayat Office, Kakumanu 522112, Guntur district	MACS	SHG
104	Ujwala MACTS	Kondapi(V)&(M), PRAKASAM DT	MACS	SHG
105	Vanitha Bharathi Mahila MACTS	D.No.7-2-28, Bhimavari Street, Bapatla	MACS	Individual
106	Vasavi Vanitha MACTS	D.No.24-12-7, Brahmin Street, Ponnur 522124	MACS	Individual
107	Vasundhara MACTS	Inkollu (V)&(M), PRAKASAM DT	MACS	SHG
108	Vedantham Kamalamma Society	WMAC , Nagulapali, EAST GODAVARI DT	MACS	SHG /Individual
109	Velugu MACTS	Ulichi (V) Ongle (M), PRAKASAM DT	MACS	SHG
110	Vikasa Bharathi Mahila MACTS	Inturu Village, Amarthaluru Mandal 522341	MACS	Individual
111	Vimochana MACTS	Venkataramaiah Street, New Earlspet, Madanapalle 517325, Chittoor District	MACS	SHG
112	Viswanadha Mahila MACTS	D No.5-15-4, Railpeta, Nidubrolu, Ponnur Mandal	MACS	Individual

Source: MICD, NABARD 2012

Notes

1. Eastern region consists of Bihar, Jharkhand, Orissa, West Bengal, and Andaman and Nicobor island.
2. West region consists of Goa, Gujurat, Maharashtra and Rajasthan.
3. Central region includes Madhya Pradesh and Chhattisgarh.
4. North region is consists of Chandigarh, Delhi, Haryana, Himachal Pradesh, Jammu and Kashmir, Punjab, Uttar Pradesh and Uttarakhand.
5. Northeast region includes Arunachal Pradesh, Assam, Manipu, Meghalaya, Mizoram, Nagaland, Sikkim and Tripura.
6. It is possible that the outreach in northeast is low because of the relatively small size of this region as compared to the other large size regions, so it would be appropriate to look at the penetration level relative to the population of the region.
7. India: MFIs caught between RBI and A.P Government Norms, Microfinance: Models, Policies and Experiences Africa Filed Under Microfinance: Models, Policies and Experiences Around the World. http://microfinanceafrica.net/tag/microfinance-institutions-network/page/3/
8. 'Access to Finance in Andhra Pradesh 2009', by Centre for Micro Finance, Chennai, commissioned by Centre for Microfinance: Models, Policies and Experiences Research, BIRD, Lucknow.
9. The Mission for Elimination of Poverty in Municipal Areas (MEPMA) is a Government of AP Society forming part of Department of Municipal Administration and Urban Development. It was launched on 1st September, 2007 by Smt. Sonia Gandhi, UPA Chairperson. Sri N. Kiran Kumar Reddy, Hon'ble Chief Minister of Andhra Pradesh is the Chairman of Governing body whereas Sri M.Maheedhar Reddy, Hon'ble Minister for Municipal Administration and Urban Development Department is the Vice Chairman. The Principal Secretary, Municipal Administration, is the chairman of MEPMA's Executive Committee. Mission Director, MEPMA, and his team implement the urban poverty alleviation schemes. The district units are headed by Projects Directors. Empowerment of urban poor women, especially those residing in slums, is the main objective of MEPMA. It is a platform of Urban (Women) Self Help Groups (SHGs) for Empowerment (see the organization chart-fig-6.11).

References

1. Lascelles. D (2012): Microfinance: Models, Policies and Experiences-A Risky Business A Time for Strong Leadership, Centre for Financial Inclusion, Publication 17, May.
2. Banking with the poor Network (2009): Microfinance: Models, Policies and Experiences Industry Report: Bangladesh 2009, Banking with the Poor Network in Collaboration with SEEP.

3. Kevany, S (2010): Debt trends in Peru's Personal, Microfinance: Models, Policies and Experiences Sectors Spark, Warnings Wall Street Journal, 16 December.
4. Bateman, M (2011): Confronting Microfinance: Models, Policies and Experiences: Undermining Sustainable Development. Sterling VA: Kumarian Press.
5. Cain, P (2010): Microfinance: Models, Policies and Experiences Meltdown in Bosnia, Al Jazeera (English), 4 January.
6. Chakrabarti. R (2004): The Indian Microfinance: Models, Policies and Experiences Experience- Accomplishments and Challenges, India Development Foundation, Gurgaon.
7. Businessmaps of India.com/finance/top-finance companies –in-India.
8. Sinha Anand (2012): Strengthening Microfinance: Models, Policies and Experiences Institutions (MFIs): good Governance and Strategic People Practices, Mumbai, April.
9. Lokpal Capital (2010): Microfinance: Models, Policies and Experiences Industry in India, taken from NCAER-The great Indian market.
10. The Social Performance report (2011): "Microfinance: Models, Policies and Experiences funders and Social Performance Management", Microfinance: Models, Policies and Experiences India, ACCESS Knowledge Series, New Delhi.
11. Sinha.F (2009): State of Microfinance: Models, Policies and Experiences in India, Institute of microfinance (InM).
12. Champatiray, Agrawal and Sadhu (2011): Map of Microfinnace Distribution in India, IFMR, Centre for Microfinance: Models, Policies and Experiences research.
13. M-CRIL, 2011: Microfinance: Models, Policies and Experiences Review 2011 Anatomy of a crisis- a financial and social analysis.
14. Department of Rural Employment and Poverty Alleviation, Annual Report: 1998-99, Ministry of Rural Development, New Delhi, 1999.
15. Fernandez, Aloysius P (2007): "History and Spread of the Self-Help Affinity Movement in India: the Role Played by IFAD." Occasional Paper Series, IFAD, July 2007.

    http://www.ifad.org/operations/projects/regions/pi/paper/3.pdf.
16. Srinivasan (2011): Microfinnace India, State of the Sector Report 2011, Sage Publication, New Delhi.
17. Joseph. P (2007): www.indiareports.com/business writters/ indian microfinance sector.
18. Lauer, K(2008): Transforming NGO MFIs: Critical Ownership issues to Consider, Occasional Paper, CGAP, No.13, June.
19. Status of Micro finance in India, 2009-10, NABARD, Mumbai.

Chapter 5

MFI's in Agency Areas' of AP: A Pragmatic Analytical View

5.0. Introduction

India is a democratic country, so are the microfinance institutions. There are two major models for microfinance delivery *i.e.* the SHG Bank Linkage Programme and the MFI model in operation in the country. While the MFI model is growing rapidly the SBLP is by far the more dominant model in terms of outreach. Both these models are very distinct from each other in terms of methodologies adopted and legal forms of institution involved in service delivery - SBLP is promoted by the apex agricultural bank (NABARD) through the commercial banks while the MFI model is privately managed with some institutions being regulated by the Reserve Bank of India.

SBLP, initially promoted by MYRADA (in Karnataka) and PRADAN (in Tamil Nadu and in Bihar) and formally launched in 1992 by NABARD began to scale-up in the mid-1990s. Under this programme, NABARD refinances bank loans to SHGs through commercial banks but the credit risk is carried by the banks. This has stimulated the growth of SHGs nationwide making these a dominant model in the delivery of microfinance.

The MFI model uses a variety of methodologies ranging from the very popular SHG methodology to Grameen and joint liability groups, as well as individual banking arrangements. Over the years, through an ongoing process of experimentation and innovation, Indian MFIs now largely follow a mixed approach customised to their target segment and area of operation. In addition to the microfinance delivery methodologies, MFI management approach, services and performance are influenced by their legal form - ranging from not-for-profit Societies/Trusts and not-for-profit Companies registered under Section 25 of the Companies Act to NBFCs licensed by RBI. The microfinance methodologies and institutional framework adopted by MFIs are illustrated below.

Fig. 5.1

Note: Figure reproduced from "State of Microfinance in India",2009

5.1. MFIs Development and Growth in ITDA's

MFIs in the state of Andhra Pradesh are as old as the Self-help groups (SHGs) in the country. There are nearly 525 MFIs operating in the state having head quarter of operation within the state (see table-5.6). Interestingly, there are 38 MFIs operating in the state with administrative head office as well as branch offices in the state. MFIs in the state and across the 29 districts are catering to the financial need of 3653118 clients which constitutes nearly 25.96 per cent. Out of the 29 districts under MFI service coverage, 19 are recognised to be poor districts. The portfolio outstanding across the districts constitutes 31.75 per cent *i.e.* 19445023878.

History goes that, tribals of agency areas otherwise known as ITDA areas are quite self-sufficient so far as the consumptive needs are concerned. Almost everything required for the day -to-day living and survival are met from the natural resources available. Paderu ITDA is home to as many as 11 tribal communities and of them three sects of PTGs are residing over in it. PTGs (Kondhs, Gadava and Paroja) are residing on the top of the hills and have very little access to the modern avenues and services (Pamela Mackenzie, 2006). The means of survival is through the age of traditional activity surrounding to the available resources. Life is simple- production-consumption- reproduction-regeneration. Life is beautiful amidst nature.

Modern economy could not have its reach to these pious places and people. Education is still an alien subject for many and hence, is their participation in common day-to-day activity in a market economy. The years of intensive effort of tribal welfare department of the Government have changed the institutional and infrastructural status quo of the area and made the injected services access to people.

Paroja Tribe (PTG)

The welfarestic activities have long reaching impacts in terms of creation of awareness and making the services reach at the door step of the tribal houses irrespective the geographical cut-off-ness. The tribals of two Mandals have been organised and determined to form a MACS under the technical support of AASSAV-an NGO working for past four decades. The MACS is taking care of financial needs of tribals and have its operating as well as administrative office at Araku valley. It is run by a group of tribal adhering all technical parameters of functioning of a MACS. It is the single private financial institution serving the tribals in Paderu agency area.

Gadava PTG

5.2. Various Forms of MFIs

A range of institutions in public sector as well as private sector offers the microfinance services in Visakhapatnam district of Andhra Pradesh. Based on asset sizes, MFIs can be divided into three categories:

(i) 5-6 institutions which have attracted commercial capital and scaled up dramatically within last five years. The MFIs which include SKS, SHARE and Grameen Style program but after 2000, converted into for-profit, regulated entities mostly Non-Banking Finance Companies (NBFCs).

(ii) Around 10-15 institutions with high growth rate, including both news and recently form for-profit MFIs. Some of MFIs are Grameen Koota, Bandhan and ESAF.

(iii) The bulk of MFIs are NGOs struggling to achieve significant growth. Most continues to offer multiple developmental activities in addition to microfinance and have difficulty accessing growth trends.

Private MFIs barring a few exceptions are still fledging efforts and are therefore unregulated. They secure microfinance clients with varying quality and using different operating models. Regulatory framework should be considered only after the sustainability of MFIs Model as a banking enterprise for the poor is clearly established.

5.3. MFIs Spread in ITDA Areas

The number and volume of spread of MFIs across the districts and across the tribal concentrated districts varies significantly. The operation of microfinance institutions are though guided by the inherent form, experimentation of different modus operandi is also experienced to be the guiding principle for many. Location of operation as well as circulation of M3 is the central to the operation of MFIs irrespective of tribal and non-tribal districts consideration. Nevertheless, presence of other MFIs and NGOs has considered being a factor for the operation of the MFIs.

There are 198 microfinance institutions (MFIs) in operation at ITDA districts of the state of Andhra Pradesh (see table-5.1). The noted MFIs among them are Asmitha Microfin Limited, Bhartiya Samruddhi Finance Limited (BSFL), Cresa Financial Services Private Limited, Fullerton India Credit Company Limited, L and T Finance Limited, Share Microfin Limited, SKS Microfinance Limited and Spandana Spoorthy Financial Limited etc. Their origin and lateral spread occurred over the continuum of two decades. Though they are present in the tribal districts of the state but their presence in all ITDAs of the state is questionable.

Table 5.1: MFIs in ITDA Districts of Andhra Pradesh

Sl. No.	District	ITDA	No. of MFIs
1	Visakhapatnam	Paderu	24
2	Vijayanagaram	Parvathipuram	18
3	East Godavari	R.C. Varam	23
4	Srikakulam	Seethampeta	20
5	Khammam	Bhadrachalam	15
6	West Godavari	K R Puram	24
7	Warangal	Eturunagaram	22
8	Adilabad	Utnoor	14
9	Mahaboobnagar	Srisailam	15
10	Nellore	Yanadi	23
	Total		**198**

Source: Consolidated from DRDA Reports from 2010 and 2011

5.4. MFI's in Paderu Agency Area-A Thorough View of SHG Mode of MFI Operation

There are about 24 MFIs functioning in the Visakhapatnam district, with a turnover of ₹ 260 crores . Nearly one lakh people took loans from them, 65,000 of them in rural areas. Paderu agency area of the Visakhapatnam district is the most difficult and highest populated agency in of the state of Andhra Pradesh. More than 3574 habitations over 11 Mandals are nestled across hills and valley area.

Indira Kranti Patham (IKP) has been implementing the microfinance programme in Paderu region with involvement of community-based-CBOs namely self-help groups and with the financial assistance of World Bank. Various activities undertaken in the agency area ranges from mobilization of women into self help groups (SHGs), federating them into Village Organization (VO), Mandal Mahila Samakhyas (MMS) and Girijana Maha Samakhya (GMS) for all 11 Mandals, providing financial access through Bank linkage (BL), Community Investment Fund (CIF), Revolving Fund (RF), Grain Banks (GB) etc. Facilitating social security through insurance and pension schemes and new pilot activities like Health and Nutrition Centers (HNC), Early Child Education Centers (ECEC) are being implemented in all the 11 Mandals and 3574 habitations of Paderu agency area (see table 5.2).

Table 5.2: Status of Self-help Groups in Paderu Agency Area

Mandal	Total HHs	BPL HHs			Total SHGs	Total members	Total VOs	Total MMS
		Poor	POP	Total				
1	2	3	4	5	6	7	8	9
Ananthagiri	10505	4305	5722	10027	665	8484	63	1
Arakuvalley	9811	4698	4460	9158	843	9092	48	1
Dumbriguda	9581	5015	3867	8882	779	8848	54	1
Hukumpeta	12313	6373	5307	11680	990	11415	54	1
Pedabayalu	11874	2804	8478	11282	906	11357	59	1
Munchingiput	10654	4423	5812	10235	619	7324	57	1
Paderu	11898	5747	4570	10317	881	10997	56	1
G.Madugula	12291	5374	6277	11651	871	11055	52	1
Chintapalli	14227	7250	6496	13746	840	10378	46	1
G.K. Veedhi	12310	5824	5513	11337	534	6413	42	1
Koyyuru	12348	6000	5475	11475	838	8617	52	1
Total	127812	57813	61977	119790	8766	103980	583	11

Source: IKP Annual Report 2011-12, ITDA, Paderu, Visakhapatanam, Andhra Pradesh

Mere presence of self help group does not make any difference in the financial status quo or ability of the people to take avenues of income generation and venture for exploring new ways of sustenance. The functioning and financial rotation is the prime mover of the economy, people and the institutions. Only 65.7 per cent (5757) self help groups are functioning in the agency area regularly, while the remaining are default and defunct. Of the 65.7 per cent, 72 per cent (4145 *i.e.* 47.3 per cent of the total SHGs) self help group are reported conducting monthly meetings, while the rest of them are following a meeting cycle of weekly and fortnightly. Member participation in meetings is reported to be as high as 98 per cent at the extreme to an average attendance of 72 per cent in general. Good participation ensures members involvement in day-to-day functioning of the self-help group and also taking up of firm productive decisions in fund devolvement on income generation activities. The stated participation rate is based on annual average participation across the seasons and it is morethan the average.

Out of the total functional self help group of 5757, there are about 85 per cent attend the VO meeting regularly. Of the total functioning self help groups, 47 per cent self help groups are maintaining the books, adhering book keeping practices and books are being timely updated. It is recorded that internal lending within the group is very low; it is about 43.84 per cent of the total functioning self help groups which accounts to 2524. This practice is resulting in lower coverage of fewer families and slowing down realization of economic benefits a meager few. There are about 56. 16 per cent of self help group members are devoid of getting access to the financial services.

Finance market is primarily and significantly controlled by the formal institution while role of informal and/or non-formal financial institutions in the agency area cannot be completely refuted. There are quite a good number of formal banks (11) operating in the agency area to cater the financial needs and requirement of tribal people and the community based organizations. Almost all nationalized banks have strong presence across the Mandals of the Paderu agency. However, the APGVB is single handedly catering 25.37 per cent of financial requirement of tribal people across agency during the financial year 2011-12 only (see table-5.3). APGVB is having a strong rural presence and especially in rural and tribal areas of the state of Andhra Pradesh. The other formal banks are CNB, DCCB, Indian Bank, Indian Overseas Bank, State Bank of Hyderabad, State Bank of India, Union Bank of India, UCO and Vijaya Bank.

Table 5.3: Bank and MFIs-wise Achievement for the year 2011-12

Sl. No.	Institutions	Total Target		Total Achievement		% of Achievement
		SHGs	Amount	SHGs	Amount	
1	2	3	4	5	6	7
1	APGVB	1821	1994.00	439	505.78	25.37
2	ASSAV MACS	431	422.50	45	55.41	13.11
3	CNB	24	23.00	11	9.50	41.30
4	DCCB	60	33.50	0	0.00	0.00
5	Indian Bank	183	275.00	33	43.30	15.75
6	IOB	60	33.00	0	0.00	0.00
7	SBH	234	228.50	159	224.25	98.14
8	SBI	973	913.50	260	274.41	30.04
9	UBI	1570	1930.50	369	443.00	22.95
10	UCO	55	34.00	15	12.00	35.29
11	Vijaya	95	48.00	0	0.00	0.00
	Grand Total	5506	5935.50	1331	1567.65	26.41

Source: Data Compiled from Progress Report-2011-12 of ITDA Paderu

All banks are functioning with a targeted mandate of covering 24 self help group at the minimum to 1821. The annual mandate varies every year and so also the fund injection. Banks such as, Vijaya, DCCB and IOB have not supported any self help group during the financial year 2011-12. However, there are 1331 self help groups across the district of Visakhapatnam are being supported with financial linkage of ₹ 1567.65 lakhs. This has resulted a little more than 26 per cent achievement but in terms of its turn over it has helped many self-help members to stand up their own and made them self-sustenance.

The presence as well as contribution of Non-Banking Financial Intermediaries (NBFIs) is visible in the Paderu agency area. Tribals of the agency area have floated Mutually Aided Cooperative Society (MACS) to regulate the financial transactions of the community through an institutional mode. The MACS and popularly known as AASSAV-MACS is a four-decade old NBFC working in the Araku Mandal of Paderu agency area. It is catering to the need of 431 self-help group constituting 13.11 per cent of total fund in circulation during the financial year 2011-12.

5.5. Accessing Social Safety-nets

May it be called welfarestic look to people's safety and/or livelihood security, may it be coinage as securing food and income of the people of varied age and may it be named as social safety-net, it all leads to one and the same basket of intervention to protect people from unemployment, deprivation, hunger and malnutrition. There are various programmes and schemes being crafted by central government and state government to serve the people for their survival with dignity. There are central grants, state grants and also common grants by state and central government jointly. Self help groups (SHGs) being institutions of catalystic importance are facilitating to bring the programme to people and enhance people accessibility to service delivery mechanism. The current section is viewing the contribution of MFIs in bringing welfare programmes to people.

5.5.1 SGSY

Swarnjayanti Gram Swarozgar Yojana (SGSY) is a Government of India initiative to bring the poor families (Swarozgaries) above the Poverty Line by ensuring appreciable sustained level of income over a period of time. This objective is to be achieved by inter alia organizing the rural poor into Self Help Groups (SHGs) through the process of social mobilization, training and capacity building and provision of income generating assets. The SHG approach helps the poor to build their self-confidence through community action. Interactions in group meetings and collective decision making enable them in identification and prioritization of their needs and resources. This process would ultimately lead to strengthening and socio-economic empowerment of the rural poor as well as improve their collective bargaining power.

SGSY lays stress on the cluster approach. Instead of funding diverse activities, each block should concentrate on a few select activities (key activities) and attend to all aspects of these activities, so that the Swarozgaris can draw sustainable incomes from their investments. These key activities are preferably taken up in clusters so that the backward and forward linkages can be effectively established. This would facilitate not only monitoring but more importantly provision of various services required by the Swarozgaris.

The key element of the scheme is the choice of activity and that has been selected based on the available local resources, the aptitude as well as the skill of the people. For selection of key activities, a profile of the poor families, as reflected in the BPL Census is analyzed. There will be poor families with assets, such as land. Efforts are made to see that those having a minimum extent of land are enabled to cross the poverty line by making additional investment on their lands, such as wells or other irrigation facilities, pump sets etc. The experiences over the years as well as Evaluation Studies have shown that investments on land based activities have resulted in generation of income on a more sustainable basis.

The programme has covered 3284 self help groups in agency area. During the past six years ₹ 248.97 lakhs have been injected into the tribal economy for undertaking land development activities. Self help groups are supported with a minimum of ₹ 5000 per annum to ₹ 60,000 per annum depending on the type of activity being selected and the financial requirement for that. The number of groups under coverage over the years is volatile in nature and depends on the interest as well as eligibility of the activities being under selection, screening and execution. However, fund support given is the maximum to an ever minimum number of self help groups (see fig.5.2).

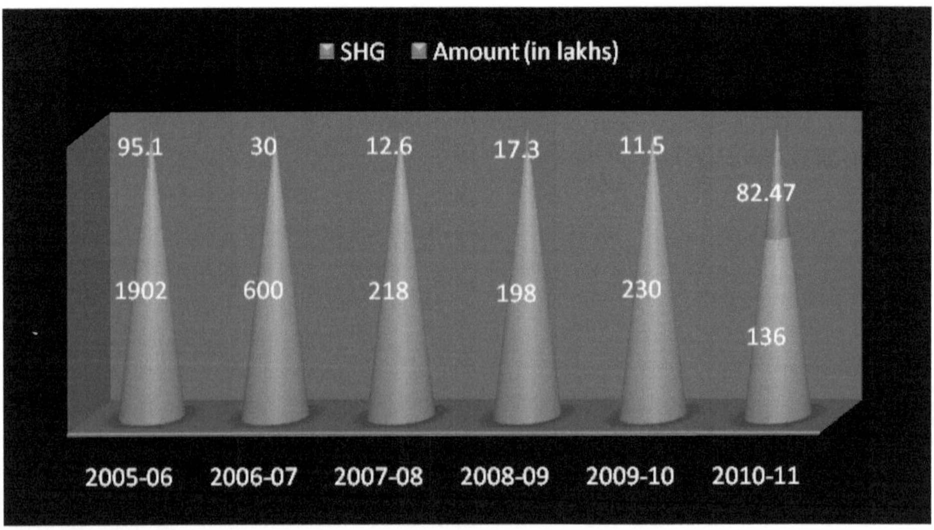

Fig. 5.2: SGSY Coverage in Paderu ITDA

The District Rural Development Agencies generally release the fund to the Zilla Mahila Samakshya (ZMS) which in turn allocate to Mandal Mahila Samakshya (MMS) and MMS in turn to Village Organization (VOs). There are 136 self help groups (SHGs) and 73 Village Organizations (VOs) of Paderu ITDA are supported with financial assistance of ₹ 108.9 lakhs. All 136 self help groups reported an expenditure of ₹ 82.47 lakhs which is about 75.73 per cent of the fund sanctioned to the ZMS. Under taking activities around land to ensure as well as enhance rural income in agency areas is a herculean task. It is guided by factors like climatic condition, type of land, location of land, land on the hill or valley, access to the site etc.

Table 5.4: SGSY Fund Disposal for FY 2010-11

Sl.No.	Mandal	Amount released from DRDA VSP	No. of VOs	No. of SHGs	Expenditure Reported	Balance available with MMS
1	2	3	4	5	6	7
1	Ananthagiri	8.90	5	5	4.5	4.40
2	Araku Valley	10.00	7	7	6.1	3.90
3	Dumbriguda	8.60	6	20	7	1.60
4	Hukkumpeta	9.80	1	3	2.8	7.00
5	Peddabayalu	9.50	8	19	9.5	0.00
6	Munchingput	8.50	4	11	8.5	0.00
7	Paderu	11.20	13	19	11.2	0.00
8	G. Madugula	9.80	9	14	9.8	0.00
9	Chintapalli	12.20	7	18	11.87	0.33
10	G.K.Veedhi	11.10	7	14	5.1	6.00
11	Koyyur	9.30	6	6	6.1	3.20
	Total	108.9	73	136	82.47	26.43

Source: Annual Report, IKP-2010-11, Paderu

Location of villages in all 11 Mandals is a challenge in itself. Mandals like Dumbriguda, Peddabailu, Paderu and Chintapalli have reported better access to the programme. Almost 56 per cent of self help groups spread across four Mandals have better access to fund in comparison to all other Mandals of the agency area. Five self help group of Hukkumpeta Mandal have accessed the programme spread across 4 Village Organizations (see table-5.4). There is no a specific pattern revealed for access of the SGSY programme and it is purely becoming an initiative of better leadership of Village Organizations and self help groups. However, evolution of self help group has to a greater extent assisted faster as well as better coverage of the programme in unreachable and most difficult tribal area. 26.43 lakhs fund still lying un-utilized across 7 MMS with a minimum of 0.33 lakhs to a maximum of ₹ 7.00 lakhs per MMS. This is due to lack of appropriate plans and insufficient avenue to invest for development of income generation activities. A good purposive planning and through scanning would entail optimum utilization of the programme and benefit the marginal and small tribal farmers.

5.5.2 Insurance - Abhaya Hastham (ABH), Aam Admi Bima Yojana (AABY) and Janasree Bima Yojana (JBY)

Delivering services to people through an institutional mode is the motto of Self help groups in rural and tribal areas. There are many schemes and programmes of state and central government designed to benefit the public and especially to the needy. Tribal areas are exceptions to it. The tribal welfare department is working dedicatedly in convergence with all other line departments following an 'umbrella approach' to provision all basic necessities of life. IKP through ITDA is serving as 'single point entry and exit' of all development programmes. All programmes are executed through ITDAs. Insurance schemes are executed with a similar approach and its impact is being captured in the subsequent sections.

5.5.2.1 Abhaya Hastham

Govt. of AP have undertaken comprehensive eradication of poverty, in rural and urban areas through the Indira Kranthi Patham, a Programme for empowerment of women, through formation, development and strengthening of women SHGs and their federations, covering 1 Crore 25 Lakh women in rural and urban areas. During the last 3 years, Government is also implementing National Rural Employment Guarantee Scheme (NREGS) to secure guaranteed wage employment to the rural poor. As a result of these measures, a large number of women and their families in rural and urban areas are enjoying stable livelihoods, increased incomes and better quality of life. While economic growth at household level is an important pre-condition for expanding the scope of Social Security, no less important is the need for adopting special social security policies. The Government regards social security as one of the important factors in improving quality of life and elimination of poverty.

The need for income security in their old age, when their ability to earn from their physical labour would have reduced considerably has been voiced by SHG members in various meetings they had with the Chief Minister of Andhra Pradesh in the month of December 2008. Responding to the demand from the SHG women, the Chief Minister of Andhra Pradesh has conceptualized Abhaya Hastham.

Abhaya Hastham is a Co-contributory Pension Scheme for the women, above the age group of 18, belonging to the Self Help Groups in Andhra Pradesh. The scheme envisages contribution of ₹ 30 per month by SHG women and Government's co contribution of ₹ 30 per month into her Pension account. The contribution of the member and the co-contribution of the Government is periodically transferred to Life Insurance Corporation of India, for investing diligently for securing better returns on the investment. The corpus thus generated till the age of 60 years, will be used for giving monthly pension amount to each women, on crossing 60 years of age.

Abhaya Hastham offers 3 kinds of benefits to its members, namely 1) Monthly pension on crossing 60 years of age, 2) Death and disability insurance cover including Scholarship to the children of the women, 3) Lump sum transfer of corpus which is accumulated in her account is transferred to the nominee in the event of her death.

The monthly pension amount payable to the women depends on the no. of years of contribution she makes till she crosses 60 years of age. In order to secure reasonable level of monthly pension, Govt. have decided to ensure that every women shall receive ₹ 500 or above pension every month, on crossing 60 years of age till her death.

A women joining Abhaya Hastham at the age of 18 will be able to draw ₹ 2,200 per month pension on crossing 60 years of age. This amount is likely to be much higher, should the level of monthly contribution / co-contribution is stepped up after every 5 or 10 years.

Every woman who joins Abhaya Hastham will have a Pass Book to record details of her contribution and she is entitled to receive an annual statement with the particulars of her contribution and co-contribution from the Govt and interest earned on the corpus. On crossing 60 years of age, the monthly pension is adjusted to her Bank Account by LIC, every month.

Abhaya Hastham offers many more benefits to the women in addition to monthly pension on retirement. The other benefits are (a) death and disability insurance cover in respect of women in the age group of 15 to 59. In the event of her natural death, ₹ 30, 000 is paid to the nominee. In the event of her accidental death, ₹ 75,000 is paid to her nominee. In the event of permanent disability ₹ 75,000 is paid to her. In case of partial disability ₹ 37,500 is paid to her. (b) The member's children studying 9th, 10th, 11th and12^{th} classes will receive a scholarship of ₹ 1,200 per annum (c) in the event of the death of the subscriber, the available contribution of the subscriber and Government's co-contribution along with accrued earnings on such contributions will be transferred to the nominee of the subscriber.

Govt. of A.P. has brought out legislation, A.P SHG Women Co-contributory Pension Act 2009. The bill was very recently passed by the A.P State Legislature and AP State Legislative Council on 10th and 11th February respectively. Government had provided budget of ₹ 365 Crores in the 2009-10. The Scheme has already been launched at the State Headquarters and also in the Districts by distributing cheques of ₹ 500 towards the monthly pension for the SHG women who joined the scheme and attained retirement age. It is expected that about 1crore 25 Lakh women are likely to join the Abhay Hastham Scheme. The AP Government has thus earned the distinction of rolling out world's largest co-contributory pension scheme for the poor in the unorganized sector.

There are 70,504 tribal women enrolled into the programme during past two years of its operation in Paderu agency area of Visakhapatnam district alone. Age group of enrolled varied from 20 years to 45 years on an average. All the young women are keen in joining the programme. The benefits of the programme are thoroughly shared with members in the monthly review meeting of Self Help Groups, Village Organisations and Mandal Mahila Samakshya (MMS). It is reported that there are 408 cases registered and 283 cases solved during the past two years. The fastness of the programme and commitment of the staff is motivating many self help group members to join the scheme. The beauty of the scheme is seen in distribution of scholarships to 4867 students to further pursuing carrier development (see table-5.5). This is serving as a moral boost to all students and enabling them to pursue their carrier even during hey-days of their life.

Table 5.5: Coverage of Abhaya Hastam in Paderu Area

Year	Members enrolled	Claims Reported	Claims Settled	Claims Pending	Scholarship distributed
1	2	3	4	5	6
2010-11	23860	328	275	53	767
2011-12	46644	80	8	72	4100
Total	**70504**	**408**	**283**	**125**	**4867**

Source: Data Compiled from IKP Annual Reports for the year 2010 and 2011

Cumulative membership of the programme across the tribal Mandals of the Paderu agency area reveals almost a similar trend for all Mandals. Almost a same number of self help group members are keen in enrolling themselves into the programme. A similar ratio is being observed for settlement of issues and claims (see table-5.6).

Table 5.6: Mandal-wise Cumulative Coverage of Abhaya Hastam till 2012

Mandal	Members enrolled	Claims Reported	Claims Settled	Claims Pending	Scholarship distributed
1	2	3	4	5	6
Ananthagiri	4895	22	20	2	402
Araku Valley	5732	22	16	6	275
Dumbriguda	5999	27	21	6	381
Hukkumpeta	6226	57	41	16	389
Pedabayalu	6761	43	27	16	483
Munchingput	5896	29	20	9	278
Paderu	7016	46	26	20	530
G. Madugula	8684	70	58	12	578
Chintapalli	5796	25	16	9	477
G.K. Veedhi	6298	27	17	10	463
Koyyur	7201	40	21	19	583
Total	**70504**	**408**	**283**	**125**	**4839**

Source: Data Compiled from IKP Annual Reports

The women members of SHGs feel that Abhaya Hastham will help them lead a dignified life in their old age, enhance their self worth and secure them greater respect from rest of the society. The SHG women are immensely thankful to the Government for the remarkable transformation Abhaya Hastham brings into their lives.

5.5.2.2 Aam Admi Bima Yojana (AABY)

Aam Admi Bima Yojana, a Social Security Scheme for rural landless household was launched on 2nd October, 2007 at the hands of the then Hon'ble Finance Minister at Shimla. The head of the family or one earning member in the family of such a household is covered under the scheme. The premium of ₹ 200 per person per annum is shared equally by the Central Government and the State Government. The member to be covered should be aged between 18 and 59 years. The scheme is executed through Life Insurance Corporation of India.

The Government of Andhra Pradesh under the dynamic leadership of Dr. Y.S. Raja Sekhar Reddy, the Hon'ble Chief Minister of Andhra Pradesh envisions that even the poorest of the poor Landless Agricultural Laborers (LAL) deserve access to relevant and affordable insurance and be protected from risks on death and disability. The Government further envisions that every person is insurable and their children are financially supported for education through scholarships. And together we can make it happen across Andhra Pradesh through "Indira Jeevitha Bima Padakam".

Response of the programme is eye appealing in tribal districts of the state of Andhra Pradesh. Beginning of the programme was made with a 30313 membership in tribal agency area of Paderu in 2008-09. This has witnessed a highest enrollment of 88,654 members in 2010-11. Cumulatively there are 1, 85, 795 beneficiaries enrolled under the programme in ITDA Paderu of Visakhapatnam districts. During the past four years 1674 claims reported and 1480 cases disposed off. Almost 88 per cent cases are settled (see fig. 7.3). This also implies for a significantly larger number of policies in force and exhibit the commitment of the people and trust on the scheme. The scheme has given away scholarships to 5970 tribal students to continue with their education (see table-5.7). All tribal Mandals have accessed the services of insurance claim and scholarship. The insurance claim is confined to two digits while scholarship is in three digits (see table-5.8).

Table 5.7: Coverage of Aam Admi Bima Yojana (AABY) in Paderu during 2008-12

Year	Members enrolled	Claims Reported	Claims Settled	Scholarship disbursed
1	2	3	4	5
2008-09	30313	309	309	1618
2009-10	39951	542	524	1313
2010-11	88654	498	443	949
2011-12	26877	325	204	2090
Total	**185795**	**1674**	**1480**	**5970**

Source: Compiled from Various IKP Reports, Paderu

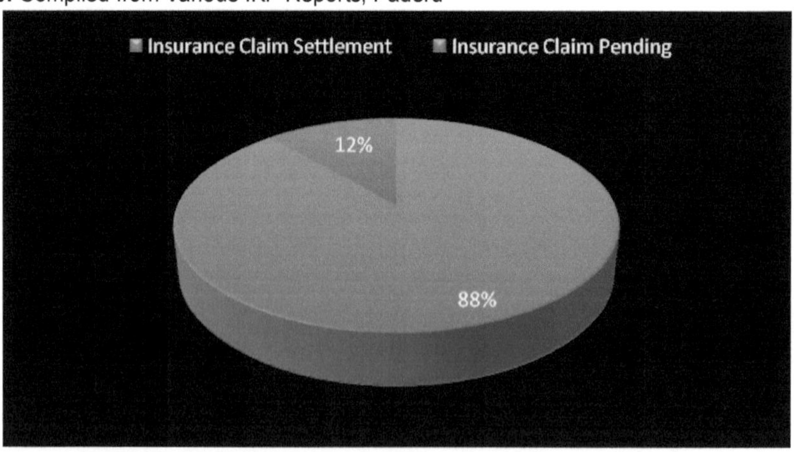

Fig. 5.3: Insurance Claim Settlement Status under AABY

Source: Various Annual Reports of IKP, Paderu

Table 5.8 : Insurance Settlement under AABY across agency Mandals

Mandal	2008-09		2009-10		2010-11		2011-12	
	Insurance claim settled	Scholarship Distributed	Insurance claim settled	Scholarship Distributed	Insurance claim settled	Scholarship Distributed	Insurance claim settled	Scholarship Distributed
1	2	3	4	5	6	7	8	9
Ananthagiri	41	103	45	34	28	55	13	158
Araku Valley	34	126	55	63	47	119	19	342
Dumbriguda	12	125	17	57	11	90	9	172
Hukkumpeta	18	84	50	11	15	13	8	59
Pedabayalu	25	211	48	93	41	125	19	154
Munchingput	24	134	50	116	49	95	21	143
Paderu	34	287	55	197	65	47	26	222
G. Madugula	13	83	55	222	53	50	21	84
Chintapalli	42	175	47	266	60	85	36	359
G.K. Veedhi	20	127	41	152	33	161	19	215
Koyyur	46	163	61	102	41	109	13	182
Total	**309**	**1618**	**524**	**1313**	**443**	**949**	**204**	**2090**

Source: Compiled from various reports of IKP

5.5.2.3 Janasree Bima Yojana (JBY)

Janasree Bima Yojana is a group insurance scheme meant for the spouses of the SHG member or any unmarred, widow or divorcee who are the SHG members. It is being implemented in collaboration with LIC of India. It is a group Insurance Scheme for the age group of 18 -58 years. The Premium paid under the Scheme is ₹ 175 per annum per member, of which 50 per cent will be contributed by the Central Government and the remaining by the individual member *i.e.* each member shall pay ₹ 75 the benefits are same as the AABY in all respects.

Beginning of JBY in agency areas of Paderu and among tribal people is very encouraging. There are about 71,162 self help group members were brought under the scheme, having participation of 24,846 Primitive Tribal families and 46,316 Non-Primitive Tribal Families. There are about 81.68 lakhs remitted to LIC by the self help groups during the period. The initiative of securing tribal through insurance coverage is on the go (See fig.5.4) and every year it is getting a forward momentum. Apart from the coverage there are about 2950 claims settled and 3391 scholarships distributed to students of both Mandals.

Fig. 5.4: Coverage of SHG under Janashree Bema Yojana in Paderu

Updation and renewal of policies are not being reported so far, rather a few policies in two Mandals are being deleted (See table-5.9) due to lack of interest of the beneficiaries in one hand and migration of families on the other. This constitutes 3 per cent of the total policies executed in the ITDA Paderu and 16.37 per cent of the policies executed in Chintapalli Mandal. It is also to be noted that maximum number of policies are being executed from the same Mandal.

This will have a little or no repercussion on the functioning of the policy in the agency area. 3 per cent with drawl is a meager number of the total policy executed.

Table 5.9: Janashree Bema Yojana in Paderu ITDA

Mandals	Total Policies	Renewed Policies	New Policies	Deleted Policies
1	2	3	4	5
Ananthagiri	40	0	40	0
Araku Valley	62	0	62	0
Dumbriguda	134	0	134	0
Hukkumpeta	112	0	112	0
Paderu	57	0	57	0
G. Madugula	111	0	110	1
Chintapalli	226	0	189	37
G.K. Veedhi	59	0	59	0
Koyyur	177	0	177	0
Total	**978**	**0**	**940**	**38**

Source: Annual ITDA Report 2012, Paderu

5.5.3 Health and Nutrition

The objective of Health and Nutrition programme in tribal agency areas is to provide balanced diet to pregnant women, lactating mothers and children by establishing Nutrition-cum-Day Care Center (NDCC). With this purpose, Fixed Nutrition Days (FNDs) are conducted in the village Anganwadi Centers (AWCs) and Growth chart of children below five years of age are monitored. The programme is limited to two critical Mandals of the ITDA *i.e.* Munchingput and Pedabayalu and would be spread on to all 11 mandals on a phased manner.

There are 166 Nutrition and Day Care Centres in operation in both Mandals with a focused approach and coverage. 75 centres in Munchingput and 91 centers in Padabayalu are functioning in remote habitations bordering neighbouring Orissa state. The ITDA action plan for the year 2010-11 shows a total number of 227 NDCC were planned for the year. Of the 166, it is reported that only 66 of them are functioning regularly to serve the pregnant women and lactating mothers.

The access to services of NDCC has been on rise. It is because of intensive awareness creation by the health workers and also due to increase in confidence as well as trust of people on the quality and timeliness of services offered (see fig.5.5). The quality of nutritional food provided is liked by all beneficiaries and in consequence resulted to increased access.

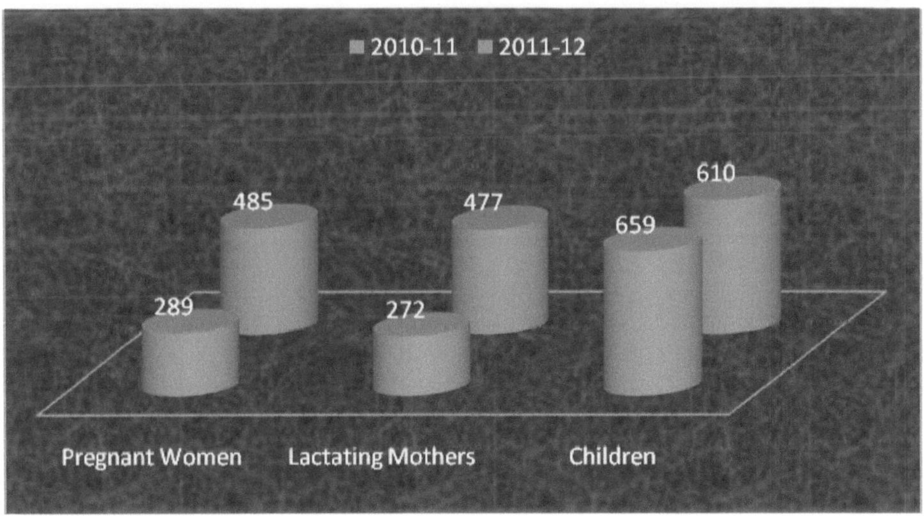

Fig. 5.5: Access to NDCC Services during 2010-12

The self-help groups have accelerated the pace of institutional deliveries. The cases of normal deliveries are usually more, while tribal women as well as their families are resorting to caesarian delivery in complicated cases. The change in mind set to trust on medical sciences is massively grown among the tribals due to the recruitment of village health workers, who are primarily hailed from the same community, to motivate and imbibe confidence in them. There are more number of healthy birth is reported over the years across the NDCC Mandals (see table-5.10).

Table 5.10: Access to and Utilization of Focused Health and Nutrition Schemes

Sl. No.	Parameters	Munchingput Mandal	Pedabayalu Mandal	Total
1	2	3	4	5
1.0	NDCC	75	91	166
2.0	Total Beneficiaries Enrolled	765	780	1545
2.1	Pregnant Women	224	234	458
2.2	Lactating Mothers	216	261	477
2.3	Children	325	285	610
3.0	Total Deliveries	148	176	324
3.1	Normal Deliveries	137	158	293
3.2	Caesarians Delivery	13	18	31
4.0	Total Children Born	149	178	327
4.1	Girls	78	97	195
4.2	Boys	71	81	152
5.0	Birth Weight of babies	149	178	327
5.1	> 3 kgs	35	48	83
5.2	2.5-3 kgs	92	97	189
5.3	< 2.5 kgs	22	33	55
6.0	NDCC to be established	6	8	14

Source: Compiled from various annual reports of IKP, Paderu

A long mile still has to tread on and there are many more mandals still to cover under the service. Intensification of activity and making people voluntarily accessing the service is a challenge in front of the Community Based Organization (CBOs) and the self-help groups, besides putting the onus on government to popularize through creation of massive awareness.

There are several flaws being noted and reported regarding the quality of food served, internal class and caste hierarchy, food served is of not to the taste and liking of the tribals, sufficiency of materials, long distance to walk on and so on and so forth. It is, in fact, a challenge for the CBOs more than the government to take the programme forward for better grounding and making its reach to the target clientele and self-help groups are working with more commitment in this direction.

5.5.4 Early Child Education (ECE)

'Charity begins at home and Mother is the first teacher of the child' are two well-known proverbs being widely used in common parlance. This is rightly applicable in case of execution as well as monitoring of "Early Child Education Centers" or Bal Badies in the villages to train the small siblings to be habituated with the school environment and provision nutrition supplement to the children. The ECE programme was initiated during 2011-12 and confined its operation to four Mandals of the agency area. 1629 children are enrolled across 90 centers (see table-5.11). Enrollment of children to the ECE is highest in Ananthagiri Mandal followed by Pedabaiylu, Munchingput and Paderu. More developed Mandals have less children enrolment ratio in comparison to less advanced Mandals of Paderu agency. The self-help groups in the villages, apart from taking care of the household chores also do take the responsibility of monitoring the activity of Bal Badies in the villages. This is emerging as an additional activity being entrusted to self help groups to look after.

Table 5.11: ECEC and Children Enrolment in Paderu Agency

Sl. No.	Mandal	No. of ECEC	No. of Children enrolled	Budget released (₹ in Lakhs)
1	2	3	4	5
1	Ananthagiri	30	503	25.00
2	Paderu	20	336	25.00
3	Munchingput	20	353	25.00
4	Pedabayalu	20	437	25.00
	Total	**90**	**1629**	**100.00**

Source: Compiled from IKP Annual Reports of 2010-11 and 2011-12, ITDA Paderu

5.5.5 Revenue Sadassulu

The land records are not accurate and available completely in the state of Andhra Pradesh. Therefore, there is a foreseen need to update the system of land records. "Revenue Sadassulu " is a means to ensure right on land and would facilitate the farmers to get loans from the banks. "Revenue Sadassulu" will act as machinery to solve land disputes at the village level. Officials of various cadres from department of revenue, Dept of Forest and Dept Rural Development are members as well as facilitators to conduct the meetings in places of convenience to people. Elected political representatives and peoples representatives use to participate in the meetings.

There are 244 gram Panchayat's in Paderu ITDA alone. During the one year period of its execution to solve peoples issues on land, there are about

9907 cases being lodged in and a few cases are resolved (see table-7.12) in the agency area. Issues are registered through petitions in the gram sabha's. There are 188 Gram Sabha's organized at different public places around the agency area. It is surprising to see from the statistics that are 9907 cases being registered averaging out about 53 cases in each Gram Sabhas.

Table 5.12: Status of Revenue Sadassulu during 2011-12

Sl. No.	Mandal	No. of GPs	Gramsabha's conducted	Petitions received	Petitions Disposed	Petitions balance
1	2	3	4	5	6	7
1	Ananthagiri	24	21	889	232	667
2	Araku Valley	14	10	115	46	69
3	Dumbriguda	18	25	391	63	328
4	Hukkumpeta	33	21	1141	119	1022
5	Pedabayalu	23	16	395	56	339
6	Munchingput	23	16	756	47	709
7	Paderu	26	19	706	55	651
8	G. Madugula	17	21	2056	221	1835
9	Chintapalli	17	5	731	24	707
10	G.K.Veedhi	16	14	1533	34	1499
11	Koyyur	33	11	1184	83	1101
	Total	**244**	**188**	**9907**	**980**	**8927**

Source: IKP Annual Report, 2011-12, ITDA Paderu

The initiative of revenue Sadassulu through conduct of Gram Sabhas is confined to accessible Mandals like Dumbriguda, Hukkumpeta Paderu, G. Madugula and Ananthagiri of the agency area. Average Gram Sabha's are being conducted for remote and Naxal prone mandals like G.K. Veedhi, Chinatalli and Koyyur. Profile of the complaint registered reveals a mixed category of people. Despite all effort, 90 per cent of petitions are to be resolved in the days to come.

5.6. Economic Impact

Rural and tribal economy is characterized with agriculture, animal husbandry, petty business and other allied activities based on abundance of alternative resources and availability of skill sets. All activities governing the economy especially of tribal are more of traditional in nature and mostly carried out with the set of skills and technologies used for generations across.

Hence, the economy as has been said elsewhere has achieved near stagnation and saturation *i.e.* almost the same level of income, equivalent volume and value of production, lack of alternative means for fulfillment of basic needs, no aspiration as well as effort to go high etc. The formation of microfinance institutions has brought a new dawn and a new goal and enables them to show new hope for sustained living. The new face of the economy and its virtual forms are being thoroughly captured in the subsequent sections.

5.6.1. Agriculture Finance

TRICOR has circulated the developmental fund through the Village Organizations where self-help groups are the beneficiaries. There are various schemes for promotion of tribal livelihood being under taken as a part of planned developmental activities of five-year plans for the tribals in agency areas of the state of Andhra Pradesh. The core segment of livelihood promotion are being revolving around three-major activities like promotion of sustainable agriculture in tribal and rural areas, promotion of animal husbandry activities to boost income of tribals and inculcation of habit of Individual Small Scale Business among tribals with a view not only to enhance family income but also enhancing the business acumen among them.

To reduce cost of cultivation and increase net income, Community Managed Sustainable Agriculture (CMSA) is encouraged, which reduces input costs by making available organic fertilizers and pesticides made from locally available material. CMSA is under implementation in 11000 villages in 653 mandals of 22 districts, in 38.71 lakh acres and with 19.67 lakh Farmers in the state of Andhra Pradesh. During current fiscal year up to end of February, 2013, 36.00 lakh acres area brought under CMSA fold by benefitting 17.07 lakhs farmers. The Flagship programmes identified under CMSA during 2010-11 are POP Strategy, 36 X 36 models and SRI Paddy.

With CMSA practices, the savings varied from ₹ 1,000 in case of Paddy to ₹ 16,000 in case of Chilies by which additional incomes to farmers ranges from ₹ 5,000 to ₹ 10,000 per Acre. POP strategy facilitates the land lease to the landless laborers and promotes CMSA practices in these lands. SRI system of paddy is cultivated in 0.25 Acre and 36 X 36 model or 7 tire model or poly crops in the remaining 0.25 Acres to earn a net income of ₹ 50,000. 14,899 PoP families adopted this strategy. Apart from this, more than 1 lakh 36*36 models have been grounded benefitting 1 lakh families with household nutritional security

When the programme got prominence as a panacea for ensuring livelihood for POPs using the means of natural resources, it has got a mixed response from the remote and tribal areas of the state. Especially in an agency like Paderu, 47.92 per cent of target 288 acres is covered under SRI, while 37.04 per cent of the targeted numbers under 36*36 models. NADEP compost pit which is firmly accepted and traditionally practiced since ages by agriculturist families has pushed through 55.56 per cent units.

Azolla in Paddy is a programme which is almost alien in nature has got a vibrant response with 91.77 per cent response from all across tribal Mandals of Paderu agency. Opening up of NPM Shops and Custom Hiring Centers have proven mixed progress during the reporting period. 135 PoP strategies had been planned for Paderu, while 52 have been crafted and executed. Membership registration under the scheme is not very encouraging, which is making the field acceptance of the scheme in tribal areas skeptical (see table 5.13).

Table 5.13: Performance of Community Managed Sustainable Agriculture during 2011-12

Sl. No.	Parameters	Units	Target	Achievement	% to cover
1	2	3	4	5	6
1	System of Rice Intensification (SRI)	Acres	288	138	52.08
2	36*36 Models	No's	2700	1000	62.96
3	NADEP Compost Pit	No's	2700	1200	55.56
4	Azolla in Paddy	Acres	3000	2753	8.23
5	NPM Shops	No's	90	77	14.44
6	Custom Hiring Centres	No's	30	20	33.33
7	PoP Strategy	Coverage	135	52	61.48
8	Members Registration	Members	7552	5800	23.20

Source: IKP Annual Report, 2012, ITDA Paderu

The Economic Support Scheme (ESS) to take the agriculture programme forward have well executed with 3513 beneficiaries with injection of ₹ 186.89 lakhs in the agency area. The fund support is extended from TRICOR for past couple of decades *i.e.* 1972 onwards to help enable tribal taking up agri operation persistently without feeling the adverse impact of fund crunch. Utilization of fund for agriculture promotion is not an universal affair in tribal agency area (see table 5.14). Areas located in the valley have possessed with a larger chunk of plain land for cultivation and hence, is the access of ESS fund in comparison to all other Mandals of the agency area. Hence, access of the ESS fund across the Mandals of the Paderu agency area is regulated not by choice of the farmers but by the availability of land for under taking agriculture programme.

Table 5.14: Mandal-Wise access to Agriculture Finance under ESS

Sl. No.	Mandals	No. of Beneficiaries	Amount (₹)
1	2	3	4
1	Ananthagiri	397	24.87
2	Arakuvalley	352	21.37
3	Dumbriguda	553	30.98
4	Hukumpeta	252	16.68
5	Pedabayalu	232	9.46
6	Munchingiput	129	7.58
7	Paderu	353	15.65
8	G.Madugula	236	14.92
9	Chintapalli	544	25.18
10	GKVeedhi	299	11.26
11	Koyyuru	166	8.94
	Total	**3513**	**186.89**

Source: IKP Annual Report 2011-12, ITDA, Paderu, Visakhapatnam, Andhra Pradesh

'Srinidhi ' is a welfare scheme dedicated for economic empowerment of women and women headed organizations. The scheme was floated with an initial corpus of ₹ 1000 crores. The primary focus of the new initiative is to make women empower to undertake activities and/or income generation avenues and make themselves self sufficient. Activities around land and agriculture is being massively supported and promoted with twin objective of income generation as well as generation of employment opportunities in rural villages of the state. The scheme categorically defined the nature of activities to be funded with and supported for and made a favoritism provision of sanction of fund to women exclusively. Men are devoid of getting the benefit. The scheme is in operation across the districts of the state of Andhra Pradesh.

Tribal areas and especially the ITDAs are getting a preferential benefit of having larger number of self-help groups exclusively formed, managed and run by tribal women. The scheme is being routed through self help groups in tribal areas. Applications for support are submitted through the Village Organizations (VOs) and Self-help groups (SHGs). Almost 83 per cent of applications for enrollment under the scheme are submitted through the self-help groups in tribal mandals of Paderu agency area while 17 per cent of applications are directly submitted through the Village Organizations. This reveals the vigor of the involvement of microfinance institutions in accelerations of schemes among the peoples (see fig. 5.6). However, grounding of the scheme is very poor. The scheme had a targeted disbursement of ₹ 600 lakhs for all 11 mandals of the Paderu agency area for the year 2010-11, while it has a limited reach of ₹ 5.02 lakhs. Interest for the scheme and execution has been limited to a few tribal mandals viz. Araku valley, Hukkumpeta, Paderu and Munchingput. Self help groups of the Srinidhi covered Mandals have proven the capability and efficiency in execution of government schemes and programmes. It may be the reason for having trial of the scheme in these mandals of agency area. Further, Paderu Mandal has a better grounding of ₹ 3.62 lakhs, which is more than half of the fund deployed in rest of the three Srinidhi programming mandals.

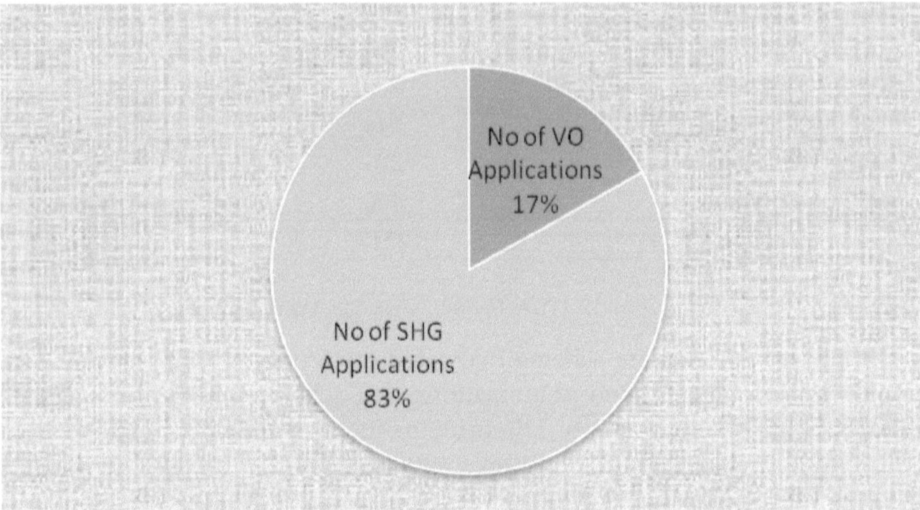

Fig. 5.5: Scenario of SHG and VO application for Shreenidhi, 2011

Poor access to scheme and utilization of fund dedicated under the scheme is a challenge for the time to come for the government departments and the rural financial institutions working in the tribal ITDAs. Access to and utilization of the dedicated fund of ₹ 600 lakhs may have enabled many thousands of self-help groups to explore new and profitable activities for promotion of economic condition of the people and region at the large (see table-5.15). There are ₹ 594.98 lakhs of fund still lying unutilized. The self-help groups would have accessed the fund provided, they could have come out with a viable and economically profitable micro plan on and around land and natural resources available in the region. A dedicated micro plan would only ensure appropriate utilization of fund and creation of adequate income for the tribal people.

Table 5.15 : Access and Utilization of Shreenidhi Scheme in Paderu during 2011-12

Sl. No.	Mandal	Target (₹ in lakhs)	Achievement (₹ in lakhs)	Balance (₹ in lakhs)	% of VOs covered
1	2	3	4	5	6
1	Ananthagiri	100.00	0.00	100	0.00
2	Araku Valley	100.00	0.85	99.15	3.64
3	Dumbriguda	0.00	0.00	0.00	0.00
4	Hukkumpeta	25.00	0.25	24.75	0.43
5	Paderu	100.00	3.62	96.38	7.55
6	G. Madugula	50.00	0.00	50.00	0.00
7	Munchingput	25.00	0.30	24.70	1.61
8	Pedabayalu	0.00	0.00	0.00	0.00
9	Chintapalli	100.00	0.00	100.00	0.00
10	Koyyur	50.00	0.00	50.00	0.00
11	G. K.Veedhi	50.00	0.00	50.00	0.00
	Total	**600.00**	**5.02**	**594.98**	**13.23**

Source: IKP Annual Report, 2012, ITDA, Paderu

5.6.2. Animal Husbandry Finance

Animal always forms a significant part rural and tribal economy. Rearing animal is not only conceived as an assured means of income to rural and tribal livelihood rather signifies the social esteem, value and position of household in the society. ITDAs therefore give focus on promotion of animal husbandry as one among the livelihood interventions to be executed in every annual plan. And it is being under execution from inception of ITDAs in the state of Andhra Pradesh.

The common animal husbandry interventions undertaken in Paderu agency area are Goat and Ship rearing, Plough bullocks, Milch animals and Backyard Poultry. These are the standard initiatives under practice since ages by the tribals. Under the four interventions local breeds use to be supplied to the tribal for rearing. Hi-breed variety of animal is proven to be a failure and especially in ITDA like Paderu where temperature is unusually very low and have a different climatic condition. As a precautionary measure, Goats, Sheep, Chicken and Bullocks and in fact all animal supplied to the beneficiaries under the scheme are of local origin.

Self help groups in tribal areas and Paderu agency in particular is serving as an effective bridge to meet the local demand of the people and enabling government to reach to the needy. It is provisioning an effective platform for the people and the government and is removing the barrier of geographic undulation and cut-off-ness for an effective programme delivery.

4585 tribal beneficiaries, in fact households, have been benefited under the scheme and there are about ₹ 417.91 lakhs of fund being injected to Paderu economy during the financial year 2011-12. All self-help groups have facilitated for developing the MCP and mobilizing the fund through the Mandal Mahila Samakshyas (MMS). All mandals in the agency area have been benefitted from the scheme (see table-5.16). However, bordering mandals of agency area such as Pedabayalu, Dumbriguda and Ananthagiri have lesser access and utilization of the programme in comparison to the rest of the cental Mandals of Paderu agency. The volume of fund injected is also comparatively less. IKP need to give close focus for better coverage of the programme and self-help groups need to be encouraged to come up with executable plans for benefitting maximum number of beneficiaries.

Table 5.16: Mandal-Wise access to Animal Husbandry Finance under ESS

Sl. No.	Mandals	No. of Beneficiaries	Amount (₹)
1	2	3	4
1	Ananthagiri	299	24.22
2	Arakuvalley	366	53.29
3	Dumbriguda	253	42.61
4	Hukumpeta	377	34.6
5	Pedabayalu	246	13.95
6	Munchingiput	463	44.00
7	Paderu	670	49.39
8	G.Madugula	358	26.86
9	Chintapalli	427	33.03
10	GKVeedhi	639	48.93
11	Koyyuru	487	47.03
	Total	**4585**	**417.91**

Source: IKP Annual Report 2011-12, ITDA, Paderu, Visakhapatnam, Andhra Pradesh

5.6.3. Individual Small Scale Business

Unlike animal husbandry, promotion of Individual Small Scale Business (ISB) has been forming an important activity of ITDA in every Annual Plan. The merit of the schemes is visualized through promotion of small scale business in rural villages with assistance of government. It is meeting twin objectives of provisioning basic household requirement of the villagers at door-step and providing employment to villagers in the village.

Activities encouraged under the Individual Small Scale Business includes Kirana Shop, Shandy, Fancy shop, Cattle Business, Cycle shop and Tent house (Catering Services) etc. The amount of fund supported depends on the type of the ISB under selection.

Self-help groups are the facilitators for submission of application, processing and finalization of financial assistance. Process for submission of application follows the MCP rout and MMS pays a dominant role in finalization.

3264 tribal households have been benefitted under the scheme during 2011-12 alone. Almost ₹ 284.28 lakhs has been pumped into the economy of Paderu for under taking petty business. Access to and utilization of the scheme is not equal in all mandals (see table-5.17). Mandals like Koyyur reported for accessing 92 units at the minimum to Mandal Pedabayalu accessing 752 units at the maximum.

Table 5.17: Mandal-Wise Access to Individual Small Scale Business Finance under ESS

Sl. No.	Mandals	No. of Beneficiaries	Amount (₹)
1	2	3	4
1	Ananthagiri	296	26.44
2	Arakuvalley	231	32.89
3	Dumbriguda	258	43.4
4	Hukumpeta	343	31.78
5	Pedabayalu	752	33.95
6	Munchingiput	385	25.95
7	Paderu	187	24.02
8	G.Madugula	203	13.57
9	Chintapalli	291	26.93
10	GKVeedhi	226	16.62
11	Koyyuru	92	8.73
	Total	**3264**	**284.28**

Source: IKP Annual Report 2011-12, ITDA, Paderu, Visakhapatnam, Andhra Pradesh

5.6.4. Pavalavaddi

The Government of Andhra Pradesh introduced Pavalavaddi (PV) Scheme during the year 2004-05 with an objective to provide interest subsidy on the bank loans taken by the Self Help Groups (SHGs) in Andhra Pradesh to reduce the financial burden on them. The Scheme is applicable to all loans extended by banks on or after 01.07.2004, under SHG Bank Linkage (SHG-BL) Programme.

The Govt. of Andhra Pradesh both in 2009-10 and 2010-11 has budgeted ₹ 200 crores each to give interest subsidy under PV scheme to SHGs. The Govt. of Andhra Pradesh has released an amount of ₹ 564.25 crore up to March 2010. Of the total amount released, ₹ 548.02 crore were disbursed covering 25.8 lakh SHGs.

The scheme has a greater impact in tribal areas of the state. The amount of money pumped in and number of self help groups covered has followed an increasingly decreasing trend (see fig. 5.7) for ITDA Paderu. However, the amount of fund injected is increasing year after the year of its execution. Year 2008-09 has been recorded highest access to the scheme.

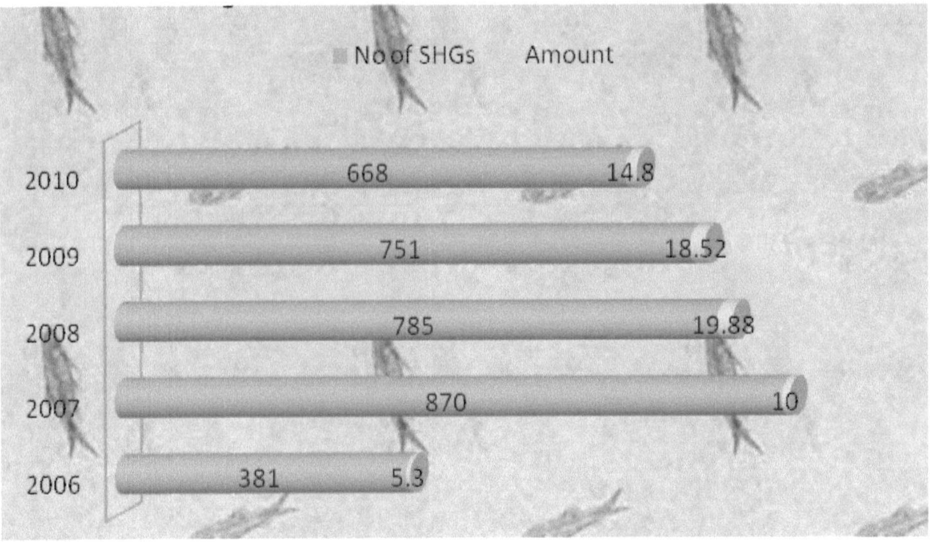

Fig. 5.7: Access and Utilisation of Pavalavaddi

Besides, highest number of self group from mandals like Arakuvalley, Pedabayalu and Koyyur have increasingly better access to fund in comparison to all mandals of ITDA. Self help groups of mandals like Munchingput and Hukkumpeta have very minimal access to the scheme (see table-5.18) during 2006-10. The access to the scheme is varying in nature and varies from year to year. However, response to the programme was negligible during its days of initiation in 2004-05. The overall response to the pavalavaddi scheme in the presence of voluminous self-help groups in the Paderu agency area is not overwhelming.

Many self help groups did not get pavalavaddi because of two reasons; firstly, poor recovery and secondly, mistakes in data entry. For many data entry work, special support may be provided as the present APMs-MS coordinators are not equipped with computer skills. Either support may be given or they may be trained on basic computer skills. Sufficient time may be given, especially to TPMUS considering the geographical challenges. Additional time may be permitted to prepare proposals for genuine cases.

Table 5.18: Access to Pavaladdi by Mandals of Paderu Agency

Mandal	2006-07		2007-08		2008-09		2009-10		2010-11		Total (₹ in Lakhs)	
	No. of SHG	Amt (₹)	No. of SHG	Amt (₹)	No. of SHG	Amt (₹)	No. of SHG	Amt (₹)	No. of SHG	Amt (₹)	No. of SHG	Amt (₹)
1	2	3	4	5	6	7	8	9	10	11	12	13
Ananthagiri	0	0	54	0.55	14	0.28	25	0.36	117	1.93	210	3.12
Arakuvalley	116	2.08	50	0.85	81	2.02	86	3.86	137	2.56	470	11.37
Dumbriguda	0	0	53	0.80	45	1.42	48	1.49	57	0.94	203	4.65
Hukumpeta	13	0.15	14	0.21	35	1.26	101	2.19	10	0.78	173	4.59
Pedabayalu	6	0.06	186	2.41	66	1.55	114	1.44	47	0.47	419	5.93
Munchingiput	0	0	43	0.32	27	0.57	25	0.33	13	0.42	108	1.64
Paderu	0	0	131	0.81	147	5.44	103	2.25	47	2.64	428	11.14
G.Madugula	0	0	9	0.04	75	1.81	82	2.03	22	0.39	188	4.27
Chintapalli	0	0	196	1.88	87	2.35	5	0.06	16	0.42	304	4.71
GKVeedhi	113	1.12	92	1.09	83	1.3	58	1.05	49	0.87	395	5.43
Koyyuru	133	1.89	42	1.04	125	1.88	104	3.46	153	3.38	557	11.65
Total	**381**	**5.3**	**870**	**10.00**	**785**	**19.88**	**751**	**18.52**	**668**	**14.8**	**3455**	**68.50**

Source: IKP Annual Report 2011-12, ITDA, Paderu, Visakhapatnam, Andhra Pradesh

5.7. Institutional Impact

Institutions are created to regulate and exercise strategies towards goal achievement. The self-help group formation, nurturance and capacity building for a self-propelling development, is an endeavour towards this direction. Several activities are grounded in this direction and maasostvalu is one among them. This has not only identified the families without interest payment but also enabled formation of self-help groups of people with different abilities (PWD).

During the process mentor system is seeing getting prominence and becoming popular among the strata's of microfinance institutions created for the tribal involving the tribal in the agency area of Paderu. Each VO is nurtured by some staffs appointed by the Indira Kranthi Pratham (IKP). The performances of the VOs are measured over a period of time. Focused attention has been given to develop social capital potentiality within the institutions. Community Resource Persons (CRPs) are identified from villages and among the tribals with good potentiality and understanding about the microfinance programme and basically the operation of self-help groups and preparation of micro credit plans and execution of livelihood enhancement programmes. Currently, as appearing from IKP data base, there are 54 CRPs have been screened and selected to serve the SHGs and VOs across the mandals of the agency area.

M-Book Keeping is entirely a new effort to regulate functioning of the self-help groups and equip institutions with easy and efficient means to document as well as expedite the functional ability as well as agility in a time bound manner. Agencies involved in the management as well as execution of the microfinance programme in tribal areas and especially in the agency area are attempting for introduction of MBook Keeping for a faster and speedier passing of information to help enable for quicker decision making. Majority of self-help groups are trained on the functioning of the Mbook keeping and it is attempted to launch the programme on the mandals and villages having mobile connectivity in the beginning and to scale up to other areas in the course of time.

5.7.1 Institutional Linkage and Money Circulation

Financial inclusion is conceived as a major driving force to achieve self-sustained inclusive economic growth. Financial inclusion can be defined as the process of ensuring access to financial services and timely availability of adequate credit where needed by vulnerable groups such as weaker sections and low income groups at an affordable cost (Report of the Committee on Financial Inclusion in India, 2008). Achieving financial inclusion through formal banking system is a cumbersome task . Unavailability of adequate financial services like credit, insurances, and remittances to majority population at an affordable cost is a major roadblock for the growth of financial sectors.

In this context, Microfinance approach can be considered as an alternative solution to provide financial services to common section of the society. Micro-finance is the provision of thrift, credit and other financial services and products of very small amounts to the poor for enabling them to raise their income levels and improve their living standards (RBI, 2005). Micro-finance has profound implications not just from a finance perspective, but also from the perspective of economic development. Microfinance consisting of micro credit, micro savings and microinsurance, is regarded as an important tool to reduce risk, poverty and vulnerability of common people. Microfinance is one of the most remarkable socio-economic developments in the present era.

Self Help group linkage model is one of the indigenously developed and successfully operated models of Microfinance in India. Under this model, the SHGs are financed by bank without any collateral securities, peer group pressure is considered as collateral by the lenders. SHG led microfinance approach also helps to reduce the burden of heavy transaction cost faced by formal financial institution in India. Further, various empirical studies also found that microfinance through SHG bank linkage programme has enabled the SHG members to improve their socio economic status through the availability of various microfinance services. Thus, Micro-finance through self help group bank linkage model can provide sustainable mechanism to meet the unmet financial needs of un-banked poor.

8779 self help groups in Paderu agency area are linked to formal financial institutions to access credit and 47.85 crores worth of finance being injected to the tribal economy of Paderu averaging out 5.45 lakhs per self help group during 2005-11. This is also confirming a per annum 7.975 crores inflow of capital across tribal mandals of Visakha agency area. There are about 1500 self help groups linked to financial institutions depending to the level of access in different mandals of agency area (see table-5.19).

Table 5.19: Status of SHG Finance in Paderu Agency Area

Sl. No.	Year	No. of SHGs	Amount (₹ in Cr)	% of SHG linked
1	2	3	4	5
1	2005-06	1582	4.60	53.12
2	2006-07	1412	5.35	56.43
3	2007-08	1780	8.20	58.11
4	2008-09	1584	9.03	52.23
5	2009-10	1245	9.08	52.78
6	2010-11	1176	11.59	56.69
	Total	**8779**	**47.85**	

Source: Compiled from Various Annual Reports of IKP, Paderu

The Village Organization in the SHG-Microfinance model is the conglomeration of self help groups in terms of hierarchy. 131 village organizations have also been linked to formal financial institutions to access credit. A volume of 5.74 crores have been leveraged and flowed in to the tribal village organizations during 2005-11. The money circulated has been utilized for creation of IGPs of various magnitudes in different mandals of Visakha agency area. It is surprising to note that all 131 VOs are linked during the financial year 2006-07 and thereafter no linkages has been recorded with village organizations (see table-5.20).

Table 5.20: Status of Village Organization Finance in Paderu Agency Area

Sl. No.	Year	No. of VOs	Amount (₹ in Cr)
1	2	3	4
1	2005-06	0	0
2	2006-07	131	5.74
3	2007-08	0	0
4	2008-09	0	0
5	2009-10	0	0
6	2010-11	0	0
	Total	**131**	**5.74**

Source: Compiled from Various Annual Reports of IKP, Paderu

Provision of bridge loans to the self help groups is in place based on the performance and as per the banking norms. There are special back up initiatives being in operation to give institutional financial support to the self help groups with good ratings to access bridge loans besides the formal linkages. 818 Self help groups covering 6327 beneficiaries are connected to formal financial institutions and accessing bridge loan facilities to the tune of 8.19 crores during 2005-11 (see table-5.21). Peculiarity is reported on the use of the facility and access to the services. Self help groups have accessed the services once in the six years of study period in the agency area. This may be due to the focused attention being given by the leader of IKP during the financial year 2008-09.

Table 5.21: Status of Bridge Loans in Paderu Agency Area

Sl. No.	Year	No. of SHGs	Amount (₹ in Cr)
1	**2**	**3**	**4**
1	2005-06	0	0
2	2006-07	0	0
3	2007-08	0	0
4	2008-09	818	8.19
5	2009-10	0	0
6	2010-11	0	0
	Total	**818**	**8.19**

Source: Compiled from Various Annual Reports of IKP, Paderu

Besides, total financial inclusion is the overall goal of the self help initiative through the self help groups. Cent per cent of self help groups are tried to be linked to the banks and other financial institutions functioning in the agency area. The effort for inclusion of the excluded is to bring them up 'below poverty line' through triggering development within the community. The footprints are being evidenced trough the enhanced access to the services extended by the Government and private institutions, participation in community level activities and political participation in Panchayat Raj Institutions (PRIs). The movement of Self help groups are establishing 'Access to', 'power within' and 'power over' and is enhancing the livelihood purview for the excluded .

Banking institutions and MFI's within the organization functional framework is contributing significantly to achieve the targeted inclusion of vulnerable mandated in every annual work plan. Despite all efforts and due to the coercive organizational operational policy, Total Financial Inclusion (TFI) initiative is failing to include all marginalized targeted in every annual plan. There are 42 village Organizations included in the TFI in Paderu during 2007-08 and ₹ 15.59 crores have been mobilized. Surprisingly, the initiative of TFI for VO is not taken forward. Immediately, subsequent year had recorded TFI for 117 self help groups and a volume of ₹ 2.76 crores had been mobilized. This initiative was also for the financial year 2008-09. In totality, there are 159 institutions in Paderu agency area included in TFI and mobilized ₹ 18.35 crores during financial year 2007-09 (see table-5.22).

Table 5.22: Status of TFI in Paderu Agency Area

Sl. No.	Year	No. of SHGs/VOs	Amount (₹ in Cr)
1	2	3	4
1	2005-06	0	0
2	2006-07	0	0
3	2007-08	42 (VOs)	15.59
4	2008-09	117 (SHGs)	2.76
5	2009-10	0	0
6	2010-11	0	0
	Total	159	18.35

Source: Compiled from Various Annual Reports of IKP, Paderu

Reasons for failure of TFI as revealed from the field survey includes limited mandate of the financial institutions for liquidation, operational stress in tribal area, poor network and communication, illiteracy and ignorance of women, credit worthiness of the SHGs/VOs, timely functioning of the SHGs in terms of credit rotation within the SHG members etc. A system of flexibility in credit disbursement to Self-help groups, especially the SHGs in tribal area, and introduction of institutional operational flexibility to branches of financial institutions operating in tribal areas may work out better. The TFI and Bank linkage target mandated by the Village Organization/SHGs needs a prior discussion of the concerned institutions at the local level followed by an intuitional discussion of the concerned financial institutions at the senior management level for addressing the target. Hence, requirement of 'structural change' within the financial institutions placed and functioning in agency area to deal with issues of tribal separately and 'operational changes' to curb the credit need of the VO/SHGs to cater to the need of the excluded is realized.

Despite the operational flaws, self help movement in tribal areas and especially in Paderu ITDA of Visakhapatnam district has got a proven track record. The self help groups have been financed by the formal and informal financial institutions, Village Organizations are also supported with a good volume of finance. Access to self help group finance is ever increasing *i.e.*

from ₹ 4.60 crores during 2005-06 to ₹ 11.59 crores during 2010-11. Besides linkages with banks and informal financial institutions, bridge loans have also been accessed by self help groups and village organizations (see table-5.23) though it is a single time operation during 2005-11. All sources in totality have amassed finance of ₹ 80.13 crores during the period and the entire money is under circulation in the agency area of Paderu.

Table 5.23: Total Institutional Finance Linkage

Sl. No.	Year	SHG Finance	VO Finance	Bridge Loans	TFI	Total (₹ in Cr)
1	**2**	**3**	**4**	**5**	**6**	**7**
1	2005-06	4.60	0	0	0	4.60
2	2006-07	5.35	5.74	0	0	11.09
3	2007-08	8.20	0	0	15.59	23.79
4	2008-09	9.03	0	8.19	2.76	19.98
5	2009-10	9.08	0	0	0	9.08
6	2010-11	11.59	0	0	0	11.59
	Total	**47.85**	**5.74**	**8.19**	**18.35**	**80.13**

Source: Compiled from Various Annual Reports of IKP, Paderu

A revolving fund is a fund or account that remains available to finance an organization's continuing operations without any fiscal year limitation, because the organization replenishes the fund by repaying money used from the account. Revolving funds have been used to support both government and non-profit operations.

In the case or revolving funds for a government project whose budget goes through annual parliamentary or other legislative appropriations that relate to a fiscal year then the unutilized balance may lapse after the close of the financial year. However, it is restored for the next year provided the agency concerned includes the amount in next year's appropriation.

Within state governments, law establishes revolving funds. Revolving funds, established for the purpose of carrying out specific activities, institute a basis under which financing for the cost of goods or services furnished to or by a government agency originate. Revolving funds are to be replenished through charges made for such goods or services. Revolving funds can be especially valuable for non-profit organizations because they afford both the donor and the non-profit significant advantages. Often a non-profit must announce a program, engage personnel, extend invitations or sign contracts well ahead of raising the donations or receiving the program revenue that will cover the costs. In such cases a revolving fund allows the non-profit to commit to programs early so that it can ensure professional execution and the program's success. The non-profit then can work to generate the revenue, donations, or other support that will repay the money spent.

It is of paramount importance for self help group to be linked with revolving fund support. The greater the volume of linkage, the better is the prospect for self help groups and the members there of to widen the avenues for taking of income generation activities for the group. Year 2006 recorded for availing ₹ 95.2 crores revolving support for 1904 self help groups. Revolving support for the subsequent years have been on decline viz. year 2007 with RF of ₹ 30 crores, year 2010 with RF of ₹ 11.5 crores and year 2011 with RF of ₹ 108.9 crores (see fig.5.8).

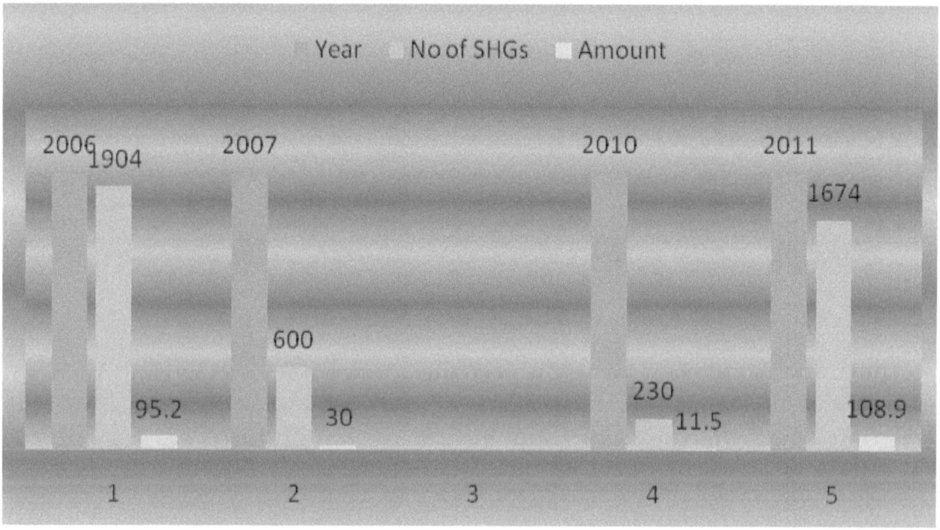

Fig. 5.8: Comparative View of RF Support to SHGs at Paderu

A comparative study of the revolving support during 2005-06, 2006-07 and 2009-10, 2010-11 reveals that RF support during the former two years is of more or less the same irrespective of the mandals, while it is varying in the later two years across the mandals. RF support to each Mandal was 8.65 crores in 2005-06. Chintapalli Mandal of Paderu ITDA has access highest RF support followed by G. K. Veedhi, Paderu, Hukkumpeta, Peddabailu and Koyyur. Mandal like Munchingput has least RF linkage (see table-5.24).

A revolving fund offers several advantages to the donor. The donor's commitment is limited and the gift can be restricted for the sole purpose of supporting the program(s) for which it is intended. At the same time the funds committed to a revolving fund may be multiplied several fold as the non-profit repays money it may have taken from the fund. The non-profit will have a great incentive to repay the money to the revolving fund, in order to continue to enjoy the advantages of the revolving fund for future years.

Table 5.24: Comparative Revolving Fund Support to SHGs during 2005-2011

Mandals	2005-06		2006-07		2009-10		2010-11		Total	
	SHGs	Amt(₹)	SHGs	Amt(₹)	SHGs	Amt(₹)	SHGs	Amt(₹)	SHGs	Amt(₹)
1	2	3	4	5	6	7	8	9	10	11
Anantagiri	173	8.65	58	2.90	21	1.05	178	8.90	430	22.00
Araku Valley	173	8.65	55	2.75	21	1.05	20	10.00	269	22.00
Dumbriguda	173	8.65	54	2.70	21	1.05	172	8.60	420	21.00
Hukumpeta	173	8.65	54	2.70	21	1.05	196	9.80	444	22.00
Pedabayalu	173	8.65	54	2.70	20	1.00	19	9.50	267	22.00
Munchingput	173	8.65	54	2.70	21	1.05	17	8.50	265	21.00
Paderu	173	8.65	55	2.75	21	1.05	224	11.20	473	24.00
G. Madugula	173	8.65	54	2.70	21	1.05	196	9.80	444	22.00
Chintapalli	173	8.65	54	2.70	21	1.05	244	12.20	492	25.00
G.K.Veedhi	173	8.65	54	2.70	21	1.05	222	11.10	470	24.00
Koyyur	173	8.65	54	2.70	21	1.05	186	9.30	434	22.00
Total	**1904**	**95.20**	**600**	**30.00**	**230**	**11.50**	**1674**	**108.90**	**4408**	**245.60**

Source: IKP Annual Report 2012

5.7.2. Legal Compliance: Audit, Registration and Return

Functioning of microfinance institutions is becoming more and more complex over the years. Formation of groups and nurturing them for accessing fund, utilization of fund and repayment of borrowed fund is not as easy as it appeals to the eye. It is the tedious procedure as ever thought up. There are many people involved in it and different processes and approaches it has to go through till it reaches the needy for realization of benefit.

The increase in base of operation and close penetration of financial services in rural and semi-urban areas is often a challenge to stand up to the expectation of the people. Disbursement of fund is as difficult as receipt of payment from the clients. Diversification of needs and designing as well as customizing loan packages as per consumer need is becoming a tedious task every now and then. There are several coercive methods being under adoption for recovery of loan. To curb, the unethical means adopted for loan recovery, malpractices adopted for loan disbursement, and to remove un-healthy completion of microfinance institutions, government and especially the Government of Andhra Pradesh has developed and executed Microfinance Ordinance 2010.

With this pretext, self help group financing is evolving as a peoples' friendly initiative to work with people for economic upgradation of the people. Mandated to work with the women group of the rural and tribal folk, it is accepted as a better means of financing for income generation programmes in rural and semi-urban areas. Each self help group is formed with women participation of the same locality and people of same and/or similar financial background and with similar anticipation of stabilization of income base for the family. The interest is no more of the individual interest rather it is the interest of the community to supplement and complement the initiative to reach the individual target with a 'comprehensive community effort'.

Table 5.25: Registration Status of VOs in Paderu

Sl. No.	Mandals	Total No. of VO	Registration completed	Registration not completed
1	2	3	4	5
1	Ananthagiri	63	18	45
2	Arakuvalley	48	48	0
3	Dumbriguda	54	54	0
4	Hukumpeta	54	35	19
5	Pedabayalu	59	59	0
6	Munchingiput	57	57	0
7	Paderu	56	53	3
8	G.Madugula	52	52	0
9	Chintapalli	46	10	36
10	GKVeedhi	42	42	0
11	Koyyuru	52	51	1
	Total	**583**	**479**	**104**

Source: IKP Paderu Progress Report 2011-12

Formation of self help groups, nurturing and bringing them up to a level for accessing bank linkages is the preliminary challenge to the promoters let it be government and/or NGOs. Accrediting institutional status to self help groups is a milestone to be achieved through registration, structuring and provisioning legal sanctity. Transparency in functioning is viewed as per prescribed government rule and regulation for self help groups and Village Organizations. It is a better parameter for judging efficient functioning of microfinance institutions working in rural and semi-urban areas. Following the principle, 583 village organizations in Paderu agency area have opted for registering themselves and 479 village organizations have successfully registered. About 17.83 per cent of village organization has to get through the registration process (see table-5.25). It has been recorded that registration document for all enlisted Village Organizations have been submitted to the DCO office. Inadequate follow up from the VOs and non-cooperation of officials have resulted to delay in registration of village organizations.

Of the total number of registered Village Organizations, about 96.22 per cent village organizations have completed internal audit and remaining VOs have not initiated the audit process. Spread of Village Organizations and audited status across the mandals of Paderu agency area reveals that 14.03 per cent of Village Organizations of Munchingput, 15.21 per cent Village Organisations of Chinatapalli, 4.76 per cent Village Organization of G.K.Veedhi and 9.61 per cent Village Originations of Koyyur Mandal have not under taken the audit as of the reporting period (see table-5.26). It is the first attempt of its kind for institutionalizing audit for Village Organizations.

Table 5.26: Audit Status of VOs in Paderu

Sl. No.	Mandals	Total No. of VO	Audit completed	Audit not undertaken
1	2	3	4	5
1	Ananthagiri	63	63	0
2	Arakuvalley	48	48	0
3	Dumbriguda	54	54	0
4	Hukumpeta	54	54	0
5	Pedabayalu	59	59	0
6	Munchingiput	57	49	8
7	Paderu	56	56	0
8	G.Madugula	52	52	0
9	Chintapalli	46	39	7
10	GKVeedhi	42	40	2
11	Koyyuru	52	47	5
	Total	**583**	**561**	**22**

Source: IKP Paderu Progress Report 2011-12

479 Village Organizations have filed the return. No clearance has been accorded to any Village Organization as of now. About 18.35 per cent village organizations have yet to file the return for the internal audit. Of the eleven mandals, 6 mandals have filled the return fully. Rate of return in the rest of the five mandals varies from a minimum of 1.92 per cent to a maximum of 78.26 per cent (see table-5.27).

Table 5.27: Status Returns of VOs in Paderu

Sl. No.	Mandals	Total No. of VO	Return Filed	Return not filed
1	2	3	4	5
1	Ananthagiri	63	18	45
2	Arakuvalley	48	48	0
3	Dumbriguda	54	54	0
4	Hukumpeta	54	35	19
5	Pedabayalu	59	59	0
6	Munchingiput	57	57	0
7	Paderu	56	53	3
8	G.Madugula	52	52	0
9	Chintapalli	46	10	36
10	GKVeedhi	42	42	0
11	Koyyuru	52	51	1
	Total	583	479	104

Source: IKP Paderu Progress Report 2011-12

5.7.3. Leadership Rotation

One member of the group needs to take the lead. How is this person identified? The best way is to ask the following questions within the group:

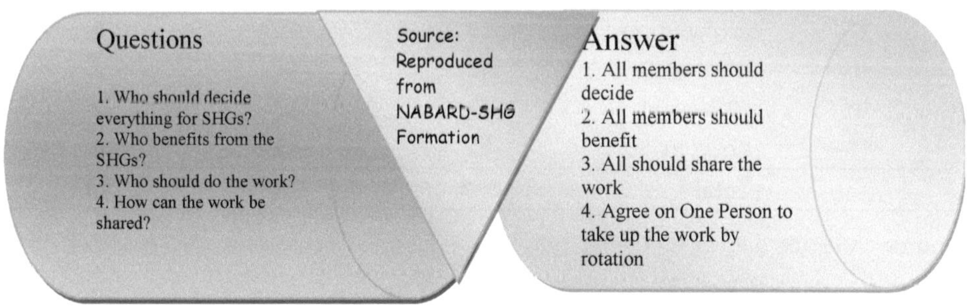

A leader is identified unanimously by the members of a group. Being the representative of the group to lead the group in terms of financial planning, indulgent in income generation programme, keeping the books and accounts up to date etc are primarily the basic business and responsibilities assigned on the position. The leader has to create another leader and give scope to all other members to serve the group. Normally, a leader should hold the position for a financial year and another leader need to replace him through people's choice. Change in leadership is viewed as a positive sign for better growth and functioning of institution. It gives scope for the other team members to complement and supplement the livelihood initiative and venture out for new outlooks.

Hence, leadership change is a major institutional issue for sustainability as well as growth of the institutions. The same is being viewed critically for the growth and diversification of VOs and self-help groups. It is reported that the office bearer of VO s has not been changed for the past 5-7 years and in most of the cases from the date of formation of Village Organizations' in the agency area. It is being reported that 95.02 per cent VOs has witnessed change in leadership officially (see table-5.28). There are about two mandals *i.e.* Hukkumpeta and Arakuvalley recorded for 100 per cent leadership rotation.

Table 5.28: Mandal-Wise Leadership Rotation Status

Sl. No.	Mandals	Total VO	Change-in Leadership	No Change-in Leadership
1	2	3	4	5
1	Ananthagiri	63	61	2
2	Arakuvalley	48	48	0
3	Dumbriguda	54	53	1
4	Hukumpeta	54	54	0
5	Pedabayalu	59	54	5
6	Munchingiput	57	52	5
7	Paderu	56	53	3
8	G.Madugula	52	47	5
9	Chintapalli	46	44	2
10	GKVeedhi	42	40	2
11	Koyyuru	52	48	4
	Total	**583**	**554**	**29**

Source: IKP Paderu Progress Report 2011-12

5.7.4. Bank and NBFI-wise Operational Issues

Self help groups in agency areas of state of Andhra Pradesh are mooted with a view to give financial and economic self-sufficiency to tribals and multiply the skills and resources for better sustainable living. With this pursuit tribal women are well organized into groups and several thousands of self-help groups are formed across villages of Paderu agency area. The self-help groups formed are initially financially supported by the World Bank supported Velugu Programme.

The financial self sufficiency of the self-help groups are provided with linking them with formal banks and informal financial institutions. All

SHGs created are groomed to a stage to be linked to the formal financial institutions. 11 mandals of Paderu agency area are served with morethan a dozen of banking institutions and hence, the corresponding operational and capable SHGs are attempted to be linked with them for larger financial assistance for under taking economically gainful livelihood activities based on available natural resources.

The mandate of IKP to link all self help groups to formal financial institution is just little away from the goal. There are many financial institutions supporting financial assistance without any pinch while others are encountering difficulties of managerial as well as operational in nature (see table-5.29). To cite an example APGVB has no operational issue in any of the operating branches and mandals with self help groups, except reporting of inadequate staff in K.D.Peta brach of G. Madugula Mandal.

Table 5.29: Consolidated Status of Issues of Bank/NBFI in Paderu Agency Area

Sl. No.	Mandal	Bank/NBFIs	Operating Branch	Issues in Operation
1	2	3	4	5
1	Ananthagiri	SBH	Kasipatnam	No Issue
		APGVB	Ananthagiri	No Issue
		SBI	Kinchumanda	Banker not supporting for linkage
		UCO	Devarapalli	No Issue
2	Arakuvalley	ASSAV	Arakuvalley	No Issue
		APGVB	Arakuvalley	No Issue
		CANARA BANK	Yendapallivalsa	Banker not supporting for linkage
		SBI	Arakuvalley	No Issue
3	Dumbriguda	SBI	Kinchumanda	Banker not supporting for linkage
		APGVB	Arakuvalley	No Issue
		ASSAV	Arakuvalley	No Issue
4	Hukumpeta	SBI	Paderu	Banker not supporting for linkage
		VIJAYA BANK	Hukumpeta	No Issue
		SBI	Kinchumanda	Banker not supporting for linkage
		UBI	Paderu	No Issue
5	Paderu	UBI	Paderu	No Issue
		APGVB	Paderu	No Issue
		IOB	Vaddadi	No Issue
		DCCB	Paderu	No Issue

Contd...

Table 5.29: Contd...

Sl. No.	Mandal	Bank/NBFIs	Operating Branch	Issues in Operation
1	2	3	4	5
6	Pedabayalu	UBI	Paderu	No Issue
		APGVB	Peddabayalu	No Issue
		SBI	Rudakota	Issuing less than the eligibility
7	Munchingput	APGVB	Peddabayalu	No Issue
		SBI	Peddabayalu	No Issue
		SBI	Paderu	Banker not supporting for linkage
8	G. Madugula	APGVB	G. Madugula	No Issue
		UBI	Paderu	No Issue
9	Chintapalli	UBI	Chintapalli	No Issue
		APGVB	Chintapalli	No Issue
10	G.K. Veedhi	UBI	Chintapalli	No Issue
		APGVB	Chintapalli	No Issue
		UBI	Seleru	No Issue
11	Koyyuru	ANDHRA BANK	Narsipatnam	No Issue
		APGVB	Koyyuru	No Issue
		SBI	K.D. Peta	No Issue
		APGVB	Nathavaram	No Issue
		SBH	Golugonda	No Issue
		INDIAN BANK	K.D Peta	Issue less than the eligibility
		APGVB	K.D Peta	Inadequate Staff to run the branch

Source: Data Collected from Bank/NBFIs Operating in Paderu ITDA

AASSAV-the only tribal run MFI has not witnessed any operational flaws in lending finance to self help groups of Arakuvalley and Dumbriguda Mandal. It is running in two mandals viz. Arakuvalley and Dumbriguda of Paderu agency area and with staff of tribals. It is the single MFI working at par with all other formal financial institutions and catering to the financial need of a handful of self help groups.

Similarly, UBI, UCO, IOB, DCCB and Vijaya Banks have not reported of any operational difficulty in self help group financing and its recovery from all operating branches of Paderu ITDA. The smooth functioning of all branches across the agency mandals is the strength for tribal self-help groups.

However, operational difficulties are reported by 3 nationalized banks for financing the self help groups. These are Canara Bank, State Bank of India and State Bank of Hyderabad. The common issues reported are (a) Bankers are not supporting for bank linkage, and (b) Banks issuing less amount to self help groups that of their eligibility. All three banks are operating in the well-connected mandals like Hukkumpeta, Paderu, Dumbriguda and Arakuvalley, where maximum number of self-help groups are well aware of the banking norms and functioning.

5.8. Viewing IKP's Role Vis-à-Vis ITDA

Integrated Tribal Development Agencies (ITDAs) in Andhra Pradesh are morethan four decades old. These are institutions created in tribal concentrated areas to conserve the hinterland, natural resources, and rich culture as well as tradition of tribal people. These agencies are given with special administrative powers and privileges. All programmes and schemes of the government are executed under one-roof following the umbrella approach.

The prime responsibility of ITDA goes towards development and welfare of the aboriginal people living over the contiguous land. A concentrated approach is followed to execute the development of programme in good coordination and liaison with all other line departments.

Indira Kranti Pratham (IKP) is a World Bank supported programme to empower women of the state of Andhra Pradesh through formation of self-help groups and engaging them in ICPs of various natures. It has a special recognition in tribal areas of the state. IKP, as popularly known among the people, is the largest women empowerment microfinance programme in the world. IKP is executed with active support of the ITDAs. The staff and building have been widely engaged to popularize the programme and ensure peoples acceptance.

It has been viewed as the real executor of government schemes devoted for ITDAs with people's engagement and participation. All activities requiring community support for execution is executed through self-help groups of IKP. Schemes and programmes of all line departments functioning in the agency area are tuned up with functioning of IKP for execution. The progresses of schemes are also measured through the IKP lenses in the state.

ITDA and IKP though are two institutions working in agency area; the scope of operation is very much limited for IKP. IKP takes the responsibility of physical execution of activities while technical support for execution used to be extended from the core department. IKP being perceived as an institution having deep rooted base at each and every habitation and coming up with 'community-involved-micro- credit-plans' is probably the most viable institution to establish a good linkage with community and all other government departments functioning in remote and tribal areas.

The Micro Credit Plans (MCPs) are developed with active participation of women self-help group members. All MCPs are developed to meet the immediate and long-term need of the members of the groups. Each MCP gets financial support either from the specific department and/or division of the ITDA or directly assisted by IKP-RF. 3167 self-help groups are linked to formal and informal financial institutions for financial assistance. About 60 per cent self help groups are eligible for renewal of bank linkage (see table-5.30). SHGs across all 11 mandals have linked to financial institutions. All 3167 SHGs are eligible to access ₹ 3316 lakhs; however a minimum of ₹ 2795.51 lakhs is planned under SBLP for self-help groups of Paderu.

Table 5.30: Mandal-wise SHG-Bank Linkage Credit Plan for Paderu during 2012-13

Sl. No.	Mandal	No. of SHG Linked	No of SHG Eligible for Renewal in 2012-13	Loan amount entitled as per SLBC Guideline (Max)	Loan Amount Planned (Min)
1	2	3	4	5	6
1	Ananthagiri	266	135	265	215.95
2	Arakuvalley	351	267	389	344.50
3	Dumbriguda	252	153	233	201.00
4	Hukumpeta	334	144	226	195.75
5	Pedabayalu	409	265	321	300.15
6	Munchingiput	269	179	215	201.25
7	Paderu	435	275	660	527.23
8	G.Madugula	425	216	434	351.93
9	Chintapalli	72	44	90	73.00
10	GKVeedhi	85	64	120	98.50
11	Koyyuru	269	163	363	286.25
	Total	**3167**	**1895**	**3316**	**2795.51**

Source: IKP-TPMU, Paderu

5.9. NGOs and MFI Programme

The Non Government Organizations involved in promoting SHGs and linking them with the Formal Financial Agencies (FFAs) perform the following functions:

(i) Organizing the poor people into groups,

(ii) Training and helping them in the organizational, managerial and financial matters,

(iii) Helping them access more credit and linkage with formal financial agencies,

(iv) Channelizing the group effort for various development activities,

(v) Helping them in availing opportunities, widening the options available for economic development,

(vi) Helping them in sustaining the group effort independently even after withdrawal of the NGO.

There are 11 NGOs involved in formation of self help groups across 11 mandals of Paderu agency area under the STEP programme. All self-help groups formed are trained on various parameters of SHG functioning, and administration. Training module and materials are developed hiring consultants and trainings are being imparted using simple materials. This has helped IKP in multiplying the effort and executing the livelihood programmes more vigorously and aggressively. The IKP programme has also yielded back end support from the DWCRA initiatives for women empowerment; where women headed institutions are imparted with training skill and financial assistance to become self-sufficient.

Besides, many NGOs have accessed fund support from the Department of Women and Child Development, GoI and assisted self-help groups with financial support and imbibed them with technical support to identify and execute livelihood activities of their choice for family sustenance.

It is the STEP programme which has persistently supported formation, nurturance, and management, enabling SHGs to rotate RF, and mobilization of CIF and enabling SHGs for bank linkage. It helped self help groups to pass through various tests prescribed by financial institutions for bank linkage. It is recorded that about 1023 groups are enabled through NGOs like VJNNS, Kovel, AASSAV and Nature to access revolving support of ₹ 8,37,450 during 2003 to 2007 across 8 mandals out of the 11 mandals. The RF support extended to members of self help groups have supported to 407 members in the second round of rotation (see table-5.31). AASSAV has supported access of maximum bank linkage of ₹ 5,17,000 during the period and it is due to the sole running of AASSAV Mutually Aided Cooperative

Society in the tribal area. The initiative of NGOs have mobilized community investment fund (CIF) of ₹ 2,63,475 during the period.

Self-help groups in rest of the 4 mandals are supported by some community based organizations like Adivasi Mitra, TERDS, ORRC, and GYSS. Self help groups are formed, trained and assigned to IKP for future nurturance.

Table 5.31: NGOs and SHG Functioning during 2003 to 2007 in Paderu Agency Area

NGOs	Mandals Covered	Groups Supported	Disbursed RF Fund	Rotation of RF among members	BL amount	CIF harnessed
1	2	3	4	5	6	7
VJNNS	Chintapalli, Koyyur	412	217200	216	330100	103475
Kovel	Paderu, G. Madugula	173	215000	67	0	100000
AASSAV	Dumbriguda, Araku valley	210	214000	64	517000	35000
Nature	Araku Valley, Ananthagiri, Dumbriguda	228	191250	60	15000	25000
	Total	**1023**	**837450**	**407**	**826100**	**263475**

Source: Consolidated from various Annual STEP Reports of 2003, 2004, 2005, 2006 and 2007, Paderu, CARE-India, STEP

5.10. MFIs and Administrative Regulations

After the State brought in the ordinance regulating the MFIs, the sector which used to lend ₹ 7,000 crore, had not been able to lend anything now to the poor who had no access to bank loans. The MFIs had registered according to the stipulations brought in, but it was not translating into lending process.

Though the banks through SHG linkage were extending loans to some extent, there existed a credit gap of ₹ 5,000 crore. As a result the poor were left with no option but to approach moneylenders paying exorbitant interest. The sector had offered loan restructuring package, interest reduction etc., to the tune of ₹ 1,800 crore, but the State did not respond formally so far. The SIDBI too modified the package further and submitted it to the government.

"The entire client base of nine million is being treated as defaulter borrowers by credit bureaux. It is time they were turned into current borrowers with credit restructuring," he said Mr. Prasad said, CEO, MFIN.

While the Reserve Bank of India was the regulator for the Non-Banking Finance Companies and MFIs, the State government came with an ordinance equating MFIs with moneylenders. "This duality of regulation was not helping the sector or the poor". He hoped the Microfinance Institutions Bill 2012 introduced in Parliament would soon put a full framework of regulations in place for the entire country (The Hindu, 22 August, 2012).

Notes

1. The MACs is running uninteruptedly from the date of its inception. It is meeting financial needs of Araku Valley and Dumbriguda Mandals.
2. http://www.hindu.com/2010/10/19/stories/2010101959900300.htm
3. AASSAV is a three-decade old Non-Governmental Organization working for development of tribal people and the region. It had been institutionalized and promoted by Gustafson- a Canadian by origin. There are several livelihood activities promoted for tribals and many small and cottage industries being institutionalized across the area. This organization though formed by a foreign national has really been handed over to the tribals for nurture and development. ASSAV-MACS is one among the many institutions created by the organizations. Besides, there are units like coffee cultivation and processing, Dairy, Carpentry unit, poultry, Bee Keeping, Printing Press, Organic Manure production and marketing, Orchards, Organic Vegetable Cultivation etc. promoted by the NGO.
4. **Swarnajayanti Gram Swarojgar Yojana (SGSY)** is an initiative launched by the Government of India to provide sustainable income to poor people living in rural areas of the country. The scheme was launched on April 1, 1999. The SGSY aims at providing self-employment to villagers through the establishment of Self-help groups. Activity clusters are established based on the aptitude and skill of the people which are nurtured to their maximum potential. Funds are provided by NGOs, banks and financial institutions. Since its inception, over 2.25 million Self-help groups have been established with an investment of ₹ 14,403 crores, profiting over 6.697 million people. The Swarnajayanti Gram Swarojgar Yojana (SGSY) was launched as an integrated programme for self-employment of the rural poor with effect from April 1, 1999.
5. "ABHAYA HASTAM" a co-contributory pension and Insurance Scheme for IKP SHG women members to provide pension in the old age and to create social security to the family members. The Government has appointed SERP as the Nodal Implementing Agency to implement the scheme for the benefit of the SHG women. Lr. No. 775/RD-SHG.I(P)/2009, Dt. 10.02.2009, www.rd.ap.gov.in
6. Department of Financial Services, Ministry of Finance, Government of India, www.financialservices.gov.in
7. The mission of Andhra Pradesh is to empower the Community (SHG - Federations) to play an active role in insurance operations through SERP, to reduce people's vulnerabilities through social security measures like Aam Aadmi Bima Yojana (AABY), Jansree Bima Yojana (JBY), Other Group Insurance

(OGI), Credit Insurance (CI) etc. It also implements the Cattle Insurance and any other scheme they consider to be relevant to protect their income and assets base of the poor. www.rd.ap.gov.in/insurance/brief_note

8. "Revenue Sadassulu" is primarily an initiative to resolve land disputes by the people. This body designed to act as facilitator to settles issues on land amicably in the village. There are "3,89,389 land disputes were resolved through "Revenue Sadassulu" during January - March, 2012," said Mr. N. Raghuveera Reddy, Hon'ble Minister for Revenue. More details can be found out at www.aplandrights.org.
9. The 'Srinidhi' scheme with a corpus fund of ₹ 1,000 crore was launched by the chief minister Kiran Kumar Reddy on 15th September 2011. This welfare scheme is meant for empowering women and the fund will be given to women organizations like SHGs, VOs and Rytu Mahila Sadassu.
10. The Banks are giving loans under SHG Bank Linkage programme with different rates of interest ranging from 8% to 12%. The groups are facing difficulty in paying such rates of interest. Whatever the interest rates charged more than 3% by the banks will be reimbursed to SHGs by the Dept. of Rural Development, Govt. of Andhra Pradesh. The DRDA will obtain the details of SHG repayment for the purpose of interest subsidy twice a year through the Mandal Samakhyas. DRDA will calculate the 'pavalavaddi to be given to each SHG basing on the bank statement. The DRDA will give the 'pavalavaddi' cheques to the Mandal Samakhya along with the statement of particulars accruing to each SHG. The Mandal Samakhya in turn will distribute the cheques to the SHGs through the village organizations (VO) in the presence of non-officials.

References

1. Pamela Mackenzie (2006): Keeping it LOCAL, Change and Development in an Indian Tribal Community, APH, Publishing House, New Delhi
2. NTFP and Tribal Livelihood:
3. Reserve Bank of India (2008): Banking statistical Return Reports, RBI, Speeches Published on Financial Inclusion
4. Badajena, S.N and Gundimeda, H (2009): Self Help Group Bank Linkage Model and financial Inclusion in India, www.skoch.in
5. Report of the Internal Group to Examine Issues Relating to Rural Credit and Microfinance: Models, Policies and Experiences (2005), Reserve Bank of India
6. Barik, S.K (2009): Policy of Total Financial Inclusion and Access to Finance by the Tribal Self-help Groups: Re-working on the Modality of Service Delivery, paper presented at the National Seminar on Dynamics of Banking Sector Reforms: Emerging Issues, Challenges and Strategies, Centre for Banking Studies and Research, in Association with IBA Ball 19-21 July 2009.
7. http://en.wikipedia.org/wiki/Revolving_fund

Chapter 6

Understanding the Development Potency, Measuring Impacts and Realizing Competing Edge of SHG Mode of MFI Operations in ITDA Paderu

6.0. Introduction

Finance is a key factor to economic development. It's availability in volume, value and timeliness significantly regulate the pace of economic activities spelt out for a region and people. The concept of SHG provides the benefits of economies of scale, cost effective alternatives for financial services, collective learning, and a democratic and participatory culture (Surender, Kumari and Sehrawat, 2011) .Thus access to finance, especially by the poor and vulnerable groups is a prerequisite for employment generation, economic growth, poverty reduction and social cohesion. Further, access to finance will empower the vulnerable groups by giving them an opportunity to own a bank account, to save and invest, to insure their homes, thereby facilitating them to break the chain of poverty. India is lagging in this and let's explore out the fate of tribal's of agency areas of Andhra Pradesh.

SNAPSHOT VIEW OF DATA COVERAGE:

1. Paderu ITDA of Visakhapatnam District is consisting of 11 Mandals
2. 8766 self help groups formed by IKP and NGOs and are functioning
3. 583 Village Organizations (VOs) formed across Mandals and functioning
4. All members are from different tribal groups
5. There is no bias in getting membership to self help groups
6. All mandals are homogeneous in almost all aspects
7. Economic condition of the people across mandals is almost similar.

ITDA Profile	Data Coverage
11 Mandals	✓ 3 mandals selected for the study representing three different regions of the agency area ✓ It is 27.27 % of the mandals of agency
8766 SHGs	❖ SHGs of three mandals constitutes 29% of the total SHGs of ITDA ❖ 620 SHGs out of 2556 is covered under the study i.e. 214 out of 779 in Dumbriguda, 224 out of 871 in G. Madugula and 182 out of 906 in Padabailu ❖ About 24.25% of beneficiaries of SHG being interviewed
583 VOs	❖ 53 VOs out of 165 is interviewed i.e. 10 out of 54 from Dumbriguda, 27 out of 52 from G. Madugula and 16 out of 59 from Padabayalu ❖ 32.12 % of village organizations have been studied.

6.1. Microfinance and Tribal Women Empowerment

Women empowerment is not giving women power. Women already have plenty of power, in the wealth of their knowledge and motivation, to do their jobs magnificently. Empowerment is defined as letting this power out (Blanchard, K, 1996). It encourages people to gain the skills and knowledge that will allow them to overcome obstacles in life or work environment and ultimately, help them to develop within themselves or in the society. Women's empowerment has five components, these are:

(*i*) Women's' sense of self-worth;

(ii) their right to have and to determine choices;

(iii) their right to have access to opportunities and resources;

(iv) their right to have power to control their own lives, both within and outside home; and

(v) their ability to influence the direction of social change to create a more just social and economic order, nationally and internationally.

Women Empowerment refers to an increase in the strength of women such as spiritual, political, social or economic. The most common explanation of "women's Empowerment" is the ability to exercise full control over one's actions. Thus, women empowerment occurs in real sense when achieve increased control and participation in decision making that leads to their better access to resources, it often involves the empowered developing confidence in their own capacities (Narang, 2012).Under the stated pretext , the microfinance interventions in tribal areas of Paderu are assumed to have strong control over the livelihood, institutional functioning, women involvement in shaping of tribal economy and bringing in empowerment for all section of women in the agency area by and large.

Hypothesis-I: Microfinance contributes to women empowerment

It is assumed that institutions play significant and positive role in shaping the life and livelihood of the members. The regulatory mechanism in a form induces new areas of thought and thereby enhances the scope of outlook of its members as well as encourages for a new means of living. Indeed, this has anticipated empowering people to have access to services, improves the living conditions and gives impetus to steer the group activity for betterment of the society. However, there are many questions raised on the meaning of tribal women empowerment and its scope?

Though empowerment, let it be for tribal and/or non-tribal, got different connotation to different people and has been interpreted mostly depending on the situation and context. The present study, in the wake of microfinance institution (MFIs) functioning to empowerment, has specifically given focus on analyzing the accrued benefit of MFI towards improving the household facilities and assurance of a decent living, women involvement in decision making, nature of decisions being taken in meetings, benefits of trainings on women and access and leverage of government schemes and programmes. Can this be called a successful and sufficient analysis of empowerment in the context of MFI operation?

Tribal in general and tribal of Paderu agency area in particular got a peculiar characteristic in terms of habitation pattern. Gross variation is noticed in the housing pattern and living conditions between PTGs and non-PTGs. Among the varied groups some tribal got access to better housing facilities and services and are pretty advanced in terms of economic condition in comparison to others. The study recorded a varied pattern of residential status across the studied mandals (see fig.6.1). About 91.61 per cent of self help group members proudly resides in own house, 5.16 per cent resides in rented house and nearly 3.2 per cent lives neither in own house nor in rented house but in a house of neighbors' and village common property places. Though non-residential member's forms a smaller portion, but they will all own a house with the persisted effort of self-help groups.

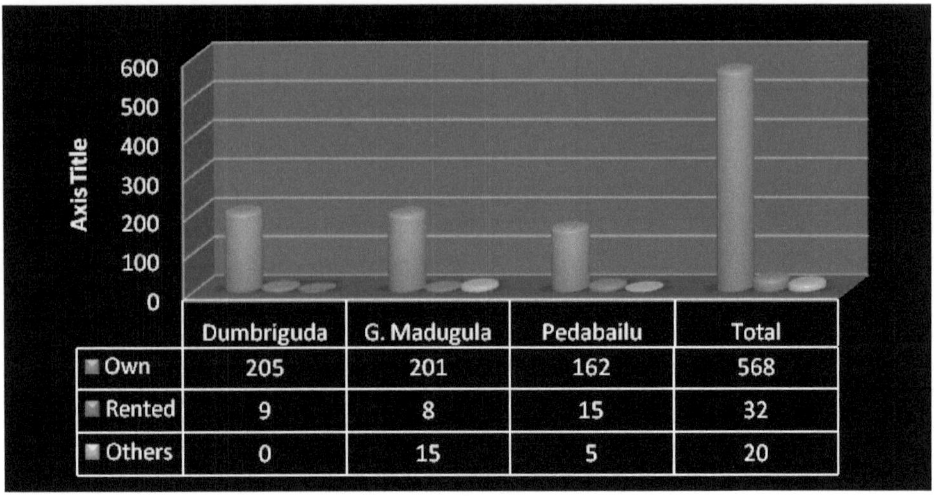

	Dumbriguda	G. Madugula	Pedabailu	Total
Own	205	201	162	568
Rented	9	8	15	32
Others	0	15	5	20

Fig. 6.1: Residential Status

The other basic necessity of life for a decent living is electricity. It is now felt as an inevitable part of life. It not only regulates children education but also eliminates issues of health, promotes creative thoughts within youngsters in a village atmosphere. The study reveals that close to 75 per cent self-help group members are accessing electricity facilities at home across the studied mandals of Paderu agency area (see table-6.1). However, 25.48 per cent of self-help group members are devoid of accessing electricity facilities in the houses. The poor state of electricity facility provision for all members of SHGs is due to the location of the villages. Some of the families are living in scattered places and close to the place of their agriculture which is mostly on the top of the hills. It is too difficult for any service to reach the people, let alone electricity. The existing electricity coverage is the sole effort of electricity department and some villages are getting access to electricity supply due to the effort of self-help groups of their villages.

Table 6.1: Access to Electricity Facility by SHG Members

Sl. No.	Name of the Mandals	% of families having electricity	% of families without electricity
1	Dumbriguda	74.29	25.70
2	G. Madugula	75.44	24.55
3	Pedabailu	73.62	26.37
	Total	**74.51**	**25.48**

Source: Village Survey

The other parameter which is basically a facility to help tribal women and men together keep healthy and keep away from occurrences of many deadly diseases is the provision of toiletry facilities in the houses. Though government is creating awareness on the use of toilets, its impact is being monitored through the use of toilets by the people. And moreover, self-help programme in the state of Andhra Pradesh is primarily supported by the World Bank which is in strong favour of promotion of use of toilets in villages and grounding the programmes at village level has strongly been viewed through self-help institutions. Promotional measures has adequately been taken to ensure that some government representative to participate in the monthly meeting of the self-help groups and the Village Organizations. The issues of the self-help group on access to toiletry services are further discussed at Mandal Mahila Samakshya (MMS) level.

Table 6.2: Access to Toilatary Facility by Tribal Women

Sl. No.	Mandal	Open Defication	Toilets
1	Dumbriguda	196	18
2	G. Madugula	216	8
3	Pedabailu	164	18
4	Total	576	44

Results:

1. **Co-efficient correlation-(r): -0.79241**
2. **STDEV : 98.60**

Source: Survey Data

The study reveals that about 92.90 per cent of self help group members reported of using open defecation across the studied mandals of the agency area. Only a small percentage *i.e.* 7.09 per cent access toilets. The reasons cited are lack of water supply in the toilets, taboo of having toilet near the house, odious smell, not habituated of using toilets etc. Access to toilets is more or less the same for Dumbriguda and Padabailu (see table-6.2). The access and use of toilets by the self-help group families is very limited revealed from the correlation *i.e.* (r...: -0.79241) and the STDEV value of 98.60. This is also giving rise to draw the conclusion that self-help as an instrument for development and change among the tribal areas could not withstand to the expected targets.

Can access to cooking devices form a part of women empowerment? To say yes or no from the point of view of women empowerment is sometimes looks shy and often sidelined with laugh that it's a silly question. However, it matters much from women's empowerment perspective. The study has recorded uses of means of cooking *i.e.* Gas, Kerosene, Wood and Others (agriculture residues, dung-cake, dried leaves, twigs and barks etc.) by the tribal women of the ITDA Paderu. Use of better cooking means is considered as one among the parameters of economic well-being. Families having access to and use of gas and/or kerosene for food preparation is said to be in better up economic categories.

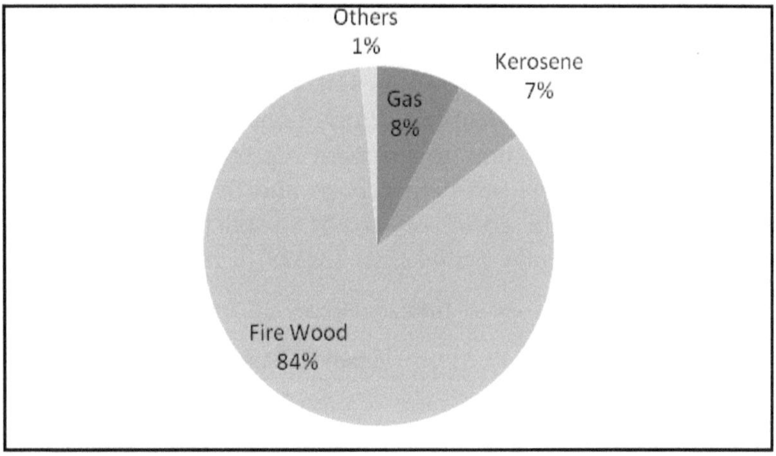

Fig. 6.2: Use of alternative Cooking Means by Tribals

The study recorded that almost 83.87 per cent of beneficiaries across the mandals are using firewood as the means of cooking. 7.74 per cent beneficiaries reported of using gas while 6.77 per cent families reported for using kerosene. About 1.62 per cent beneficiaries reported accessing other means to prepare food (see fig-6.2). Use of means of cooking and family economic condition do have a close bearing revealed from the correlation established among average monthly income and the adoption of means (see table-6.3). Hence, it can be safely deduced that economic condition does have a strong bearing on the selection of means of cooking in tribal areas and self-help group as grass root institutions have contributed significantly to empower people to have greater access to better facilities.

Table 6.3: SHG Members Access to Means of Cooking

Sl. No.	Parameters	Dumbriguda	G. Madugula	Pedabailu	Total
1	Gas	9 (4.21)	12 (5.35)	27 (14.83)	48 (7.74)
2	Kerosene	18 (8.42)	0 (0.00)	24 (13.20)	42 (6.77)
3	Fire Wood	186 (86.91)	203 (90.62)	131 (71.97)	520 (83.87)
4	Others	1 (0.46)	9 (4.03)	0 (0.00)	10 (1.62)
	Total	**214 (100.00)**	**224 (100.00)**	**182 (100.00)**	**620 (100.00)**

Results:

1. Monthly Family Income and Access to Gas Connection: correlation (r) is 0.185
2. Monthly Family Income and Access to Firewood : correlation (r) is 0.196
3. Monthly Family Income and Access to Kerosene: correlation (r) is - 0.699
4. Monthly Family Income and Other Sources of cooking : correlation (r) is 0.792

Source: Survey Data

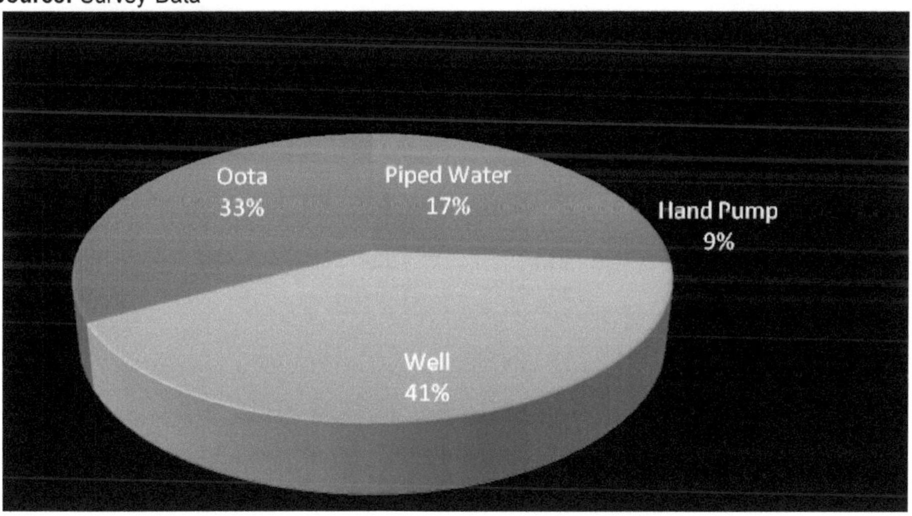

Fig. 5.3: Access to portable Drinking Water

The study considered access to portable drinking water is one among the vibrant parameter to be considered for measuring empowerment. Fetching of water is directly related to the working time available for women in the villages while consumption of portable water is the basic key to family health. Besides, health and economic well-being of a village is mostly judged from the availability of portable water. Under the circumstance, the study has recorded that about 41 per cent respondents are consuming water from dug-well, 33 per cent still depends on Oota (a perennial spring), 17 per cent got access to piped water supply and about 9 per cent are using water from the hand pumps (see fig.6.3).

The agency areas in general are undulated in nature and habitations are settled across the hillocks. There is a popular saying that tribal habitations emerged from the places where there is water. Hence, judging access to water facilities through the microfinance institution lenses is somewhat erroneous and not justifiable to suffice that areas that are not adequately provisioned with water are less developed. Because, water supply is regulated by many factors of course location may be one among them. While considering the fact, it has been somehow or other accepted that 'better accessed mandals may have better access to piped water supply facilities' and 'mandals with more number of microfinance institutions have comparative advantage of better water supply provisions'.

Table 6.4: Access to Portable Drinking Water by SHGs

Sl. No.	Parameters	Dumbriguda	G. Madugula	Pedabailu	Total
1	Piped Water	49 (22.89)	33 (14.74)	26 (14.28)	108 (17.41)
2	Hand Pump	23 (10.74)	7 (3.12)	25 (13.74)	55 (8.88)
3	Well	87 (40.65)	113 (50.44)	54 (29.68)	254 (40.97)
4	Oota	55 (25.72)	71 (31.70)	77 (42.30)	203 (32.74)
	Total	**214 (100.00)**	**224 (100.00)**	**182 (100.00)**	**620 (100.00)**

Source: Survey Data

Of the three mandals under study, Dumbriguda mandals is better positioned with more than average number *i.e.* 22.89% of self help group members accessing piped water supply and also reported with more than average number of beneficiaries accessing hand pump water. The statistics in a way confirms to the fact that bettered accessed mandals do access better water supply services (see table-6.4). However, while examining the other consideration *i.e.* better MFI positioned Mandals do have better access to advanced water supply, it is found that the consideration is adversely supported with co-efficient of correlation (r) result -0.99951 for piped water supply and r - 0.15143 for access to hand pump . Hence, access to better water supply services is independent of the number of MFIs in operation.

Decision making is an area mostly left on with male especially in rural villages. It is often assumed accrued to male by birth. This attitude hampers optimal decision making for appropriate work. To err is human. Hence, there is every possibility of committing mistakes while taking decisions of any sort. This practice is in vogue since dawn of human society and running generation across. In the process, it is in turn completely neglecting one segment of the society.

In contrast, a judicious decision may possibly reduce the risk of committing mistakes and in turn maximize benefits. The current study is attempting to give an overview of the decision making process being adopted by tribals in ITDAs and involvement of women in decision making. Study of the decision making practice being in operation in tribal areas is leading to several questions. These are:

(1) Is decision making guided by the income contribution of people and especially by women?

(2) Does decision making positively influence the level of education? And

(3) Is it guided by the types of employment women of tribal families engaged in?

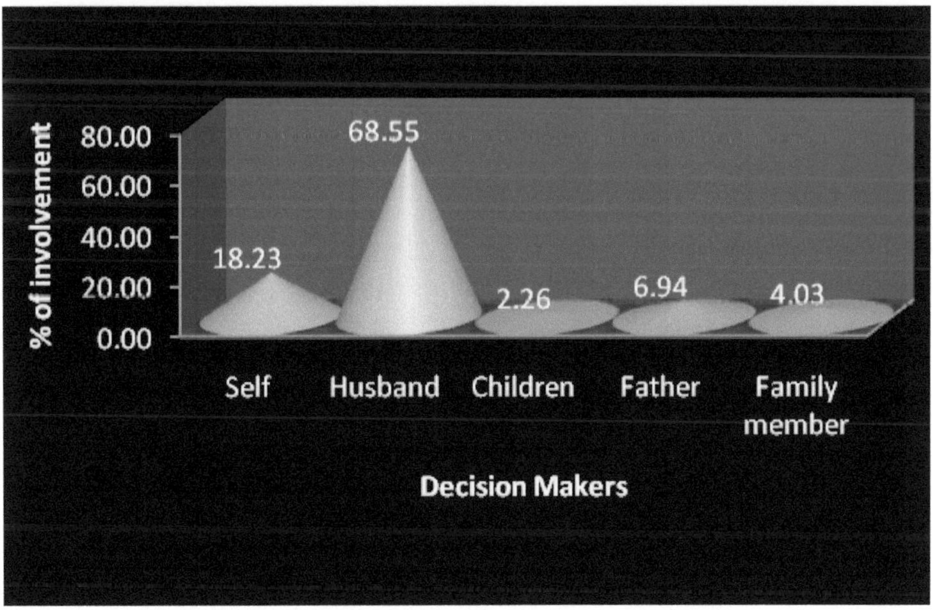

Fig. 6.4: Decision-making on Family Spending

It is a very trivial question to answer the degree of involvement of people in decision making. Indeed, the study captured that about 18.23 per cent tribal women across the PTG and Non-PTG category of the selected mandals reported of self-decision making in spending up of family income. However, 68.55 per cent women self-help group members reported of male and especially husband decision making in spending of family income. It is not surprising that father and /or in-laws in tribal families are also involved in decision making on spending because tribal families are mostly joint families and about 10.97 per cent of respondents are supporting the argument (see fig. 6.4).

Over the studied regions in the agency area, 25 per cent of self help group (SHG) members of G. Madugula Mandal and 21.43 per cent of Pedabailu Mandal reported of self-decision making which is more than the aggregate average of the Paderu agency area. Dominance of male in income spending is reported across the studied regions. Involvement father and/ or in-laws in family income spending decision making is relatively high in G.Madugula Mandal (see table-6.5). It is a good sign of progress that children and especially educated children are consulted with and involved in spending of family income.

Table 6.5: Decision on Spending of Family Income

Sl. No.	Parameters	Dumbriguda	G. Madugula	Pedabailu	Total
1	Self	18	56	39	113
		(8.41)	(25.00)	(21.43)	(18.23)
2	Husband	184	125	116	425
		(85.98)	(55.80)	(63.74)	(68.55)
3	Children	3	5	6	14
		(1.40)	(2.23)	(3.30)	(2.26)
4	Father/In-laws	4	31	8	43
		(1.87)	(13.84)	(4.40)	(6.94)
5	Family member	5	7	13	25
		(2.34)	(3.13)	(7.14)	(4.03)
	Total	**214**	**224**	**182**	**620**
		(100.00)	**(100.00)**	**(100.00)**	**(100.00)**

Source: Survey Data

The study recorded that tribal woman involvement in income spending is positively correlated to their volume of income contribution to the family income kitty. Examination of women contribution against the decision making arrives at a coefficient correlation of 0.95 (see table-6.6). The larger is the contribution the better is the stake of women involvement in decision and hence, better is the participation in spending. In the context of tribal women of Paderu agency area, women voice is increasingly recognized.

Table 6.6: Income Contribution of Women and Involvement in Decision Making

Sl. No.	Mandal	Avg Income (₹)	Self Decision
1	Dumbriguda	592	8.41
2	G. Madugula	951	25
3	Pedabailu	781	21.43
Results:	Correlation coefficient between average income contribution to family income and self decision making on spending is 0.959045906		

Source: Survey Data

Education has been identified and accepted by tribal as the only panacea of development and there is a positive response from tribal in sending children to schools and also educating themselves under the adult education programme to run the institutions and especially the self-help groups. Rising education level has a strong bearing on decision making of income spending. The study established that educated women have a better stake in decision making and they are actively involved (see table-6.7). It has positively influenced the expenditure pattern of many tribal families.

Table 6.7: Education of Women and Decision Making on Spending

Sl. No.	Mandal	Educated SHG Members	Self Decision
1	Dumbriguda	49	8.41
2	G. Madugula	102	25
3	Pedabailu	120	21.43
Results:	Correlation coefficient between education of women self help group members and self decision making on spending is 0.89949183		

Source: Survey Data

Examination of the fact that, self employed women have a positive involvement in family decision making on income spending, has been proved correct. Self-employed women and educated professionals have exhorted greater influence on income spending pattern at the household level. The co-efficient correlation (r) between employment of self-help group women and the decision making is arrived at 0.92, which reveals a high degree of causal relation among the factors (see table-6.8).

Table 6.8: Status of Women Employment and Involvement in Income Spending

Sl. No.	Mandal	Employed Women Members	Self Decision
1	Dumbriguda	94	8.41
2	G. Madugula	139	25
3	Pedabailu	149	21.43
Result:	Correlation coefficient between employment status of self help group members and self decision making on spending is 0.92963553		

Source: Survey Data

Besides the involvement of women in household spending, they too participate in group decision making in self help group meetings. The Self-help group members used to meet at least once in a month regularly to decide the functioning of the group. A point to be noted here is that group functioning and so also the decision making are influenced by many internal as well as external factors. IKP and NGOs operating in the studied regions do exhort influence and conversely members do sought their assistance in taking group related decisions. The study reported that irrespective of the regions 44.19 per cent of group president's use to take group functioning decisions and about 39.68 per cent group function decision used to be taken up by all members jointly (see table-6.9). The influence of NGOs and IKP staff in decision making across the studied regions is minimal *i.e.* 4.19% and 7.42% respectively.

Table 6.9: Decision Making in Self-help Group Meeting

Sl. No.	Parameters	Dumbriguda	G.Madugula	Pedabailu	Total
1	All Members	100	78	68	246
		(46.73)	(34.82)	(37.36)	(39.68)
2	President	74	129	71	274
		(34.58)	(57.59)	(39.01)	(44.19)
3	IKP Staff	18	15	13	46
		(8.41)	(6.70)	(7.14)	(7.42)
4	NGO Facilitator	22	2	2	26
		(10.28)	(0.89)	(1.10)	(4.19)
5	Mixed of IKP & NGO	0	0	28	28
		(0.00)	(0.00)	(15.38)	(4.52)
	Total	**214**	**224**	**182**	**620**
		(100.00)	**(100.00)**	**(100.00)**	**(100.00)**

Source: Survey Data

In totality, 83.87 per cent group related decisions are reported to be taken up by group members themselves. The decisions are virgin and pious. However, this self-decision making scenario is further enabling to hypothesize that decisions taken by the presidents of SHG president are due to their education level. In simple terminology, self-decision making on functioning of self-help groups is a direct function of education status of presidents. This fact has been thoroughly supported by the statistical results of value (4.33E-08) and the standard deviation value of 25.9416 (see table-6.10).

Table 6.10: Decision making in SHGs meeting and education

Sl. No.	Mandals	Educated SHG Members	Joint Decision	Total
1	Dumbriguda	49	100	149
2	G Madugula	102	78	180
3	Pedabailu	120	68	188
	Total	271	246	517

Results: χ^2 is 4.33E-08

STDEV is 25.9416

Source: Survey Data

Besides the inherent education, income contribution and employment status of tribal women persistent effort is being made by self-help groups to enhance the skill and potency of its members through training and capacity building which will in turn capacitate them to compete for living and retaining identity at the family and the society at large. Various forms of training have been organized and imparted from time-to-time to enhance the functional capacity of institutions through individuals. The cascading effect is broadly aiming at a larger change at the society level. A thorough examination of impact of training programmes being conducted for self help groups reveals a glaring impact on individuals in some areas while low or no impact in certain areas.

The ranking exercise conducted to assess the visible impact of training on the stipulated 12 parameters reveals that all members strongly represent for increase in level of confidence to face any situation and handling up of group activities, realized a better improvement in skill and many of them are equipped with business skill to run petty business of their own, besides many reporting for benefits that is not captured in the list (see table-6.11). Hands on experience on bank linkages, marketing linkages, linkage with government programmes, and enhancement of managerial skills are some of the important visible changes reported among the members of self-help groups across the studied regions of Paderu agency area.

Table 6.11: Impact of Training on Tribal Women Empowerment

Sl. No.	Parameters	Ranks	Assigned Ranks	Ranks for Dumbriguda	Ranks for G. Madugula	Ranks for Pedabailu
1	Confidence Building	1	145	106	27	12
2	Skill Development	2	141	57	30	54
3	Marketing Linkage	3	33	12	10	11
4	Bank Linkages	4	40	20	12	8
5	Linkages with Govt Officials	5	17	10	6	1
6	Knowledge on Rights, Entitlements and Development Programmes	6	8	4	3	1
7	Managerial Efficiency for Micro Enterprise Development	7	15	2	9	4
8	Enhanced Income and Earnings	8	26	3	0	23
9	Active Participation in Family Decision Making	9	3	0	3	0
10	Active Participation in Development Programmes	10	0	0	0	0
11	Active Decision Making Outside the Family	11	2	0	1	1
12	Others	12	190	0	123	67
	Total		**620**	**214**	**224**	**182**

Source: Survey Data

Simultaneously, the study recorded that capacity building programmes failed to break the silence of tribal women to participate in broader issues of community development and bringing tribal women out of home issues to address issues of greater magnitude. Hence, though the effort has leveraged some change in some areas but still failed in many areas.

The study has further gone ahead to examine the access of tribal self help group women to government schemes and programmes. Access to

Government Schemes and Programmes and realization of benefits by tribal women self-help group members is validated against a set of parameters such as;

(i) Universalized access to government schemes and programmes across self help groups;

(ii) Access to schemes and programmes is region neutral;

(iii) Income generation schemes do have better access than the welfarestic schemes;

(iv) Location of habitations do play a deciding role in accessing government schemes;

(v) Education and access to government schemes are positively correlated.

Universalized access to government schemes is assessed against 13 major schemes of Central and State Government under execution in tribal areas. It is reported that all schemes in execution are accessed by the tribals from a minimum of 4.52 per cent to a maximum of 41.45 per cent (see fig.6.5). The wide variation in accessing of Government schemes has further given rise to several questions like why such variation is reported in accessing schemes and what would be the causal factors associated. And further, it has been presumed that is it the regional variation within the tribal areas that secluded many tribal from access? Is it the education level which made the schemes accessed by a limited few? And the judgmental view on the very nature of the schemes at offer.

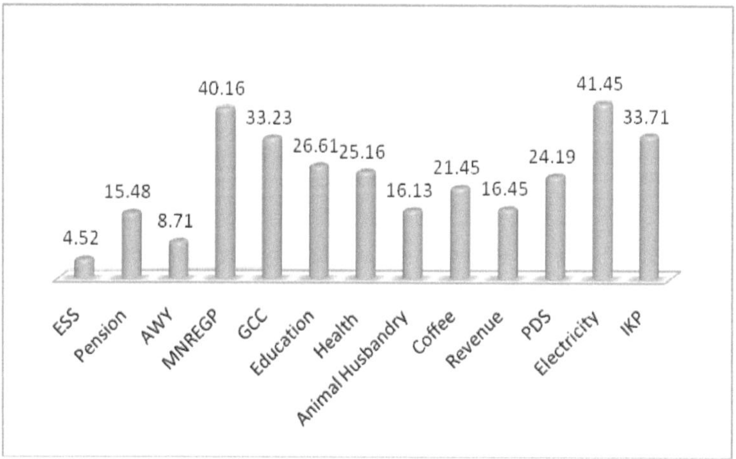

Fig. 6.5: SHG Members Access to Govt. Schemes

Tribal agency areas by and large are cut off areas and areas poised with many hillocks. They are reach in natural resources and poor in infrastructures. All contrasting features of a developing economy can be witnessed in almost all ITDAs and hence, ITDA Paderu is not an exception.

However, entire Paderu agency area does not portray a similar feature all across. The selected 3 regions represent 3 different characters and behavior and this probably has significant bearing on the life, livelihood, institutional functioning and economic well-being of the resident's over here. This can be visualized from access of Government schemes across the regions (see table-6.12). The average deviation in access of government schemes by the self-help group members among the regions is glaring for regions located away from the urban centers and vice-versa. Besides, focused execution of IKP and consequently performance of self-help groups out weighted the access of Government schemes by some tribal self-help group members across the studied regions.

Table 6.12: Access to Government Schemes Across Studied Regions

Sl. No.	Schemes/Departments	Dumbriguda	G.Madugula	Pedabailu	% Total
1	ESS	1.61	1.45	1.45	4.52
2	Pension	2.74	10.65	2.10	15.48
3	AWY	1.77	3.87	3.06	8.71
4	MNREGP	16.13	18.06	5.97	40.16
5	GCC	4.00	21.13	3.23	33.23
6	Education	14.52	8.23	3.87	26.61
7	Health	15.65	6.29	3.23	25.16
8	Animal Husbandry	4.68	7.26	4.19	16.13
9	Coffee	10.65	7.42	3.39	21.45
10	Revenue	5.81	5.00	5.65	16.45
11	PDS	14.03	5.97	4.19	24.19
12	Electricity	26.13	10.65	4.68	41.45
13	IKP	25.00	5.16	3.55	33.71
	Total	**142.71**	**111.13**	**48.55**	**307.26**

Results:

1. Average Deviation between Dumbriguda and G Madugula is 5.78
2. Average Deviation between Dumbriguda and Pedabailu is 5.43
3. Average Deviation between G.Madugula and Pedabailu is 3.11

Source: Survey Data

Access and utilization of government schemes and/or programmes gives an un-even view when viewed from the welfarestic and economic contribution. A comparative view of access to welfarestic schemes against the economic schemes reveals a wide degree of variation across the regions under study (see table-6.13). Economic schemes by its nature contribute visibly to upgrade the income kitty of the tribal families and hence, better accessibility is reported while welfarestic schemes are more of generalist in nature and contribution is limited to eligible clients.

Table 6.13: Access to IGP Schemes Vis-à-vis Welfarestic Schemes in Tribal Agency Area

Sl. No.	Income generation schemes	% of beneficiaries	Sl. No.	Welfarestic Schemes	% of beneficiaries
1	ESS	4.52	1	Pension	15.48
2	MNERGS	40.16	2	AWY	8.71
3	GCC	33.23	3	Education	26.61
4	Animal Husbandry	16.13	4	Health	25.16
5	Coffee	21.45	5	PDS	24.19
6	IKP	33.71	6	Electricity	41.45
7	Revenue	16.45	7	0	0.00
	Total	**165.65**		**Total**	**141.6**
Results:		**Standard Deviation is 11.41223803**			
		VAR is 130.2391769			

Source: Survey Data

Finally, the study is establishing that the accrued benefit of MFI for improving the household facilities and assurance of a decent living, women involvement in decision making, nature of decisions being taken in meetings, benefits of trainings on women and access and leverage of government schemes and programmes for tribal women members in self-help groups of Paderu agency area is significant. There is a glaring change among the women of the agency area and more focused intervention by self-help group is the call of the day to further improve women potency.

6.2. Microfinance and Sustainable Economic Development

Credit plays vital role in beginning and expanding the business, and hence, microfinance has been treated as an important tool for economic development. The SHGs offer small loans to aspiring individuals as well as group of individuals. These loans assist people in getting access to traditional financing and offer jobs to local communities. "Microfinance is about more than money. None of us in this room would accept living in a community with bad schools, little healthcare, no electricity, and filthy water - but with a great bank. Economic justice means the poor must have the power to speak up, speak out and speak for themselves (Lewis, Jonathan) ." There is an open acceptance to the catalystic role visualized for microfinance operation for bringing up people and especially peoples residing under Below Poverty Line (BPL) conditions and also in in-accessible cut off areas.

Can the same conduit of development be thought up for people residing in tribal dominated areas and also for tribals. The research in the wake of understanding and validating the notions, has believed in the fact that there is a quid pro quo relation exists between economic development of tribals and the functioning up of the so called MFIs in the agency areas or the Integrated Tribal Development Agency (ITDAs) areas of the state of Andhra Pradesh. Hence, it is hypothesized that;

Hypothesis-II:Micro Credit operation in agency areas brings in sustainable economic development.

Poor access to credit markets is the key reason why most of the economies cannot expand economically, institutionally, socially and culturally. However, microfinance can provide economic interventions, which help to improve access to financial technologies. A durable microfinance system with well-equipped resources can help to stimulate the economic growth from very basic level. This will be the primary plinth for the edifice of tribal development. Financial accessibility is serving as a platform to choose activities of interest and economic gain from. It is a source as well as inspiration for deciding the kind and type of activity to be selected for survival and the means to be adopted for better development and growth of family and the society at large. Hence, a vibrant means of access to finance determines the activities and especially the economic activities to be taken up as a measure of family sustenance.

Table 6.14: Contribution of Male and Female to Household Income Kitty

Mandals	Avg monthly income of female	Avg monthly income of male	Avg. monthly income to family	% of Growth
Dumbriguda	592	847	1439	43.14
	(41)	(59)	(100)	
G. Madugula	951	712	1663	-25.15
	(57)	(43)	(100)	
Pedabailu	781	661	1442	-15.39
	(54)	(46)	(100)	
Total	**2324**	**2220**	**4544**	**-4.47**
	(51)	**(49)**	**(100)**	

Source: Survey Data

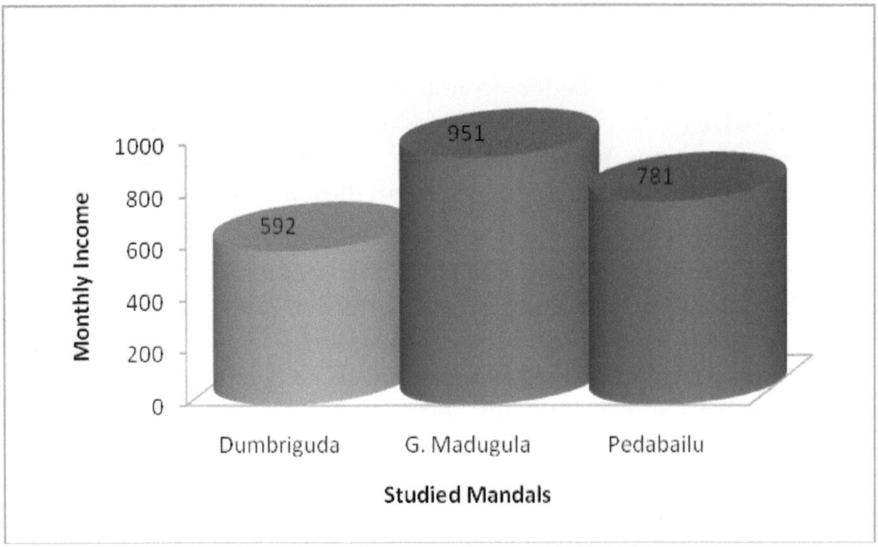

Fig. 6.6: Average Monthly Income of SHG members

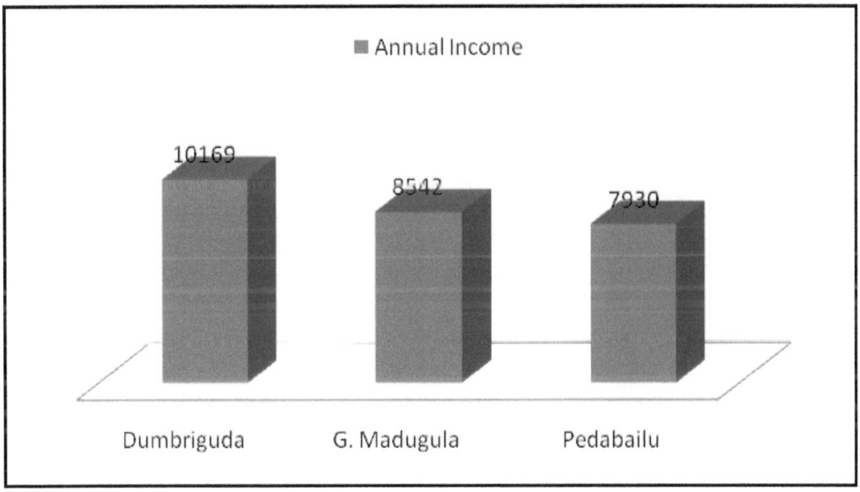

Fig. 6.7: Income Profile of Studied SHGs

It has a direct bearing on the income contribution of the family. MFIs presence and contribution in tribal area is overwhelming as seen from table 6.14. The average income contribution of male and female is not very encouraging in terms of the per capita income (PCI) as specified by the Indian economy . However, it is giving a positive hope for the family so far as the regularity of income to household is concerned. Most importantly contribution of income by female is amply visualized in comparison to male in mandals like Dumbriguda (see table-6.14). Besides, monthly income contribution to family income kitty is marginally better by male (see fig.6.6

and 6.7). Indeed, this suffices to the fact that women contribution to family income is significant and certainly a parameter worth considering instrumental in ensuring better economic condition for the family in long run. The research has considered income contribution of women as the single most dominant parameter which almost decides and determines women's involvement in the day-to-day activities as well as their institutional involvement.

Choice of activity is directly linked to the availability of capital at hand and/or the possibility of getting access to formal and informal sources of funding, besides the availability of resource base and the socio-infrastructures. Availability and accessibility to technology and skill of human resources also influence the choice of activity to be selected. Hence, in technical terminology it can be interpreted as;

Choice of Activity *i.e.* $\Delta A = f(\Delta K + \Delta L + \Delta Tech + \Delta Govt\ Policy)$

Among the factors of production, land is often considered as the single most dominant factor to determine the basket of activities to select. The type and quality of land determine the nature of activities to be taken up in an area and its suitability to the people. Detailed analysis of land in the studied regions reveals that about 8 per cent of beneficiaries interviewed are landless. Among the landless category, there are highest percentile (42%) of land less reported for Dumbriguda followed by G.Madugula (38%) and Pedabailu (20%) (see fig.6.8).

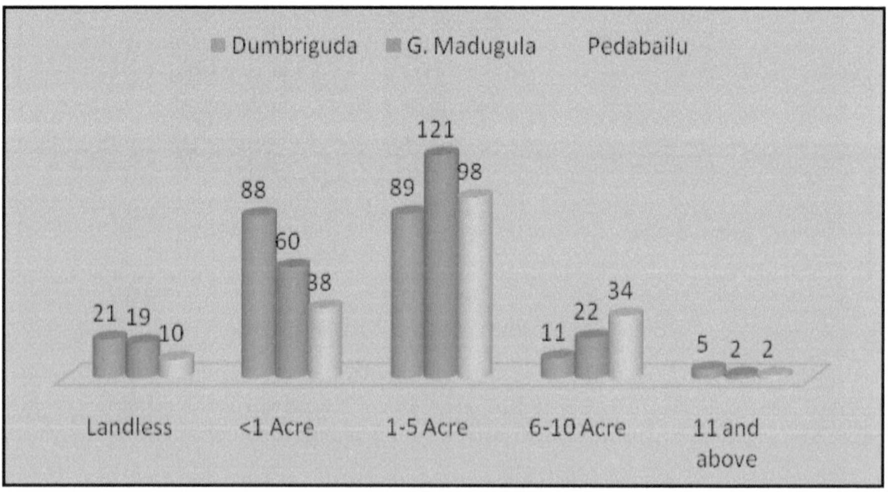

Fig. 6.8: Land Holding Position of SHG Families

The land ownership in tribal areas follows a unique pattern. Land distribution is not skewed as has been the case with general population category. There are about 30 per cent beneficiaries reported of owning less than an acre of land across the studied region. Within the category, 47.31 per cent of Dumbriguda, 32.25 per cent of G.Madugula and 20.43 per cent of Pedabailu beneficiaries own less than an acre of land.

About 49.67 per cent (rounded up to 50%) beneficiaries across the studied regions reported owning land 1 to 5 acres (see fig.6.9). Across the studied regions, G. Madugula Mandals reported highest number of beneficiaries in the category (39.28%), followed by Pedabailu (31.81%) and Dumbriguda (28.89%).

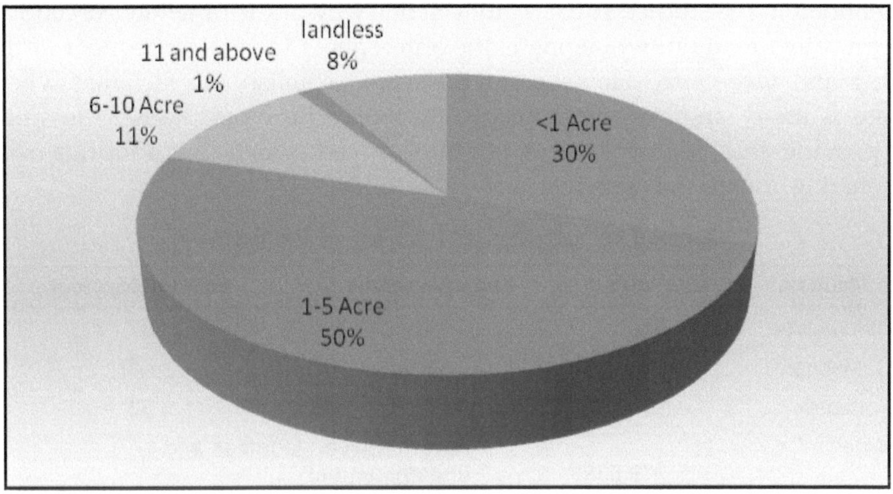

Fig. 6.9: Overall Land Holding Position of HG members

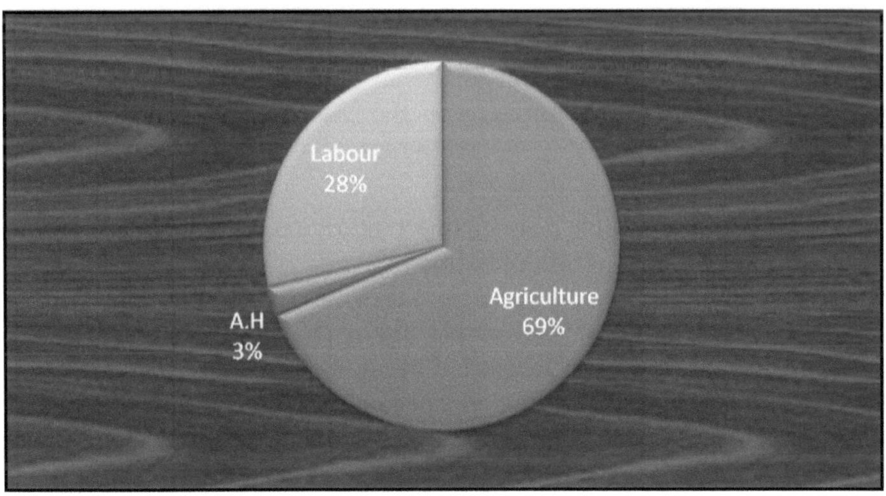

Fig. 6.10: Sources of Income to SHG Families

Employment pattern so to say engagement pattern of tribal reveals that about 69 per cent be beneficiaries across mandals are engaged in agriculture as the primary means of survival. There are about 28 per cent beneficiaries across mandals revealed adopting wage labour as the second best option of livelihood sustenance while a significantly marginal percentile beneficiary reported of being adopting animal husbandry as the 3rd best option for survival (see fig.6.10).

Pattern of landholding in one hand and the selection of means for livelihood in the studied tribal regions in one way or the other are susceptible to causative relation; while one is the antecedents for the consequent of the other and vice-versa. However, it does not establish the fact that where there is more landless employment rate would be more. Rather the study is proving that higher rate of landless forced tribals for adopting self-sustaining means for survival (see table 6.15).

Table 6.15: Landless and Choice of Employment

Mandals	Landless	Unemployment	self employment
Dumbriguda	21	116	87
G. Madugula	19	39	179
Pedabailu	10	125	33
Note:	**-79.8889**	**covar between landless and unemployment**	
	191.5556 covar between landless and selfemployment		

Source: Survey Data

Going beyond the landholding size as determinant of means of employment, the study is establishing that land as a factor of production has got little and even no influence on the choice of livelihood means for survival to be selected for. It is indifferent of the nature of engagement people of Paderu ITDA is opting for.

Irrespective of multiple avenues of employment, changing pattern of household income, recognition of women income contribution at household level, engagement of tribal people in non-agricultural activities and persistent dependency of tribal's on self-employment in Paderu agency area, an important question is yet remained unanswered. The question to be answered is has MFI led changing engagement portfolio and land ownership pattern in ITDA Paderu ensured better satisfaction in family management? The exploration to the question reveals that about 50 per cent self help group members across the studied regions are indifferent of experiencing a change in family management, 22 per cent reveled of feeling burdensome, and about 3 per cent report of non happy in family management. About 25 per cent of beneficiaries interviewed reported to be satisfied and very satisfied. The post joining self help group impact on family management has recorded a mix response (see fig.6.11).

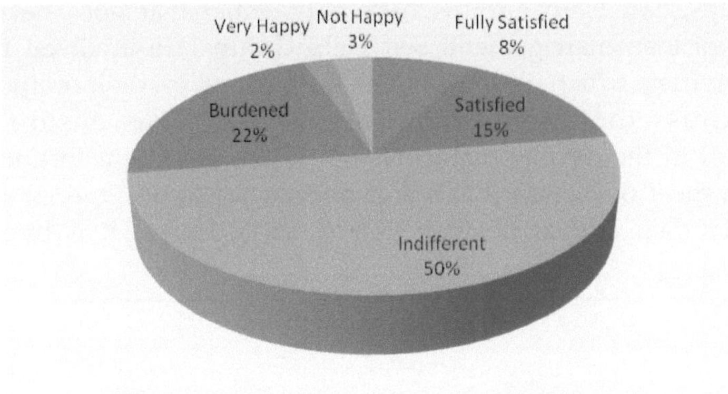

Fig. 6.11: Satisfaction on Family Management

Discussion of the previous sections and statistical analysis of the data collected over the regions though exhibit a rosy picture of empowerment of tribal women, self help contribution to family management is yet not eye appealing. Economic betterment without peace in family management would not make its contribution sustainable and persistent in long run.

6.3. Microfinance Institutions and its Use

Use value or value in use is the utility of consuming a good; the want-satisfying power of a good or service in classical political economy may be rightly used for analyzing the use value of microfinance institutions and especially the self help groups in the tribal agency areas of the state of Andhra Pradesh. Micro finance is making a significant contribution to both savings and borrowings of the poor in the country. The main use of microcredit is for direct investment. While the presence of microfinance institutions has increased the borrowing options for the poorer clients, it seems not (yet) to have significantly affected the terms and conditions of different informal credit providers .

Apart from the provision of assured supply of credit to all needy at a quicker time and without any collateral securities, self help groups do play a more competent role to shape the institutional functioning and bringing in socio-cultural changes within the tribal communities. The holistic role of MFIs and especially of self help groups as institutions of reformation is therefore not limited to finance only.

The inherent features and characteristic of the self help groups have been drawn from the very nature of family type of participants. Members from nuclear and joint families are active members of the self help groups.

The multiplicity family background draws references from the social capital structures of nuclear and joint families (see fig. 6.12). Though the access to credit has been seen as a motivational factor behind the formation of Self Help Groups, Self Help Groups have a potential that goes beyond mere economics of loan management. Self Help Groups are involved in various Social Activities which is very much important in their empowerment process. Grass root level organizations like SHGs ensure people's participation in the development process . They provide a forum in which people can meet on regular basis and discuss the issues and concerns that they face in their day to day life (Zakir, 2011). Hence, it is hypothesized that;

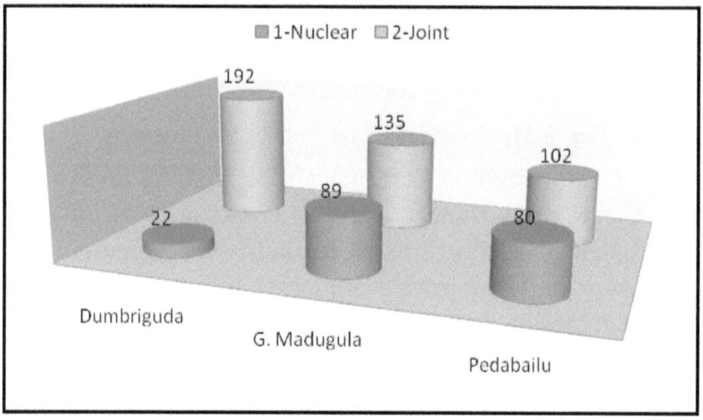

Fig. 6.12: Family Type of SHG members

Hypothesis-III: Micro Credit is used for both investment purposes as well as to meet household needs.

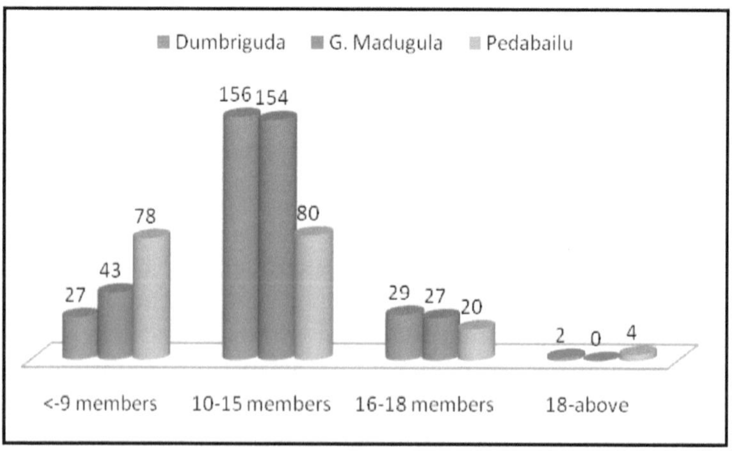

Fig. 6.13: Membership Profile of SHGs Across Regions

Self help groups across the regions of ITDA Paderu more or less exhibit similar membership pattern. However, SHG membership distribution is skewed towards the group of less than 9 and 10-15 members (see fig. 6.13). There are about 86.77 per cent self help groups are falling under both the membership categories.

Membership is as important as a factor for formation of an institution like self help group, retention of members for longer term is equally important for stable performance. Exploration of the fact reveals an unhealthy scenario. About 49.87 per cent self help group reveals stability across the studied regions. About 37.41 SHGs reported for increasing membership since its formation. The increased membership as advocated by the groups are due to better performance, financial security for families, experiencing the benefit to the neighbour, influence of neighbour etc. Besides, cases of decrease in membership are also reported by the tribal self help groups of Paderu. There are about 12.74 per cent SHGs across the ITDA area reported of decreasing membership over the years (see fig.6.14). The changing membership behavior of SHGs would have a strong bearing on its functioning and financial credibility over the period.

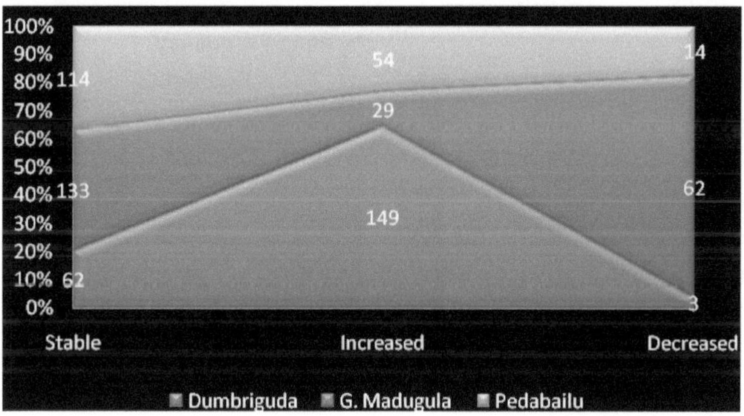

Fig. 6.14: Stability of SHG Groups Across Tribal Regions

The changing membership is linked to the motivating factors/agents encouraged for self help group participation. This has a direct bearing on the functioning of self help groups. Factors reported for motivation of joining the groups includes self decision, motivation of family members, influence of friends and/or relatives, motivation of NGOs and influence of member's of the group. Of all the factors mobilization of NGOs to join the group is by and large considered to be influential (44.19%) in comparison to all other factors. Indeed, influence of family members cannot be negated as a factor for mobilizing tribal women (25.64%) to join the self help group (see fig.6.15). Tie bout effect of neighbouring members of self help groups has moderately acted as a factor to influence members joining the groups.

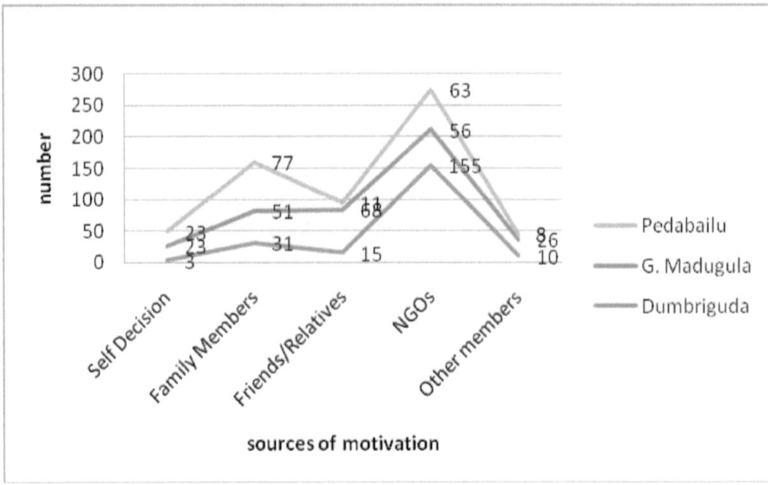

Fig. 6.15: Sources of Motivation to Join the Group

Table 6.16: SHG Meeting Frequency Across Tribal Regions

Mandal	**Meeting Frequency**		
	Weekly	**Fortnightly**	**Monthly**
Dumbriguda	58	8	148
	(41.13)	(23.53)	(33.26)
G. Madugula	54	17	153
	(38.30)	(50.00)	(34.38)
Pedabailu	29	9	144
	(20.57)	(26.47)	(32.36)
Total	**141**	**34**	**445**
	(100.00)	**(100.00)**	**(100.00)**

Source: Survey Data

Can it be said that scheduling a meeting of self help group is not a matter of serious concern? There are about 22.74 per cent self help groups reported for holding weekly meeting and 71.77 per cent reported for organizing meeting in every month (see table 6.16). Can it be concluded that holding a meeting is independent of agency of mobilization of self help groups? However, the study recorded that agenda of the meeting as well as urgency of the situation warrants holding of meeting either in weekly interval or in monthly interval. Very often than not, a close review of the scheduling of meeting establishes a correlation with the agents mobilizing the meeting. It is revealing that NGO mobilization for weekly, fortnightly and monthly meeting is not correlated, mobilization of group members for meeting is partially correlated (r=0.58 for weekly meeting), involvement of IKP in mobilization is strongly reported with positive correlation in all meetings and coordinated effort of IKP and NGOs is the only successful means (see table 6.18, 6.19 and 6.20).

Table 6.17: SHG Meeting Coordination Across Selected Tribal Regions

Sl. No.		Dumbriguda	G. Madugula	Pedabailu	Total
1	Group Members	160	151	123	434
		(74.77)	(67.41)	(67.58)	(70.00)
2	NGO	32	19	34	85
		(14.95)	(8.48)	(18.68)	(13.71)
3	IKP staff	11	39	18	68
		(5.14)	(17.41)	(9.89)	(10.97)
4	Mixed	11	15	7	33
		(5.14)	(6.70)	(3.85)	(5.32)
	Total	**214**	**224**	**182**	**620**
		(100.00)	**(100.00)**	**(100.00)**	**(100.00)**

Source: Survey Data

Table 6.18: Coefficient of Correlation between Weekly SHG Meeting and sources of coordination

Sl. No.	Parameters	Dumbriguda	G. Madugula	Pedabailu	Correlation
1	Weekly meeting	41.13	38.3	20.57	
2	Group members	74.77	67.41	67.58	0.589524
3	NGO	14.95	8.48	18.68	-0.69313
4	IKP Staff	5.14	17.41	9.89	0.002312
5	Mixed	5.14	6.70	3.85	0.762269

Source: Survey Data

Table 6.19: Coefficient of Correlation between Fortnightly SHG Meeting and sources of coordination

Sl. No.	Parameters	Dumbriguda	G. Madugula	Pedabailu	Correlation
1	Fortnightly	23.53	50	26.47	
2	Group members	74.77	67.41	67.58	-0.60163
3	NGO	14.95	8.48	18.68	-0.89109
4	IKP Staff	5.14	17.41	9.89	0.957535
5	Mixed	5.14	6.70	3.85	0.840804

Source: Survey Data

Table 6.20: Coefficient of Correlation between Monthly SHG Meeting and sources of coordination

Sl. No.	Parameters	Dumbriguda	G. Madugula	Pedabailu	Correlation
1	Monthly Meetings	33.26	34.38	32.36	
2	group members	74.77	67.41	67.58	-0.08315
3	NGO	14.95	8.48	18.68	-0.99584
4	IKP Staff	5.14	17.41	9.89	0.656395
5	Mixed	5.14	6.70	3.85	0.999953

Source: Survey Data

Logical induction to the causal relation is the decision on the agenda of the meeting of the self help groups. Who decides the agenda of the meetings? How effective is it to take a call on the socio-cultural and institutional issues? How effective is it to tackles financial issues at household level and provide seed capital for investment? A close view of the SHGs in Paderu agency area is revealing the importance of presence of these institutions in shaping the tribal economy.

Agenda of the meeting by and large is decided by the members of the self help group. It is sometimes need based and based on urgency at other. There is no laid guideline for setting up of meeting agenda, however, it has been mostly guided by a host of factors like category of members, economic condition of people in the group, cultural barriers to women's participation in meetings, available raw-materials, business opportunities in the surroundings, market availability, Govt Schemes and programmes, Panchayat schemes and programmes, seasonality in occurrence etc.

Peculiarity in agenda setting up of self help group meeting agenda is reported in Paderu agency area. Formation of agenda in some way or the other is providing impetus and influencing the decision making on the issues discussed in the self help group meeting. The study has recorded that meeting agenda of self help group are primarily taken by majority of members, some members, link workers and link workers and group members together. Level of participation in setting up of the meeting agenda varies from self help groups across the regions and also between self help groups. Self help of Paderu agency area is not an exception. About 41.29 per cent involvement of majority of self help group members in agenda setting is reported, while about 35.65 per cent of some members of SHGs, and 17.58 link workers and group members, and 5.48 per cent link worker participation is also reported across the tribal self help groups (see table-6.21).

Table 6.21: Decision on Meeting Agenda of SHGs Across Tribal Regions

Mandal	Majority of Members	Some Members	Group Members and Link Workers Jointly	Link Worker
Dumbriguda	116	49	43	6
	(45.31)	(22.17)	(39.45)	(17.65)
G. Madugula	71	92	35	26
	(27.73)	(41.63)	(32.11)	(76.47)
Pedabailu	69	80	31	2
	(26.95)	(36.20)	(28.44)	(5.880
Total	**256**	**221**	**109**	**34**
	(100.00)	**(100.00)**	**(100.00)**	**(100.00)**

Source: Survey Data

Openness in deciding agenda for the self help groups and especially in tribal areas has still a long-way to go. This has not only revealing the fact of poor participation of self help group members in effective functioning of the institutions, rather creating a visible gap in tribal participation in self-sustenance of SHGs.

Table 6.22: Procedure Adopted for Decision Making in SHG Meeting Across Tribal Regions

Mandal	By Consensus	By Voting	By Group Representative	Mixed
Dumbriguda	181	1	27	5
	(37.24)	(2.70)	(37.50)	(20.00)
G. Madugula	180	15	11	18
	(37.04)	(40.54)	(15.28)	(72.00)
Pedabailu	125	21	34	2
	(25.72)	(56.76)	(47.22)	(8.00)
Total	**486**	**37**	**72**	**25**
	(100.00)	**(100.00)**	**(100.00)**	**(100.00)**

Source: Survey Data

Can this pattern of agenda setting by Self Help Group bear any 'antecedent and consequent' relation in decision making at self help group level? There are primarily four major methods and means adopted for decision making by the tribal self help groups. These are (i) By Consensus, (ii) By Voting, (iii) By Group Representatives and (iv) Mixed. The overwhelming consequence is that about 78.39 per cent of decisions are being taken by consensus. It signifies the level of participation of tribal women in solving the issues of the self help groups (see fig.-6.16). Similarity in decision making is reported across regions of formation and functioning of self help groups (see table-6.22).

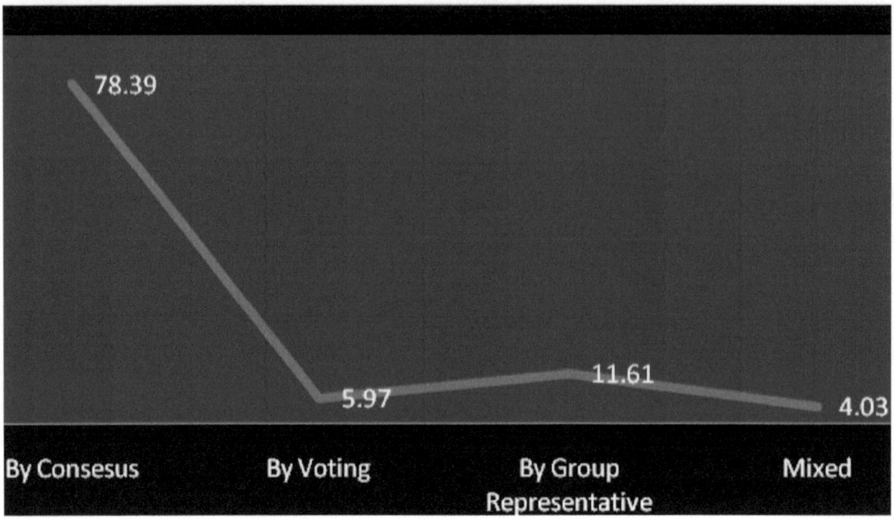

Fig. 6.12: Decisions taken by SHGs

However, it is a matter of serious concern to look into the means adopted by agencies to decide the agenda and the decision making on agenda by agency. The means adopted and decision taken are positively correlated (r=0.665456). Hence, the study is establishing that method of agenda fixation has got a positive bearing to define the pattern of decision to be arrived at.

Decisions on socio-cultural and institutional issues have irreversible impact on the life and livelihood of the tribal people. Self help groups being the institutions functioning at the community level are importantly involved in day-to-day functioning of the institution directly and /or indirectly and also influence functioning of the society at large.

Tribal society is different from non-tribals and so also the system of marriage and family type. Self help groups functioning in the regions of ITDA area apart from discussing the business also discusses on the social issues like marriage, dowry systems, consumption of liquor, reservation for women in Govt jobs, promotion of a liquor free society etc. It is worth mentioning that tribal women do not entertain inter caste marriage. About 7.74 per cent have supported for inter-caste marriage. All SHGs across the regions unanimously advocated against child marriage (90.16%), widow marriage (91.45%) and Divorce (93.23%). However, inter religion marriage has been accepted by less than 29.53 per cent (see fig.6.17). Approval for love marriage has lightly been viewed in cognizance of the education level of the children. System of dowry in marriage has negatively been viewed by 86.94 per cent of self help group members (see table-6.23). Uniform decision against the dowry system has been recorded across all studied regions of Paderu ITDA.

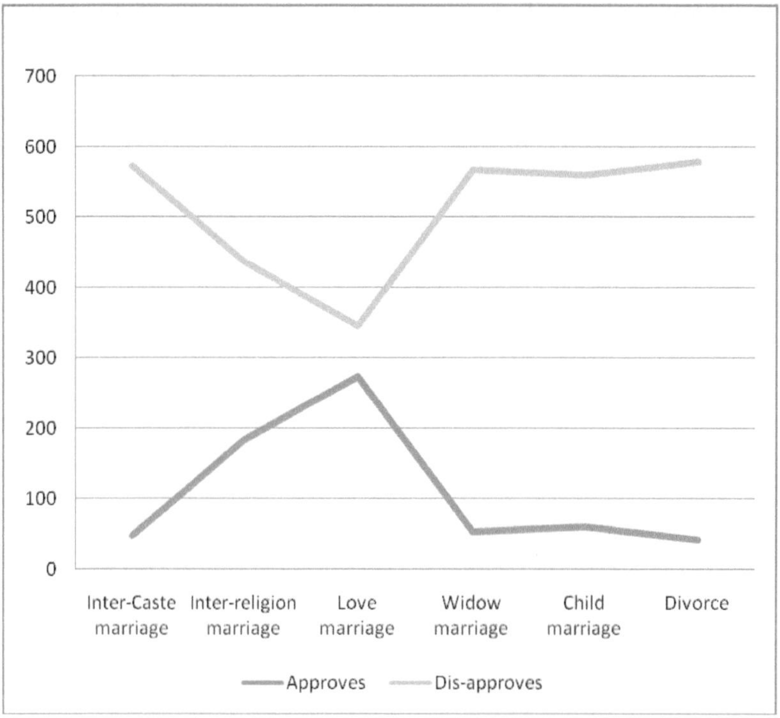

Fig. 6.17: SHG Consent on Types of Marriages

Table 6.23: Approval for Dowry System

Name of Mandal	Approves	Disapproves	Total
Dumbriguda	34	180	214
	(15.89)	(84.11)	(100.00)
G. Madugula	28	196	224
	(12.5)	(87.5)	(100.00)
Pedabailu	19	163	182
	(10.44)	(89.56)	(100.00)
Total	**81**	**539**	**620**
	(38.83)	**(261.17)**	**(300.00)**

Note: Figures in bracket is per cent to total
Source: Survey Data

Issue of family planning is a serious problem in tribal areas. This is either due to lack of health services and facilities in one hand and lack of motivation and awareness on the benefit of small family on the other. The prevailing scenario has encouraged for poor rate of adoptability of family planning by tribal community. Family planning scheme of the government is accessed by 31.93 per cent of tribal across the studied regions, while about 68.06 per cent of SHG beneficiary have not yet been accessed the scheme (see fig.6.18). Mandal Pedabailu has better accessibility to family planning programme in comparison to all other regions under study (see table-6.24).

Table 6.24: Approval of Family Planning Across the Regions

Name of Mandal	Approves	Disapproves	Total
Dumbriguda	62	152	214
	(28.97)	(71.03)	(100.00)
G. Madugula	36	188	224
	(16.08)	(83.93)	(100.00)
Pedabailu	100	82	182
	(54.95)	(45.04)	(100.00)
Total	**198**	**422**	**620**
	(100.00)	**(200.00)**	**(300.00)**

Note: figures in bracket is per cent to total
Source: Survey Data

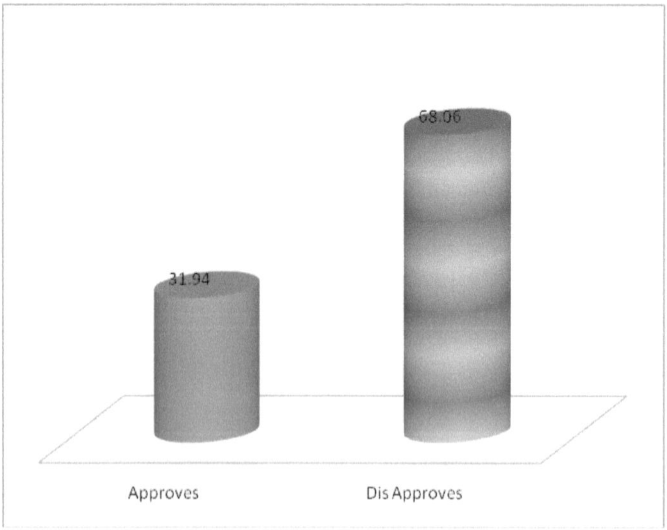

Fig. 6.18: Approval of Family Planning

The probe for reservation for tribal women in government jobs has recorded an indifferent opinion by the self help groups of Paderu agency area. About a quarter self help groups (25.12%) opined for the necessity of reservation in tribal women's life across the Paderu agency area (see table-6.25). while it is equally surprising to note that morethan 54 per cent tribal SHGs of Padabailu Mandal alone have been reported for reservation of jobs for women.

Table 6.25: Concurrence for Women Reservation Across Regions

Name of Mandal	Approves	Disapproves	Total
Dumbriguda	24	190	215
	(11.21)	(88.79)	(100.00)
G. Madugula	33	191	224
	(14.73)	(85.27)	(100.00)
Pedabailu	99	83	182
	(54.40)	(45.60)	(100.00)
Total	**156**	**465**	**621**
	(80.34)	**(219.66)**	**(300.00)**

Note: figures in bracket is per cent to total
Source: Survey Data

Table 6.26: Punishment to Wicked Husband Across Regions

Name of Mandal	Approves	Disapproves	Total
Dumbriguda	178	36	214
	(83.18)	(16.82)	(100.00)
G. Madugula	214	10	224
	(95.54)	(4.46)	(100.00)
Pedabailu	93	89	182
	(51.10)	(48.90)	(100.00)
Total	**485**	**135**	**620**
	(229.81)	**(70.19)**	**(300.00)**

Note: figures in bracket is per cent to total
Source: Survey Data

Consumption of country liquor is a common phenomenon in tribal area. It is customary for tribal and almost accepted at all household level. However, with the passage of time and improved access to education, consumption of liquor is viewed as a development stigma by the society.

The meaning and value of family is understood in a different means and have been decided to lead in a different way. Hence, SHGs (78.22%) have consented to punish the wicked husbands involved in illicit activities (see table-6.26). This decision of SHGs and the determination would certainly have positive reformation impact on the days to come.

Reform within household and the society at large is possible only when people are economically independent and self-sustainable. May be that was the long term objective of formulation of self help groups. Small savings collected from the tribal are used for the larger developmental goal. The previous sections elaborately shared the developmental impacts of the self help groups in the life and livelihood of the tribals residing all across the three studied regions of the Paderu agency area.

The SHGs across the agency area are collecting savings ranging from ₹ 12 to ₹ 36 per individual members per month. The money collected has effectively been circulated within the group to meet needs of multifarious nature. The money circulated has not only benefited the individual family and/or group of families in petty business but also made them repay back the borrowed amount in time, which in turn enabled other needy people of the group. The growth rate of savings has been enormous (see table-6.27). It has revealed a bare minimum of 2.08 times to a maximum of 55.56, leading to an average growth rate of 29.17 per cent. In absolute term, ₹ 12 became ₹ 12.50 for Pedabailu, ₹ 24 became ₹ 24.50 for G. Madugula and ₹ 36 became ₹ 56 for Dumbriguda.

Table 6.27: Scenario of SHG Saving Across Regions

Mandals	Now	At the time of Formation	Rate of Increase
Dumbriguda	56	36	55.56
G. Madugula	24.5	24	2.08
Pedabailu	12.5	12	4.17
Total	**93**	**72**	**29.17**

Source: Survey Data

This has necessitated for doubling up of overall group savings. However, persistent effort of the groups, made the saving to reach a better cumulated figure over the period. Interest rate on internal lending has contributed for a better multiplication of fund within the group. Interest rate on internal lending is about 3.46 times over the savings collected and especially, almost 10.86 times for G. Madugula mandal.

Overall bank balance with SHGs shows morethan the interest and savings collected. This is due to more inflow of IKP fund to SHGs based on their performance over the period of time. Baring the general scenario, Mandal Dumbriguda is end-up with a huge bank balance of ₹ 11,46,163.91/- which is about 31.49 times morethan the savings collected and interest rate earned. This in a way shows the volume of money being kept idle without changing hand and lacking timely liquidation. The velocity of money has been restricted and benefit of it has not been properly given to all other members of the group.

The other Mandal G. Madugula has just been endowed with 40 per cent of its collection through savings and interest earned. Mandal Padabailu looks efficient in collection, disbursement and rotation of fund. The inflow of external fund in volumes is not reported. The fund position that is collection through savings as well as interest rate is matching with the fund disbursement and hence, reported of a balance equivalent to collection (see table-6.28).

Table 6.28: Financial Turnover Across SHGs of ITDA Paderu

Mandals	Cumulative Savings	Interest earned on Savings	Saving and Interest Total	Present Bank Balance
Dumbriguda	8,993.85	27,392.67	36,386.52	11,46,163.91
G. Madugula	1,678.98	18,244.00	19,922.98	7,198.29
Pedabailu	2,958.97	1,588.40	4,547.37	4,603.41
Total	13,631.80	47,225.07	60,856.87	11,57,965.61

Source: Survey Data

Collection of fund and fund behavior is reported to be phenomenal across the regions of the agency. However, a basic question is being raised why the members contributed the fund? What was the intention of parting away a portion of monthly income? It is certainly not for charity and also not for undertaking some development work for the larger development of the tribal. However, it is certainly a case for individual growth led community development and not vice-versa.

While probing for the cause of saving by tribal women, social security objective has been reported as one among the motivational factors contributed for timely and sincere monthly savings. It is one among the factors supported by ranks (79.51%) to motivate women for saving. The intent and spirit of social security objective of saving varies across the regions of the ITDA (see fig.6.19).

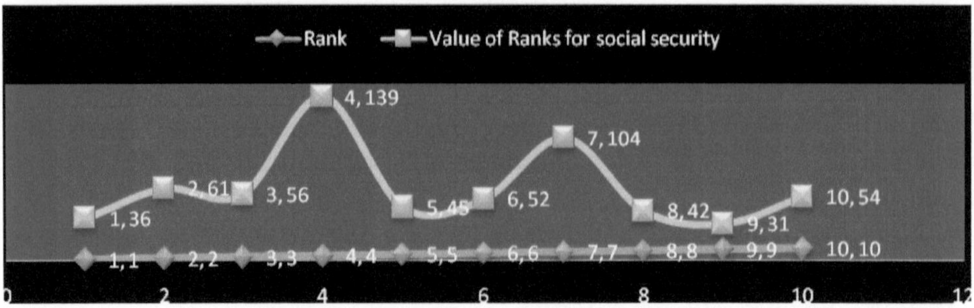

Fig. 6.19: Tribal SHGs View on Social Security

Are tribal food secure? Can it be a major source of inspiration for group saving? This is not a statement and nor even a question to be asked to tribal women, when the changing forest coverage is taking a different pattern every now and then. And so also is the climatic condition. The practice of shifting cultivation on the hills of Paderu ITDA is largely followed by many and reduction of fallow cycle which in turn reduced yield is the symptoms for strive for food. This fact has been well accepted by tribal women all cross the regions of ITDA Paderu (see fig.6.20). A ranking exercise carried out across the region noted for 89.03 per cent people assigning higher ranks to issues of food security as the major motivating factor for savings in the self help groups of their respective villages.

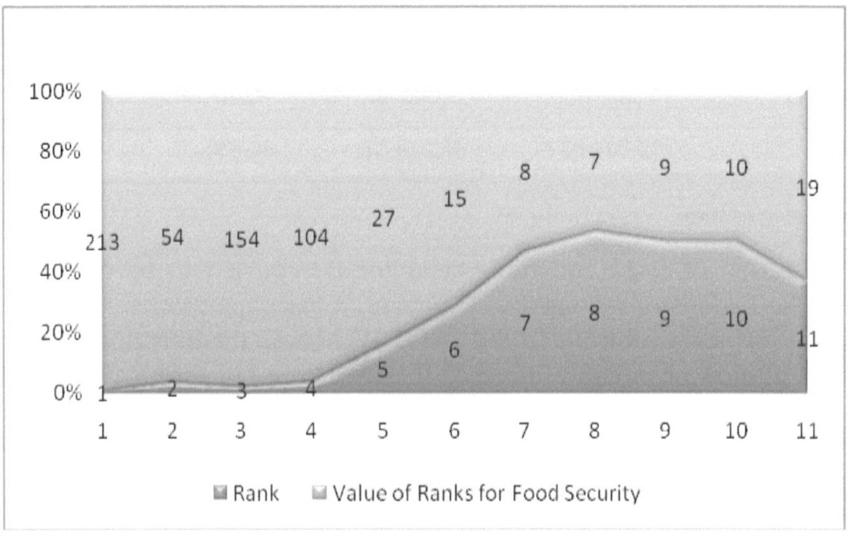

Fig. 6.20: Importance of Saving for Food Security

Access to education is increasing. There are more and more people becoming aware of the value and benefit of education. Education is gaining self-propelling momentum in agency area of Paderu. The SHG members across all regions assigned high rank to education as a major objective and motivational factor for saving with self help groups (see fig.6.21). Higher education is being accessed by tribal children. The demand for college education has forced Govt. to open technical and general colleges in different locations.

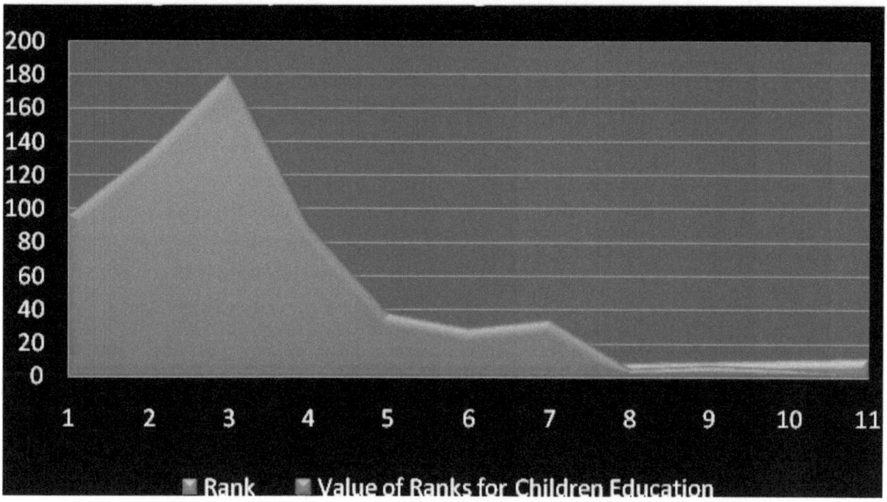

Fig. 6.21: Importance of Saving for Children Education

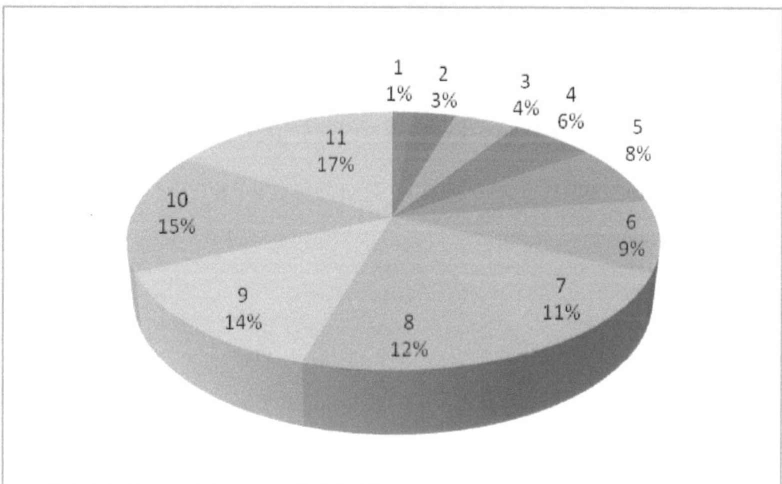

Fig. 6.22: Consideration of Saving for Medical Purposes

Objective of SHG saving for accessing medical and health services can't be ruled out. However, this has been viewed as an objective by a few and families with critical conditions. Health has viewed lightly because of availability of free medical services (see fig.6.22). All the health related expenses are borne by Tribal Welfare Department, GoAP. Similar is the argument for savings for marriage. Marriage among the tribal is celebrated following their custom and tradition. Hence, performance of marriage is not a major cause of joining the group and save (see fig.6.23). Another social commitment attached to the tribal is the performance of festivals. There is almost one festival falls in every month, however, there are two major festivals being celebrated all across the agency for weeks together. The expenditure incurred for festivals are the planned savings of each family and also collection from the people. Hence, spending for festival is not a cause for self help group savings (see fig.6.24).

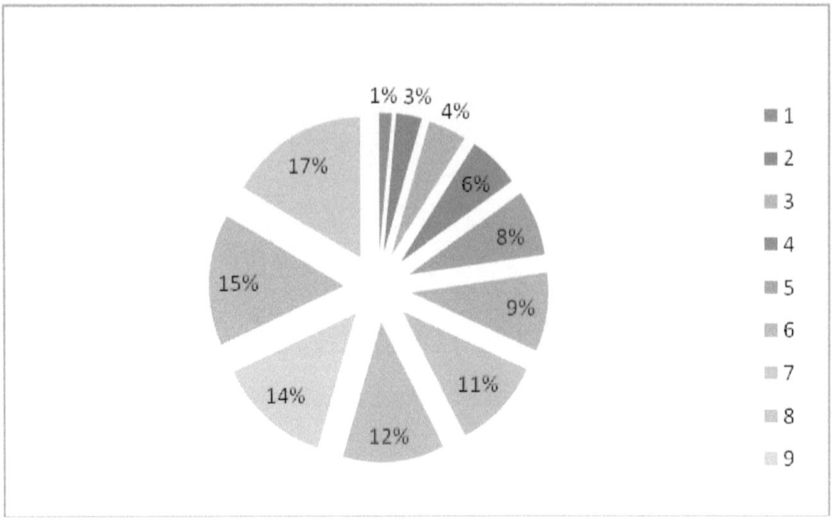

Fig. 6.23: Assigned Ranks to Savings for Marriage

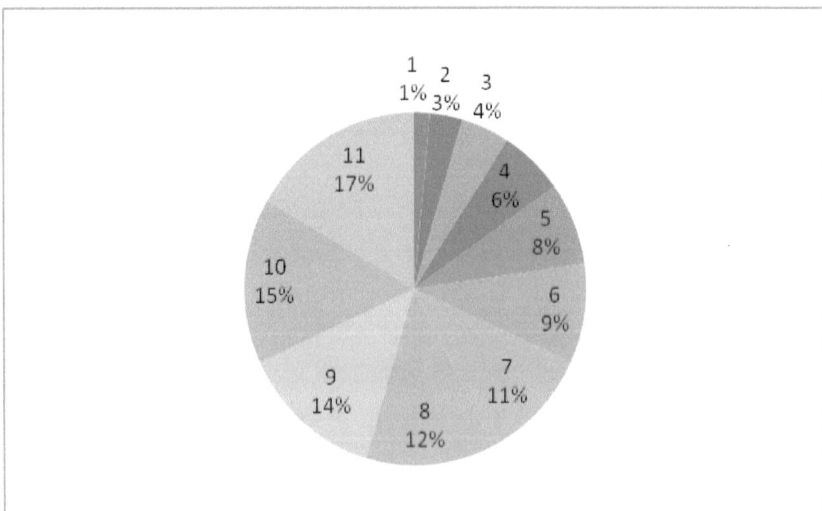

Fig. 6.24: Assigned Ranks to Savings for Festival

Emergencies and exigencies are always a part of life and tribal area and people are in fact more susceptible to this. It is un-foreseeable and unpredictable. And hence, preparedness for that kind of situation is always thought up of by the people and the institutions working in the tribal areas of Paderu ITDA. There are some insurance companies working with the self help groups and some insurance policies are in vogue. This has also corroborated by the fact of assigning higher ranks to savings for emergencies (see fig.6.25). All regions of the agency area reported for the need for saving during emergencies.

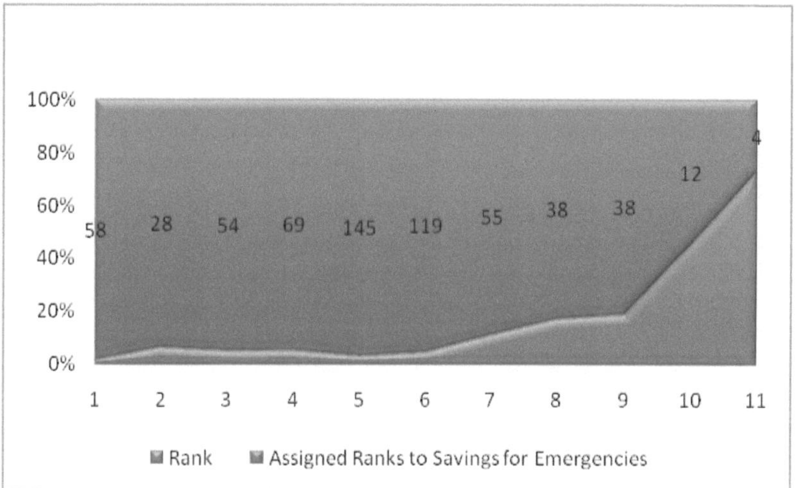

Fig. 6.25: Assigned Ranks to Savings for Emergencies

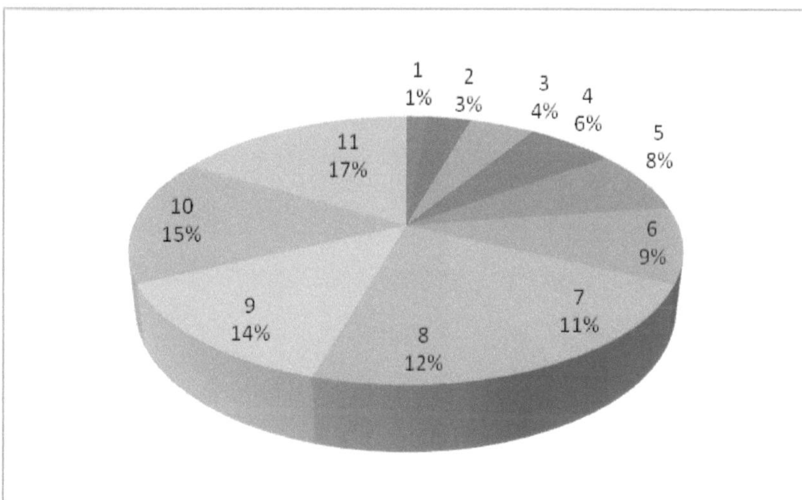

Fig. 6.26: Assigned Ranks to Savings for Agriculture

How legitimate is it to consider tribal self help group saving is to carry out agriculture and/or to meet the agricultural expenditures. Agriculture practiced in tribal dominated pockets though seems traditional and is an age old practice; the expenditure incurred is on rise over the years. Some high yielding variety of crops and modern cropping patterns are introduced in agency areas of Paderu and as a consequence the expenditure has gone up by manifold. The phenomenon of agriculture is limited to a few pockets of the agency. Hence, this has been cited as the minor motivational force for group savings (see fig.6.26). Majority of self help groups across the ITDA of Paderu assigned lower rank to savings to meet the agricultural expenditure.

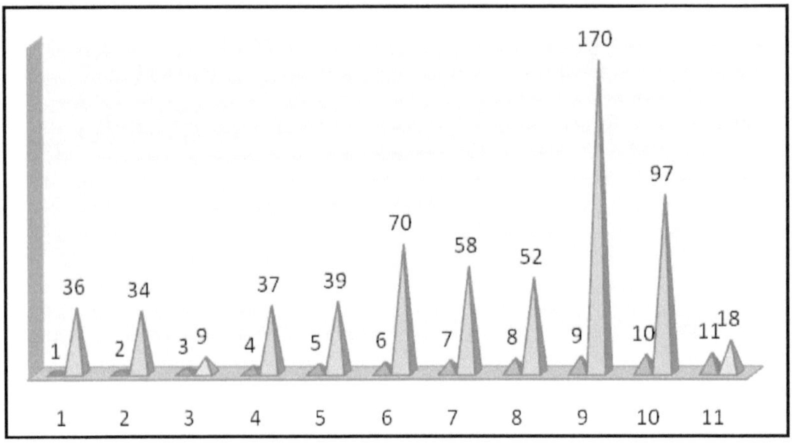

Fig. 6.27: Assigned Ranks to Savings for Asset Buildings

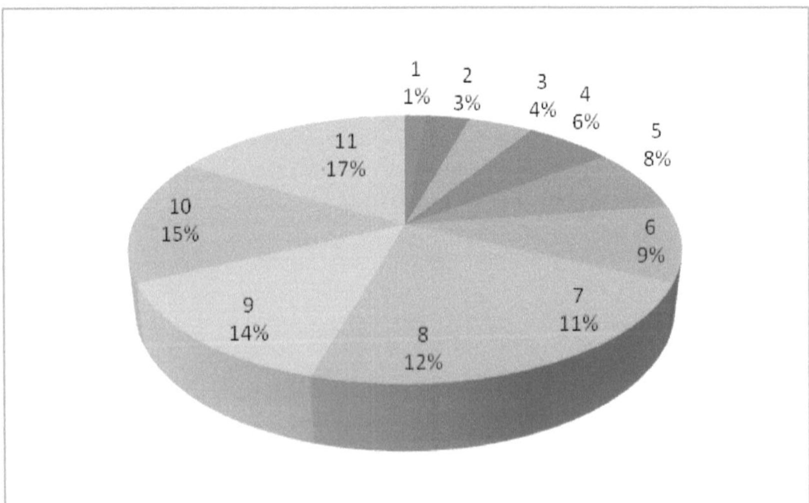

Fig. 6.28: Assigned Ranks to Savings for Self Respect

The dream for building assets, household belongings, and creation of a strong base for the future generation has been the lure for contribution of savings with self help groups in the normal area. Assured livelihood kitty and generation of income using the common pool of resources has also been viewed as an internal strength for the community and hence, this has by and large motivated many to save with the self help groups (see fig.6.27). But this is not the case in tribal dominated pockets and especially in agency area like Paderu.

Finally, the study has examined the saving as a consequence of belief on self-respect by tribal. Saving with self help groups as a matter of self-respect has not been supported by many (see fig.6.28). Tribals are indifferent of the inequalities. All are equal in the village and principle of egalitarianism is basically viewed for all.

As evident from the foregoing discussions that savings with self help groups are for multipurpose use. It is serving as a solution for long standing issues and also provisioning income earning support to many. Besides, the economic contribution, self help group institutions are also taking care of developmental needs of the household and the rural communities at large.

6.4. Microfinance and Capacity of Risk Aversion

Microfinance is the provision of basic financial services to the poor. Access to loans, savings and insurance helps poor entrepreneurs to create and grow small businesses. Initially, microfinance was largely gender neutral: it sought to provide credit to the poor who had no assets to pledge as collateral . It quickly emerged, however, that women invested their business profits in ways that would have a longer lasting impact on their families and communities. Consequently women became fundamental to the success of the microfinance model as a poverty alleviation tool. Hence it is hypothesized that;

Hypothesis-IV: Microfinance builds the capacity of individuals and households to manage risk.

What kind and type of risks tribal of Paderu agency area can face? How frequently do they come? Are these risks manmade or natural? What kind and type of preparedness is required at community level to avoid the risk? How effective are the MFIs and especially self help groups to minimize the risk associated at the individual and household level of tribal? It seems a barrel of bullets flown from the storehouse for tribal self help groups.

Tribal of ITDA Paderu unlike other tribals across India living in an area primarily cut off in nature. The geographic conditions always invite emergencies created by nature and hence, manmade risk and/or emergencies are limited. The natural calamities in the form of rain, turnado's, landslides, depletion of forest cover etc are a common feature experienced by the tribals of the Paderu agency area every now and then. This is viewed as a part of life by tribals.

In order to assess the situation and also the gravity of understanding of risks and emergencies by the tribals across the studied region a ranking exercise had been done involving 620 self help groups. All tribals irrespective of the regions have ranked high to exigencies and the need for saving in the self help groups of their respective villages (see table-6.29). This exercise confirms the preparedness of tribals to manage the risk in one hand and the acceptance of regular emergency emergent situations all around the year.

Table 6.29: Region-wise Importance of Savings for Emergencies

Mandals	**Value of Assigned Ranks**										
	11	10	9	8	7	6	5	4	3	2	1
Dumbriguda	36	13	15	29	61	36	19	5	0	0	0
G. Madugula	16	8	14	14	45	49	32	22	13	8	3
Pedabailu	6	7	25	26	39	34	4	11	25	4	1
Total	**58**	**28**	**54**	**69**	**145**	**119**	**55**	**38**	**38**	**12**	**4**

Source: Survey Data

More and more experiences shows that MFIs cannot avoid disaster risk, let alone ignore it (Pantoja, 2002). However, Government and the NGOs working over the Paderu agency villages are ready with a subsidiary plan to avoid the risk and also manage the emergent situation with the available resources based on the situation.

Table 6.30: Emergency Support Extended to SHGs

Mandals	**Required**	**Total loan received**	**Avg loan received**	**No. of Loans**
Dumbriguda	31902	31902	1077.70	5
G. Madugula	0	0	0	0
Pedabailu	20000	42300	14100.00	7
Total	**51902**	**74202**	**15177.70**	

Source: Survey Data

The savings with self help groups is reportedly used for managing the emergency situation arose. It is revealed that an amount of ₹ 20,000 for Pedabailu and ₹ 31,902 for Dumbriguda have been utilized towards managing emergency situations. Mandal G Madugula has not reported for any fund utilization towards emergencies. The average support received is ₹ 14,100 and ₹ 1,077.70 for Padaibailu and Dumbriguda respectively (see table-6.30).

It is interesting to note that the fund sanctioned is equivalent to and morethan the fund required to arrest the situation. The ITDA emergency fund has also been flown in to self help groups to make people come out of the grieve situation and preserve normalcy. Collection of fund either from the self help group's savings corpus or from the district emergency cell was a matter of calls distance from the self help groups operating in the affected areas.

Despite the fund flow from the savings groups as well as from the ITDA administration, involvement of people is important to control an emergency situation. Hence, preparing tribal community to face the challenging environment and get them out of the situation is a better challenging job for the MFIs and especially the self help groups. The self help groups have played multiple roles to help tribal imbibe with skill and technology to combat the challenging situation. The common roles played are (a) Creation of Awareness on Types of Risk, (b) Formation of Youth Groups to control the situation, (c) Provision of Supply of Necessary First Aids, (d) Keeping the PDS Stocks for Rainy season at Panchayat Depots, (e) Storing of Medicine Stocks in anticipation of rainy season, (f) Organizing Emergency Preparedness Training for SHGs, and (g) Installation of Alarm System in the Vulnerable tribal areas of the ITDA. Each activity taken up signified the importance of helping tribal to either equip with or to protect them from the emergency (see fig.6.29).

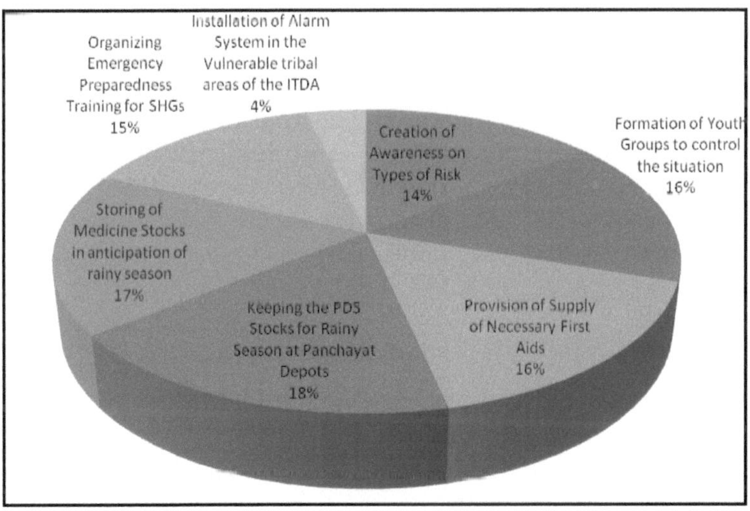

Fig. 6.29: SHG Performance Ranking to Emergency Response

Satisfaction of community on the emergency response role of SHGs working in tribal agency area of Paderu is encouraging. About 85.49 per cent of self help groups across the regions of ITDA reported satisfied with the emergency response and community preparedness initiatives of the self help groups. It has helped many families and enhanced morale of many to confront a situation like this. A meager 14.51 per cent self help groups are not satisfied on the emergency response performance (see table 6.31). 100 per cent satisfaction on emergency response role of SHGs has been reported by G. Madugula Mandal of ITDA Paderu. Level of satisfaction expressed over emergency response ability of self help groups is not welcoming by Pedabailu Mandal. About 67.03 per cent of expressed to be satisfied over the SHGs role on responding to the emergency situation.

Table 6.31: People's Satisfaction on Role of SHGs during Emergencies

Mandals	Satisfied	Not-Satisfied	Total
Dumbriguda	184	30	214
	(85.98)	(14.02)	(100)
G. Madugula	224	0	224
	(100)	(0)	(100)
Pedabailu	122	60	182
	(67.03)	(32.97)	(100)
Total	**530**	**90**	**620**

Source: Survey Data

The inter regional variation on satisfaction is primarily due to the access to the ITDA headquarters in one hand and access to better transport and communication facility on the other. The functioning capabilities of self help groups cannot be ruled out as a parameter of success of effective response to the emergency situation. The role of tribal women self help groups in capacitating households and institutions to respond to unpredictable situations is notable and recognition worthy. The developmental role of self help groups across the mandals of ITDA Paderu would have been constructively imbibed to ensure better sustainable growth in the future.

6.5. Microfinance Vis-à-Vis Tribal Employment

Self Help Groups (SHGs) and the Village Organizations (VO) in tribal areas, as in other areas, are directly and indirectly contributing towards provision of employment to the men and women. SHGs are novel and innovative organizational setup in India for the women upliftment and welfare. SHGs are promoted with a larger view of incorporating enterprise skill among the women and made them successful entrepreneurs. Apart from the mobilization of saving and distribution of corpus to needy families, the SHGs also organize training programme for the members and do make SHGs aware about the feasible marketing opportunities and scope for the business. As a social movement with government support, SHGs become more or less a part and parcel of the society (Moses, 2011). Hence, it is hypothesized that;

Hypothesis-V: Micro Credit operation in agency areas takes care off the issues of employment , under employment and unemployment

Box-6.1: Seasonal Employment Avenues for Tribals in Paderu Agency Area

Season	Major Avenues of Employment	% of engagement
Rainy	1. Collection of NTFPs (Adda leafs, roots, mushrooms, sale of firewood, bamboo roots)	40%
	2. Agriculture works (Donger Cultivation, Vegetable cultivation, Paddy in low areas)	
	3. Wage labor in Govt works (Forest Dept, and ITDA)	
Winter	1. Harvesting of agriculture and horticulture products	55%
	2. Sale of Products in local *santa-weekly market*	
	3. Preparation of field for horticultural crops	
	4. Construction work	
	5. Forest work-fire line creation, construction roads, plantation etc	
	6. Sale of firewoods	
	7. Collection and sale of forest products	
Summer	1. Wage labour (mostly in department works)	5%

Note: Paderu agency area is different from normal areas. Rainy and winter is witnessed for about 10 months while summer is just confined to one and half month. This is due to the geographic location of the Paderu Agency Area. It is located above 3500 MSL.

Tribal people anywhere in the country in general and Paderu ITDA in particular are primarily engaged in forest and agricultural activities. It is also a fact that almost all family members are engaged in the age-old traditional occupation of agriculture, forestry, rearing of domestic animals and few are engaged in activities away of traditional one (see Box-2). There are cases of a few being engaged in government jobs. The sporadic pattern of engagement has not only earning a poor visible income to the family but also creating a situation of under employment/disguised employment which is very common across the mandals of the agency. Issues of lack of avenue of employment in one hand and lack of motivational force on the other have left the tribal with family survival with the traditional sources of income. Under the stated context of tribal, and tribal areas like Paderu in particular, attention is broadly drawn towards hammering the situation to pull out tribal from the state of changelessness to utilize them in avenue of income and well-being and self help groups of the areas are considered as an effective means to break the deadlock.

Self help groups are assumed as institutions capable of showing new direction to rural life and especially to the women and thereby redefine the household well-being. It is also termed as institution par excellence to give a development boost to rural economy. Apart from the formal activities relating to finance, SHGs are also involved in activities of social and institutional importance. However, while not deviating from the prime objective of provisioning income to its members, supports have always been extended to member families to start and stabilize with a means of IGAs that not only support the member of the self help group rather the entire family.

Box-6.2: Listed Income Generation Activities of SHGs

❖ Agriculture	❖ Carpets
❖ Chicken shops	❖ Brick making
❖ Animal husbandry	❖ Khadi, leather items
❖ Mess	❖ Pot making
❖ Hosiery	❖ Spinning and weaving
❖ Mid day meals	❖ Saree business
❖ Candle preparation	❖ Basket weaving
❖ Embroidery	❖ Candle making
❖ Seeds	❖ Woolen blankets weaving
❖ Pickle making	❖ Agarbatti making
❖ Vegetable vending	❖ Sale of fruits and pongamia
❖ Phenyl making	❖ Carpets
❖ Missionary	❖ Brick making
❖ Spinning and weaving	❖ Khadi, leather items
❖ Saree business	❖ Pot making

Source: Compiled from Village Survey

In the context of rural India and tribal areas in particular, a customized approach to income generation has been crafted for the self help groups based on the resource potential of the area and the locality (see box-3). Selection of an activity is as important as a factor as its logical end to assured income generation. The self help groups are therefore generally engaged in development of a 'business plan' for each and every activity. Each business plan in turn is supported with a 'Micro Credit Plan '-popularly known as MCP for fund disbursement and group approval.

Besides, sanction of activities and allotment of funds, self help groups are arranging various training programmes to enhance the technical skill, help enable efficient running of social institutions and development institutions, create awareness in society on schemes and programmes of central and state government etc. However, people will not listen to any discussion and accept any idea-how ideological it may be- unless it is backed to benefit people in short and long-run. Self help groups being the people's institutions and local communities being united to fight back poverty, backwardness and envision for a brighter down for tomorrow, use to contribute to the SHG corpus.

The fund collected from each member as a matter of principle is saved in the bank and circulated within the group of needy to take up income generation activities to help enable family for survival. The study has recorded that SHGs across the regions have revolved the fund for 3 times from the date of formation of the self help groups. In the 1st phase of its circulation, 163 members had been benefited with ₹ 3,64,310 to take up IGA. The 2nd phase had benefited 25 families with a fund support of ₹ 1,04,921 and the 3rd phase circulated an amount of ₹ 1,69,600 over 7 families. Total money under circulation was ₹ 6,38,831 across the Mandals of the selected regions; while Mandal G. Madugula had been highly parted away with ₹ 2,29,160. In terms of family coverage, all beneficiaries are not covered with the financial benefit but the neediest families got relief from many ills (see table-6.32). However, less than half of the members of Pedabailu Mandals have enjoyed the benefit of self help groups.

Table 6.32: Status of Internal Loan for IGAs

Mandals	Not availed	1st time	1st timer Amt.	2nd time	2nd timer Amt.	3rd time	3rd Timer Amt.	Total Amt	Avg Amt.
Dumbriguda	144	57	117900	12	52020	2	39000	208920	2942.54
G. Madugula	190	25	68060	5	30500	5	130600	229160	6547.43
Pedabailu	91	81	178350	8	22401	0	0	200751	2255.63
Total	**425**	**163**	**364310**	**25**	**104921**	**7**	**169600**	**638831**	**3276.06**

Source: Village Survey

Mere financial support to self help group members would not help tribal women to deploy the fund on appropriate IGAs without provisioning handholding support through training and capacity building for marketing the product and service to realize the benefit. This kind of a financial assistance would rather aggravate the situation of poverty and may lead the family/ies to another debt burden. On the other hand families with skill and technology can do well with fund support from the self help groups.

SHGs across the Paderu agency area have conducted various training programmes of different natures such as skill building, selection and execution of income generation programmes, and also organized exposures to livelihood activities of similar in nature and to activities of interest to the self help groups at various places. Almost 46.77 per cent of members of self help groups had been imparted training on multifarious fields across the mandals of the studied regions. A higher per cent (63.55%) members of self help group of Dumbriguda Mandals has obtained trainings in comparison to rest of the two regions *i.e.* G. Madugula and Pedabailu (see fig. 6.30). There is still a long way to go. Almost 53.23 per cent of members have to be given training and orientation support to make them employable.

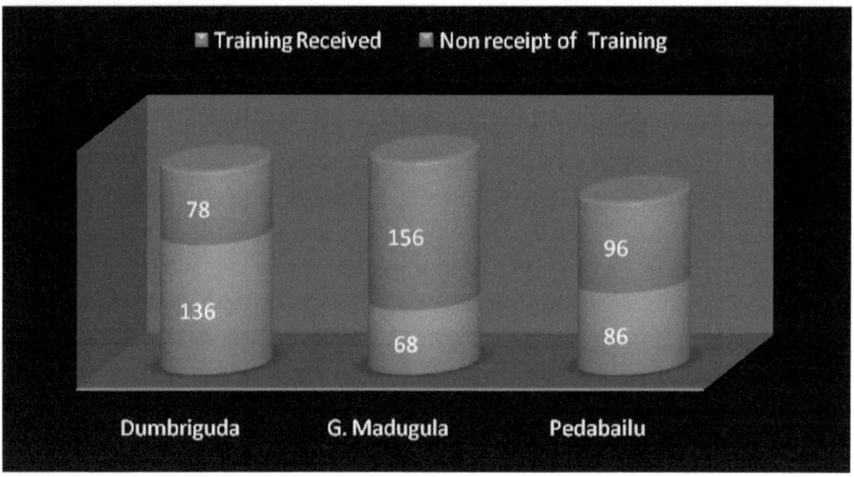

Fig. 6.30: Status of Training to SHG Members

There are families who have received all capacity building programmes organized by self help groups and to their dismay there are some families completely left with no training. Both the scenario has been recorded for the entire agency area. It is documented that out of the families trained about 87.58 per cent has also been given away with orientation programmes on multifarious activities (see fig.6.31). Inter regional comparison of orientation program is revealing overwhelming satisfaction that two regions of the Paderu agency area have organized the orientation programmes to nearly 100 per cent of its members.

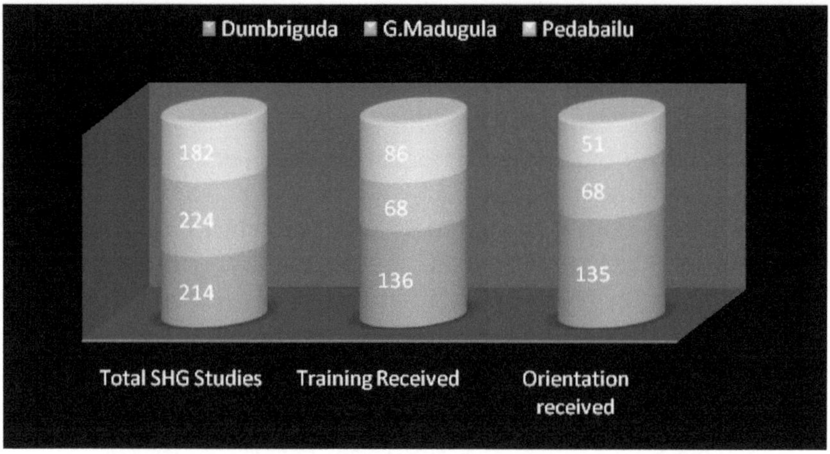

Fig. 6.31: Access to Orientation Programme on IGA

Besides training, exposure programmes have been conducted to different livelihood activities *viz.* lac cultivation, honey extraction, coffee cultivation, kitchen garden management, rearing of different varieties of country chicken, tailoring, running up of iodized salt preparation plant, firefighting, security, computer learning to work in retail marts, spoken English programmes etc. About 36.20 per cent members across the studied regions has been exposure to above said livelihood programmes (see fig. 6.32). Inter region scenario of exposure programmes varies from a minimum of 6.97 per cent in Padabailu Mandal to a maximum of 52.20 per cent for Dumbriguda Mandals. The variation in exposure program is primarily due to the locational variation of the regions in one hand and the available resource potential as well as selection of livelihood baskets on the other.

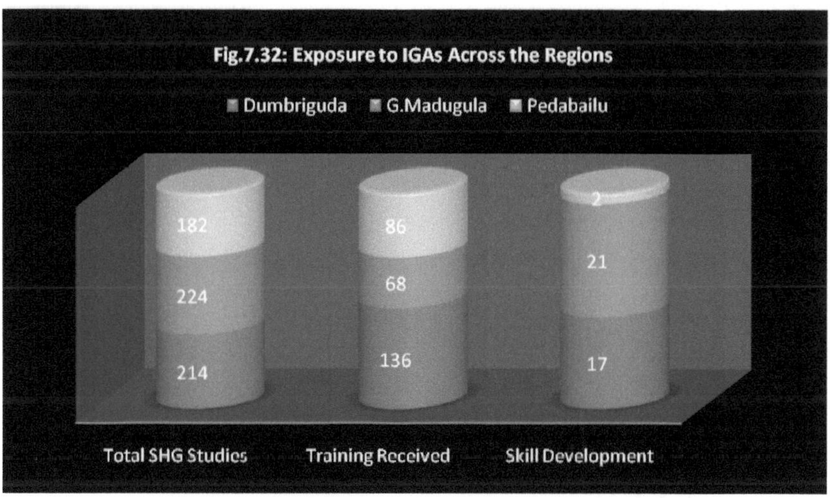

Fig. 6.32: Exposure to IGAs Across the Regions

With a view to imbibe skill in tribal women, girls and youths, various skill development programmes have been executed by self help groups from time-to-time. These includes masonry, tailoring, embroidery works, wielding, electrician, computer training, crafts making etc. About 14 per cent members across the studied regions has been exposure to above said livelihood programmes (see fig. 6.33). Inter region scenario of exposure programmes varies from a minimum of 2.32 per cent in Padabailu Mandal to a maximum of 30.88 per cent for G. Madugula Mandals. Out of the total beneficiaries, 52.50 per cent of G. Madugula, 42.50 per cent of Dumbriguda and 5 per cent of Pedabailu have been covered under the programme (see table-6.33). The variation in access to skill development program is directly proportional to the interest of the youth.

Table 6.33: Conduct of Skill Building Programmes for IGA

Mandals	Orientation on IGA	Exposed to similar activities	Skill Development	Marketing Linkage	Micro Enterprise Development	Total Trg Recepient	Sample SHGs
Dumbriguda	135	71	17	25	56	304	214
	(53.15)	(67.62)	(42.50)	(40.98)	(34.78)	(48.95)	(34.52)
G. Madugula	68	28	21	23	65	205	224
	(26.77)	(26.67)	(52.50)	(37.70)	(40.37)	(33.01)	(36.13)
Pedabailu	51	6	2	13	40	112	182
	(20.08)	(5.71)	(5.00)	(21.31)	(24.84)	(18.04)	(29.35)
Total	**254**	**105**	**40**	**61**	**161**	**621**	**620**
	(100)	**(100)**	**(100)**	**(100)**	**(100)**	**(100)**	**(100)**

Source: Survey Data

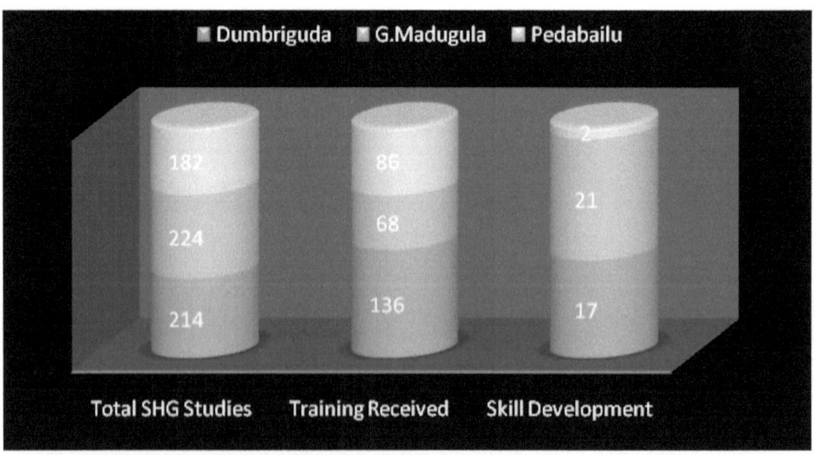

Fig. 6.33: Access toi SkillDevelopment Programme

All trained candidates are provided with a job immediately after training through campus. Companies like Birla, Peter England, Vimal, Reliance Mart etc have taken a good number of tribal youths, both boys and girls together, and posted them in cities like Gujurat, Coimbatoor, Visakhapatnam etc. Hence, marketing linkage has been viewed as a matter of serious concern for making the skill development programme fruitful and viable. The study reveals that about 21.03 per cent of skill development programmes have been linked to the market across the ITDA over the years (see fig.6.34). About 18.38 per cent of trainees of Dumbriguda, 33.82 per cent trainees of G. Madugula and 15.11 per cent trainees of Pedabailu have been directly linked to the open market. IKP being the nodal agency of implementation of the World Bank supported self help group programme has taken all precaution and care for capacity building of tribal youths along with adequate provision of marketing their skill and potentiality.

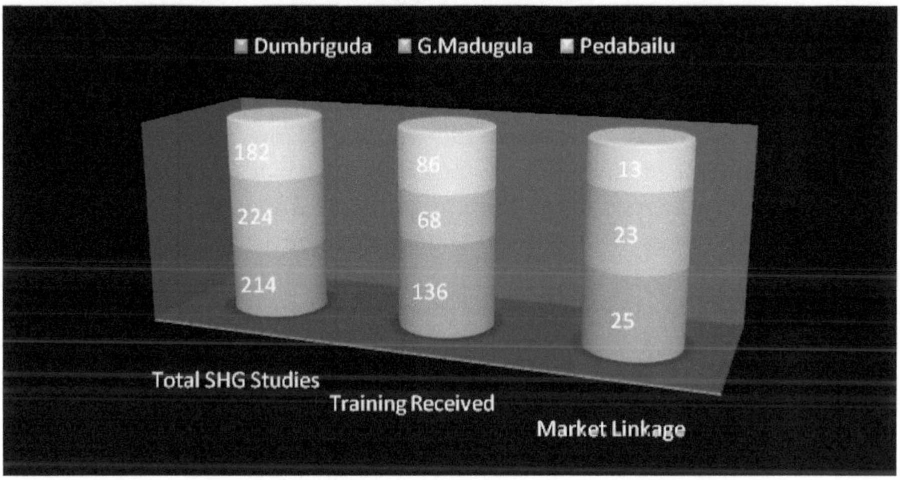

Fig. 6.34: Status of Training on Market Linkage

The training and exposure for tribal helped them to develop micro, small and medium enterprises in the region of their residence. Training programmes and exposure have firmed the commitment of the tribal to get united and set up Cashew Nut Processing Plants at Araku Valley-the tourist place. Besides, Micro Enterprise Development (MED) programmes on Turmeric Processing, Rajmah (kidney beans) Value Addition, Medicinal Plant Value Addition etc have been taken up. About 55.51 per cent tribal self help group members have been trained on various MED programmes (see fig. 6.35). It is reported that highest number of beneficiaries (95.58%) of G. Madugula have accessed to the programme followed by Pedabailu (46.51%) and Dumbriguda (41.17%).

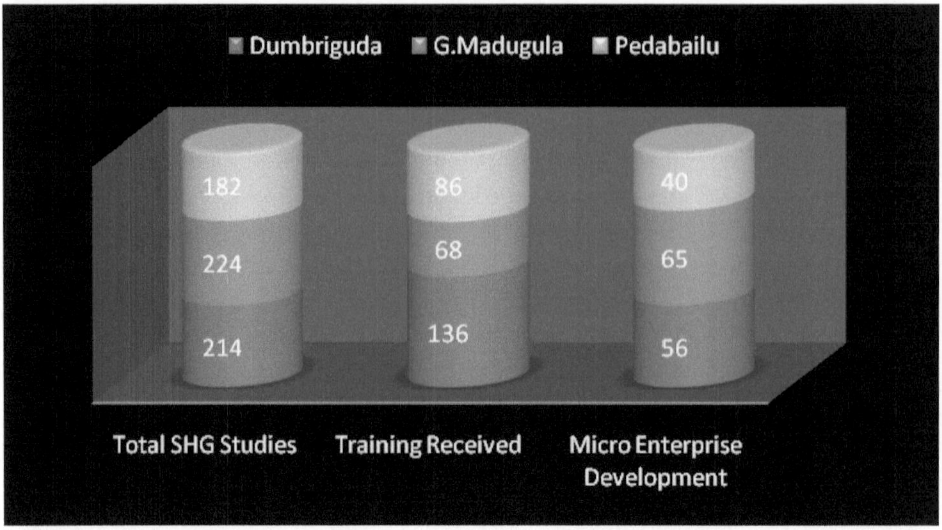

Fig. 6.35: Training on Micro Enterprise Development

Such a massive capacity building programme coverage is raising questions that who has conducted the training? How trainings, exposures and MED of such a magnitude was possible for tribal's living in forest and hillocks of Paderu agency area? Empowering tribal women, girls and youth and wiping out issues of disguised unemployment is a concerted effort of many agencies and institutions working in the tribal agency area of Paderu. Agencies like NGOs, Indira Kranthi Pratham (IKP), DRDA and ITDA etc are executing schemes and programmes to inculcate habits of self reliance and making tribals earn their living in a sustainable way, adopting a sustainable means, and without destroying the natural resources available and the natural environment at large.

There are more than hundreds of Non Governmental NGOs working in the Paderu ITDA for morethan 3-4 decades and couple of new ones being added every now and then. And all NGOs are promoting in-situ development. About 66 per cent of skill development programmes for tribals are under taken by NGO activists across the agency (see fig. 6.36). Looking at activist involvement across the Mandals a significantly larger proportion (47.07%) of their involvement is reported by G. Madugula followed by Dumbriguda (31.71%) and Pedabailu Mandals (21.22%). Besides, IKP has conducted about 23 per cent of the programmes, resource person of govt officials about 10 per cent and only 1 per cent by other than the three agencies. Apart from the NGOs, IKP has a better training coverage over the mandals (see table-6.34). In fact, role of other institutions in shaping skill of tribals cannot be refuted.

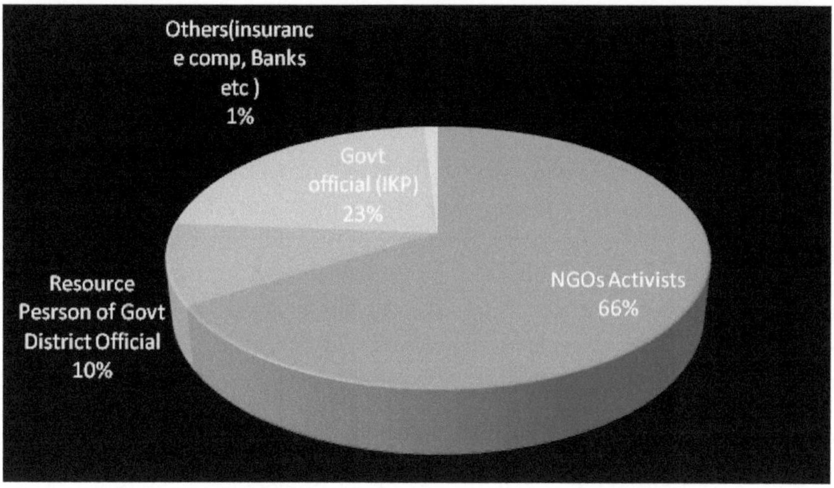

Fig. 6.36: Training Providers to Tribal SHGs

Table 6.34: Scenario of Agencies Imparted Training to SGHs

Mandals	NGOs Activists	Resource Person of Govt. District Official	Govt official (IKP)	Others (insurance comp., Banks etc.)	Total
Dumbriguda	130	36	66	1	233
	(31.71)	(58.06)	(46.48)	(16.67)	(37.58)
G. Madugula	193	15	37	0	245
	(47.07)	(24.19)	(26.06)	(0.00)	(39.52)
Pedabailu	87	11	39	5	142
	(21.22)	(17.74)	(27.46)	(83.33)	(22.90)
Total	**410**	**62**	**142**	**6**	**620**
	(100)	**(100)**	**(100)**	**(100)**	**(100)**

Source: Survey Data

The concerted effort has reverberating impact on the tribal of ITDA, Paderu. This has given boost to the confidence level of a significant number of beneficiaries (23.38%) and especially to the youths to undertake activities and income generating avenues of their choice and based on the available resources and are also confident to take the IGAs to its logical end of marketing the services and realizing the benefits (see table-6.11). About 4.19 per cent self help group members have reported enjoying enhanced income, 22.74 per cent confident of their skill development, 6.45 per cent can do bank transactions on their own, 5.32 per cent opined for linking their produce to market effectively and most importantly more than 1.29 beneficiaries claimed knows their rights, and entitlements on development programmes.

At the end what matter is how effective are these strategies to curb the employment scenario of tribals? Are there any shift in tribal engagement and/or employment pattern in an agency like Paderu? It is very easy to visualize a drastic change out of a programme and strategy basically deals with people, which is in practical not! Because after all the programme is for human beings and not for machine. The result would go with the popular saying 'slow and steady wins the race'.

Table 6.35: Current Employment Status Across Studied Regions of ITDA Paderu

Mandals	Employed	Self Employed	Professional	Labour	Total
Dumbriguda	1	84	39	90	214
	3.13	38.01	24.53	43.27	108.93
G. Madugula	10	32	97	85	224
	31.25	14.48	61.01	40.87	147.6
Pedabailu	21	105	23	33	182
	65.63	47.51	14.47	15.87	143.47
Total	**32**	**221**	**159**	**208**	**876.53**
	100.00	**100.00**	**100.00**	**100.00**	**400.00**

Source: Survey Data

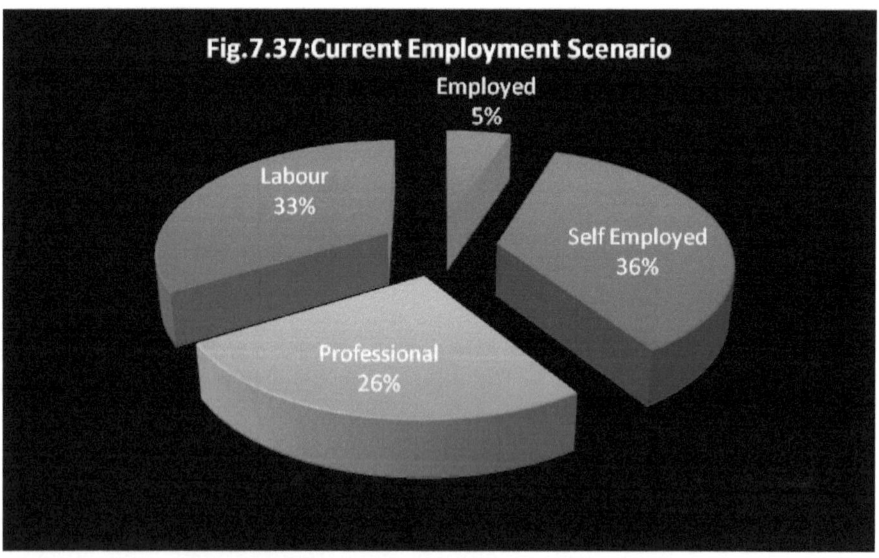

Fig. 6.37: Current Employment Scenario

The effort has resulted in creating an environment where 36 per cent claims to be self employed, 26 per cent are professionals, and about 5 per cent engaged in small government jobs (see fig. 6.37). A glaring variation in engagement pattern is reported across the studied Mandals (see table 6.35). There is visual change in employment pattern for 67 per cent of self help group members across the studied mandals. However, the woes of rest 33 per cent claimed to survive on wage labour still to be addressed and it would still be a tough job ahead.

A comparative view of employment pattern before joining the self help group and after joining the group is portrayed (see table-6.36). It is revealing that about 40 per cent of self help group members reported surviving on wage labour, which has came down to about 34 per cent after joining the group and over a period of six years. As a part of survival strategies about 59.67 per cent self help group members do collect Non-timber forest products (NTFP) all across the season. This is claimed to be the major employment provider to tribals. The scenario has completely changed. Collection of NTFP is now clubbed with wage employment. It is no more an independent source of income and employment to many.

Table 6.36: Comparative Employment Scenario Before Joining SHG and Now

	Before Joining SHG				After Joining SHG				
Mandals	**Govt. Job**	**Wage Labour**	**Collection of NTFP**	**Total**	**Employed**	**Self Employed**	**Professional**	**Labour**	**Total**
Dumbriguda	0	74	140	214	1	84	39	90	214
	(0)	(34.58)	(65.42)	(100)	(0.47)	(39.25)	(18.22)	(42.06)	(100)
G.Madugula	2	102	120	224	10	32	97	85	224
	(0.89)	(45.54)	(53.57)	(100)	(4.46)	(14.29)	(43.30)	(37.95)	(100)
Pedabailu	5	67	110	182	21	105	23	33	182
	(2.75)	(36.81)	(60.44)	(100)	(11.54)	(57.69)	(12.64)	(18.13)	(100)
Total	**7**	**243**	**370**	**620**	**32**	**221**	**159**	**208**	**620**
	(3.64)	**(116.93)**	**(179.43)**	**(300)**	**(16.47)**	**(111.23)**	**(74.17)**	**(98.13)**	**(300)**

Source: Survey Data

Finally, it is establishing a causative relation between the factors of skill building and the volume of self employed professional being created (see table-6.37). The packaged skill development programmes has imbibed many tribal with skill and potentiality to market themselves. In the attempt of understanding the correlation between self employment and the volume of skill development a poor correlation is seen, while similar is the case for MED programme and exposure. The direct trainings imparted to self help group members have a positive correlation (r=0.047833) with the creation of self employed professionals across the regions of ITDA Paderu.

Table 6.37: Skill Development Programme and Employmentability

Mandals	Self employed & Professional	Skill Dev Prog	Micro Ent. Prog
Dumbriguda	123	17	56
G.Madugula	129	21	65
Pedabailu	128	2	40
Total	380	40	161
Coefficient of correlation for Self Emp Vis-a Vis Skill Dev Prog			-0.84565
Coefficient of correlation for Self Emp Vis-a Vis MED Prog			-0.92079
Coefficient of correlation for Self Emp Vis-a Vis Training Prog			0.0478
Coefficient of correlation for Self Emp Vis-a Vis Exposure Prog			-0.0986

Source: Survey Data

It is evident from the foregoing discussions that employment and /or engagement pattern of tribal over the years have diversified. Many self employed professional have generated, dependency on forest produce has substantially reduced and eagerness for better regulated life is created. In the process, many youths finds the definition of their engagement and living. Entire family members are no more adjusting with the age-old traditional agriculture.

6.6. Microfinance Vis-à-Vis Rural Enterprises

In a world where almost half the population lives in poverty, microfinance is one of the `better tools for poverty alleviation, economic growth and development in emerging economies. These loans offer the same benefits to major world economies that face growth problems. It is very well realized that these loans can help lower-income groups setup and grow the small businesses, which generate income and employment that helps their communities and their economies . Microfinance thus comprises all segments of the rural population. In fact non-agricultural activities often complement agricultural income and labour flows and hence help build livelihood stability and growth. It is further argued that rural financial services actually become much better if we take the micro-specificity away and concentrate on sustainable financial services to a wider array of rural households and businesses. This increases institutional viability, reduces unit costs for services to the poorest and helps finance not only other poor but other important rural actors including - truckers, agribusinesses, side businesses, etc. all of which have an important role to play in rural development (Seibel, 2007). Hence it is hypothesized that;

Hypothesis-VI: Microfinance promotes micro enterprises at micro and macro level

One of the objective of self help group as a MFI is to help enable its member to pursue activities that directly and/or indirectly contribute towards building up their economic well-being. Training, financial assistance, marketing back up as well as timely catalytic support are considered as prime pillars of support to any beginner in enterprises of micro and macro form. The foregoing sections have given analytical view of the Micro Enterprise Development role of MFIs of Paderu agency area (see table-6.34) and the outcome of corresponding self-employed professionals.

Mere imparting training will not have any impact on any individual and farm. It will remain as some skill-set in the mind and heart without resulting cascading effect. To realize the full benefit of MED, it should be backed by adequate natural and financial resources to translate the skill and techniques imbibed into physical products and actions. The self help groups (SHGs) are meant for provisioning financial worthiness to the members either directly from the group savings collected or through leveraging from banks as a guarantee.

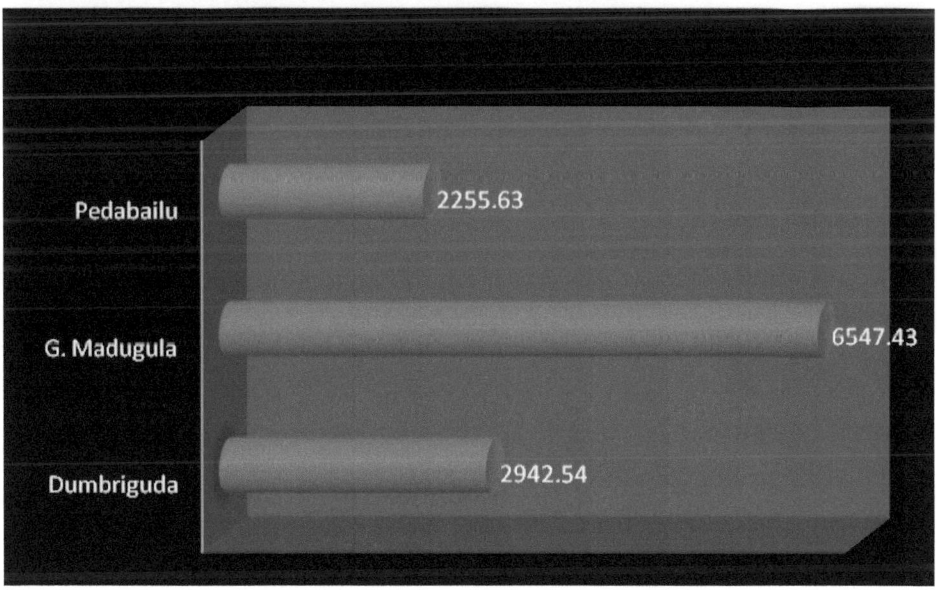

Fig. 6.38: Average Internal Borrowing Across Regions

Table 6.38: Bank Loan for Enterprises

Mandals	Not availed	1st time	1st timer Amt.	2nd time	2nd timer Amt.	3rd time	3rd Timer Amt.	Total Amt.	Avg. Amt.
Dumbriguda	92	117	1645667	5	104000	0	0	1749667	14341.53
G. Madugula	93	106	803012	20	2896800	5	388640	4088452	31209.56
Pedabailu	112	65	300100	4	14600	1	3000	317700	4538.571
Total	297	288	2748779	29	3015400	6	391640	6155819	19058.26

Source: Survey Data

The institutional worth of SHGs of ITDA Paderu is adequately visualized. They are playing all-pervasive role to capacitate tribal women to act as catalytic agent for family development (see fig.-6.38). There are 323 enterprises of various capabilities *i.e.* micro to small granted with bank finance to the worth of ₹ 6155819 with an all enterprise average of ₹ 19058.26 (see table-6.38). Access to Bank finance is more for G.Madugula Mandal with per unit bank linkage of ₹ 31209.56 and lowest for Pedabailu Mandal with per unit Bank finance support of ₹ 4538.57. The entire effort of establishment of units has taken a long way and is the result of concerted credible effort of SHGs across the agency area of Paderu (see fig.-6.39).

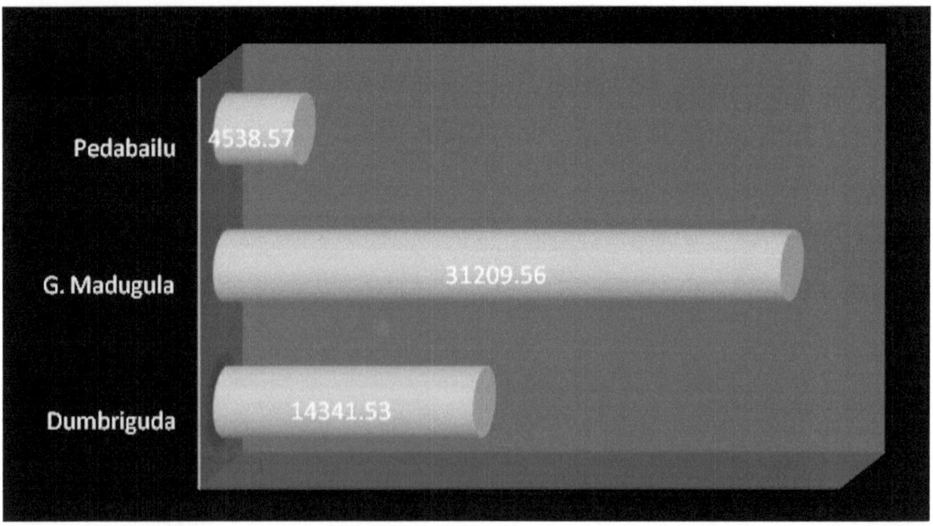

Fig. 6.39: Average Bank Loan Across teh Studied Region

The incredible role of self help groups is visible through extension of financial support as start-up capital to many enterprises. This has been extended to 140 beneficiaries all across the studied regions with an average support of ₹ 5829. This act of SHGs have not only given boost to morale of many new entrepreneurs but also made them capable to sustain with the enterprise (see table-6.39). Average financial support extended to its members is more in Dumbriguda Mandal, followed by G. Madugula and Pedabailu (see fig.-6.40).

Table 6.39: Status of Cash Credit for IGAs

Mandals	Not availed	1st time	1st timer Amt.	2nd time	2nd timer Amt.	3rd time	3rd Timer Amt.	Total Amt.	Avg. Amt.
Dumbriguda	146	65	453820	3	18000	0	0	471820	6938.529
G. Madugula	155	36	109472	3	21000	3	44600	175072	4168.381
Pedabailu	138	24	86250	5	81450	1	1500	169200	5640
Total	**439**	**125**	**649542**	**11**	**120450**	**4**	**46100**	**816092**	**5829.229**

Source: Survey Data

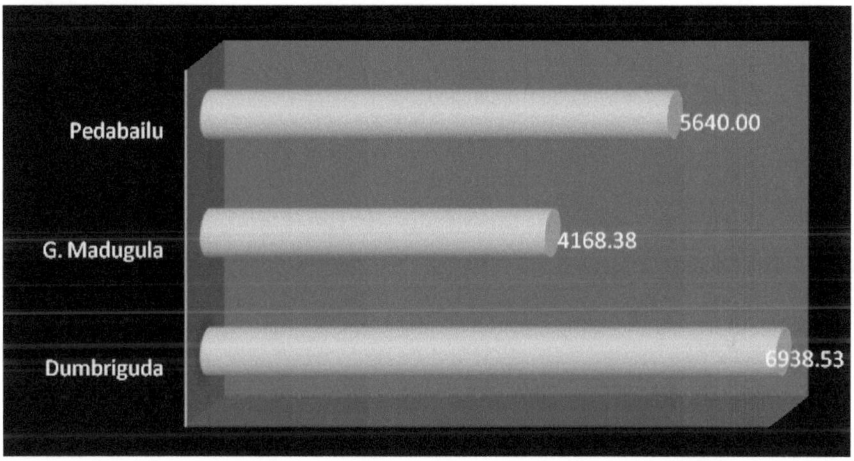

Fig. 6.40: Average Cash Credit Across Studied Regions

Table 6.40: Loan Repayment Status of SHGs Across Regions of ITDA Paderu

Mandals	Not availed	1st time	1st timer Amt.	2nd time	2nd timer Amt.	3rd time	3rd Timer Amt.	Total Amt.	Avg. Amt.
Dumbriguda	85	68	399032	19	141819	10	59533	600384	6189.526
G. Madugula	76	97	164555	34	381945	12	1149880	1696380	11862.8
Pedabailu	105	52	94680	19	70250	6	16000	180930	2349.74
Total	**266**	**217**	**658267**	**72**	**594014**	**28**	**1225413**	**2477694**	**7816.069**

Source: Survey Data

Injecting finance as and when required is not a wiser policy for any institution dealing with finance. There should have a 'heart and mind approach' to fund disbursal for better functioning. Meaning thereby, a good inflow of fund to enterprises coupled with adequate out flow of fund to pay-back the loans utilized would enhance the velocity of money to help a larger number of members of self help groups. Self help groups of Paderu agency could able to repay back about 48.82 per cent of total loan received. However, all enterprises have repaid some amounts either partly or fully (see table-6.40). The repayment in fact varies from a minimum of 0.64 per cent of loan to 63.44 per cent. In terms of repayment, all enterprises across the studied regions are with a poor repayment tract record. This may be due to low return on investment (RoI), which is leading to poor realization from business undertaken and/or indifferent attitude of members towards repayment of loans (see table-6.41).

Table 6.41: Fund Utilisation Status of Regions Across ITDA Paderu

Mandals	Avg. Internal Loan	Avg. Bank Loan	Avg. Cash Credit	Loan Repaid	Avg. Loan Rcvd.	Outstanding
Dumbriguda	2942.54	14341.53	6938.53	6189.526	24222.60	18033.07
G. Madugula	6547.43	31209.56	4168.38	11862.8	41925.37	30062.57
Pedabailu	2255.63	4538.57	5640.00	2349.74	12434.20	10084.46
Total	**3276.06**	**19058.26**	**5829.23**	**7816.069**	**28163.55**	**20347.48**

Source: Survey Data

Table 6.42: Status of Loan for Health Purpose

	Total Loan Received	Avg. Loan	Interest rate
Dumbriguda	43203	251.18	1%-2%
G. Madugula	826701	9394.32	1%-2%
Pedabailu	54969	1962.5	0.25%-2%
Total	**924873**	**11608**	

Source: Survey Data

Self help groups have given fund support for various purposes let alone development of enterprises. It is an institution of people to address many fold issues of socio-economic-institutional-cultural magnitude. This is revealing from the fund sanctioned for health purposes of tribals. An amount of ₹ 9,24,873/- had been sanctioned by the self help groups all across the studied regions towards addressing health issues of tribals (see table-6.42).

To reiterate, tribals of Paderu agency area are getting free medical treatment through the PHCs, CHCs and a Multi-Super Specialty hospital run by Tribal Welfare Department (TWD), Govt. of Andhra Pradesh. All medicines, consultation fee and food for the patients as well as for one attendant of patient are provided at free of cost to tribals of Paderu ITDA.

Table 6.43: Status of Investment on Agriculture in Studied Regions

	Required	Total Loan Received	Avg. Loan
Dumbriguda	N/A	270424	1430.81
G. Madugula	199410	39221	3920
Pedabailu	N/A	21000	2100
Total	**199410**	**330645**	**7450.81**

Source: Survey Data

Besides, members of self help groups are also supported with loans to undertake agriculture activities (see table-6.43). There are about 210 members of self help groups supported with agriculture loan across three studied Mandals to a tune of ₹ 3,30,645. Nearly an equal amount of fund support has also been given to SHG members towards undertaking animal husbandry activities. Across the regions, it is revealing that many self help group members of G. Madugula Mandal have shown interest and are benefitted with financial support to undertake animal rearing activities and primarily concentrated with Goat and Poultry rearing activities (see table-6.44). Of the three regions under consideration in ITDA Paderu, Mandal G. Madugula have accessed ₹ 2,15,010 as against ₹ 44,502 in Dumbriguda Mandal and ₹ 36,600 Pedabailu Mandal.

Table 6.44: Status of Animal Husbandry in Studied Regions

	Required	Total Loan Received	Avg. Loan
Dumbriguda		44502	298.67
G. Madugula	319298	215010	4056
Pedabailu	21800	36600	4066.66
Total	**341098**	**296112**	**8421.33**

Source: Survey Data

Paderu agency area is known for its climatic condition and environment. Tribals over here apart from concentrating on agriculture have taken up IGAs like Banana Cultivation on large scale, Opening up of Petty shops, leaf plate using siyali leaf, Petty business in weekly sandy days etc. These activities are taken up at the group level and hence, the investment is incurred at group level. The investment on these activities recorded for ₹ 2, 61,178 (see table-6.45). Once again G. Madugula Mandal of Paderu recorded for high investment on IGAs based on use of natural resources available. In terms of total investment on the micro and small enterprise development for income generation Mandal Dumbriguda has recorded significantly more investment than the other two regions *i.e.* G. Madugula and Peddabailu (see fig.6.41).

Table 6.45: Investment on IGA Across Selected Regions

	Required	Total Loan Received	Avg. Loan
Dumbriguda	N.A	228178	687.28
G. Madugula	105740	32200	4600
Pedabailu	15000	800	800
Total	**120740**	**261178**	**6087.28**

Source: Survey Data

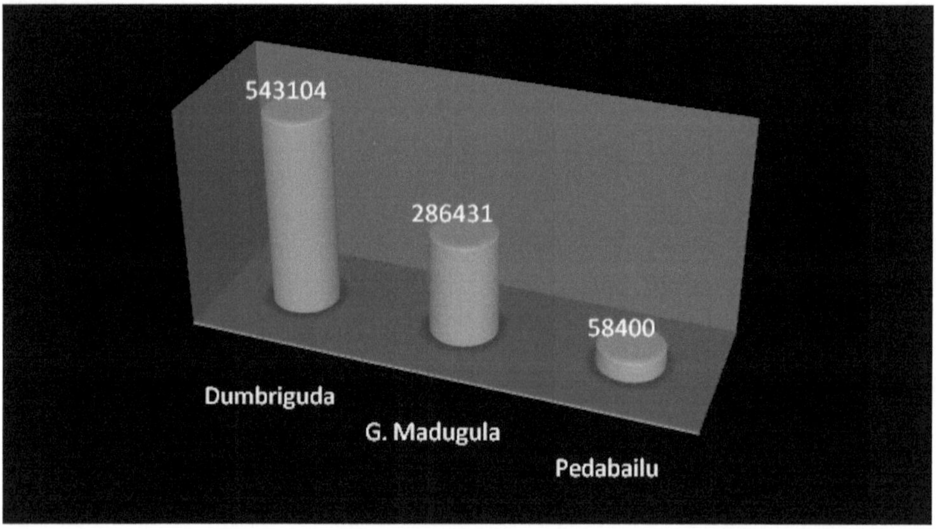

Fig. 6.41: Fund Development on IGA Across Regions

The facts and figures presented on promotion of micro and small enterprises in ITDA Paderu are not overwhelming but at the same time importance of SHGs across the regions of agency area cannot be undermined. The efficiency of tribal self help groups in promoting Micro Enterprises (MEs) is unique. Access to loan and development of enterprises across the regions are positively correlated (r=0.889907) and so also is the scenario of correlation (r= 0.984366) livestock programmes and the investment incurred across the Mandals of Paderu agency area. The contribution of self help groups towards promotion of micro enterprises is notable, visible and enterprise promotion would have not been possible without the self help groups.

6.7. Microfinance Vis-à-Vis Poverty Reduction

SHG is a group of rural poor who have volunteered to organize themselves into a group for eradication of poverty of the members. They agree to save regularly and convert their savings into a common fund known as the Group. The members of the group agree to use this common fund and such other funds that they may receive as a group through a common management (Bundela, Dubey and Pandey, 2013). In both developing and emerging economies, microfinance has vastly and increasingly been seen as one of the most important means for enhancing the lives of the poor and therefore a major tool for economic and social development mostly in rural areas (Austinice, 2010).

Does microfinance promote economic development? Does a poor person get care? It is well and good to wish that millions of microloans might create a surge of GDP growth, but that is asking for an economic macro miracle in the countries where microfinance is most needed. "Poverty is too much complex to be cured by one intervention. We don't ask if basic education alone ends poverty. We don't ask if vaccinations alone end poverty. Why should microfinance be held to the standard of single-handedly reducing poverty? (Sam Daley-Harris, 2009)." Hence, it is hypothesized that;

Hypothesis-VII: Microfinance contributes to poverty reduction

'Microfinance is a means to an end but not an end itself'. All previous sections of the chapter while witnessing positive correlation of self help group functioning in tribal areas for empowerment, promotion of sustainable economic development, injecting capacity of risk aversion, generates employment and addressing of unemployment and/or disguised employment and helps enable tribal to initiate enterprises (see table 6.37, 6.43, 6.44, and 6.45), does it require to study the poverty reduction abilities of self help groups.

The birth of self help group is to be a savior and it is a savior of poor. The intent of SHG formation is to make people get access to the basic necessities of life, get access to education, health and institutional services and amenities etc. Indeed, collection of savings, rotation of savings and further injection of money into the system back is one among the functions meant for self helps across the globe. It is a panacea to take the poor, vulnerable people along to a statute of food sufficiency, handful of income, plenty of employment, society free of disease, and a society where voices are equally viewed --- it does not mean as a society in the Utopian State.

The formation and functioning of self help groups across the ITDA Paderu made tribals, as has been seen from all previous section, to amass wealth and create assets for their own. On an average each beneficiary across the agency area is supported with ₹ 12,551 for creation of assets. Dumbriguda Mandal of Paderu agency area has extended larger average support (₹ 7141) to beneficiaries in comparison to G. Madugula (₹ 4018) and Pedabailu (₹ 1391) Mandals (see table-6.46).

Table 6.46: Investment on Development of Assets Across Mandals of ITDA Paderu

	Required	Total Loan Received	Avg Loan
Dumbriguda	N/A	2035256	7141.29
G. Madugula	108440	68279	4018.52
Pedabailu	41900	15300	1390.9
Total	**150340**	**2118835**	**12550.71**

Source: Survey Data

However, building asset is one among the parameters to ensure a healthy socio-economic and institutional status to the people and area. But there are a host of other parameters considered equally important to redefine the living of tribal people of Paderu. Some basic questions are being asked time and again on the 'intent and pace' of SHG functioning in every platform. These are as follows; i) has women came out of home? ii) Has her voiced heard in a male dominated society? iii) Has her contribution viewed and recognized at HHs and community level? iv) Has SHGs contributed for assuring a health living? v) What impact does self help groups made on tribals of Paderu at a large?

Table 6.47: Impact of SHGs on Tribal Women Mobility

Mandal	Same	Increased	Decreased	Deteriorated	Total
Dumbriguda	71	143	0	0	214
	(33.18)	(66.82)	(0.00)	(0.00)	(100.00)
G. Madugula	121	100	3	0	224
	(54.02)	(44.64)	(1.34)	(0.00)	(100.00)
Pedabailu	99	83	0	0	182
	(54.40)	(45.60)	(0.00)	(0.00)	(100.00)
Total	**291**	**326**	**3**	**0**	**620**
	(46.94)	**(52.58)**	**(0.48)**	**(0.00)**	**(100.00)**

Source: Survey Data

Answering these questions is very difficult and each question has got a strong reason to be probed on. Finding answers to these questions would give an understanding of the multi-dimensional role viewed for SHGs on a tribal area platform and the role played to meet tribal people's aspiration. Beneficiaries' satisfaction is viewed as the ultimate truth to view the 'functional ability and capability' of SHGs in tribal area to address the issues of poverty- a concept beyond income to a just society where basic amenities and services are amply available.

Tribal are self contented and strongly confine to the resources available in the vicinity of their living. Formation of self help groups has more or less created a platform for the tribal women to meet frequently-may be more than once in a month- to share and discuss social issues, and family issues. SHG meetings followed by Village Organisations (VOs), Mandal Mahila Samakshya (MMS) and Zilla Mahila Samakshya (ZMS) are the platforms created for members to meet and take their issues forward. The involvement of tribal women in SHG activities has enhanced mobility of 52.58 per cent members (see table-6.47). Participation women in SHG activities have brought the world closer reported 45.97 per cent of self help group members. The entire ITDA Paderu is the world for them. In fact, all members in any institution are not equally vocal but lack of scope is also a barrier to suppress the qualities in many. Hence, SHG has provided the platform to explore the talent of tribal women and enhanced the confidence through talks with many officers, institutions and sharing of the issues of communities, areas and finally get it addressed (see table-6.48).

Table 6.48: Women Confidence to talk to Outsiders

Mandal	Same	Increased	Decreased	Deteriorated	Total
Dumbriguda	133	79	0	2	214
	(62.15)	(36.92)	(0.00)	(0.93)	(100.00)
G. Madugula	89	129	6	0	224
	(39.73)	(57.59)	(2.68)	(0.00)	(100.00)
Pedabailu	105	77	0	0	182
	(57.69)	(42.31)	(0.00)	(0.00)	(100.00)
Total	**327**	**285**	**6**	**2**	**620**
	(52.74)	**(45.97)**	**(0.97)**	**(0.32)**	**(100.00)**

Source: Survey Data

The activities taken up by the self help groups are not individual centric at any time. Issues of poor and vulnerable are heard on priority while the commitment of development of all has never been taken the back bench. The sportive spirit of dealing with micro household/family issues to the macro level society issues and arriving at solutions have widely been recognized for 51.94 per cent of SHG members at HH level and 58.87 per cent SHGs members at the community level (see table-6.49 and 6.50). The self respect for women is an accrued benefit for their dedicated effort at the platform given by SHGs.

Table 6.49: Recognition of Women Contribution at HHs Level

Mandal	Same	Increased	Decreased	Deteriorated	Total
Dumbriguda	118	96	0	0	214
	(55.14)	(44.86)	(0.00)	(0.00)	(100.00)
G. Madugula	75	148	1	0	224
	(33.48)	(66.07)	(0.45)	(0.00)	(100.00)
Pedabailu	104	78	0	0	182
	(57.14)	(42.86)	(0.00)	(0.00)	(100.00)
Total	**297**	**322**	**1**	**0**	**620**
	(47.90)	**(51.94)**	**(0.16)**	**(0.00)**	**(100.00)**

Source: Survey Data

Table 6.50: Recognition of Women Contribution at Community Level

Mandal	Same	Increased	Decreased	Deteriorated	Total
Dumbriguda	73	139	2	0	214
	(34.11)	(64.95)	(0.93)	(0.00)	(100.00)
G. Madugula	76	137	11	0	224
	(33.93)	(61.16)	(4.91)	(0.00)	100.00
Pedabailu	93	89	0	0	182
	(51.10)	(48.90)	(0.00)	(0.00)	(100.00)
Total	**242**	**365**	**13**	**0**	**620**
	(39.03)	**(58.87)**	**(2.10)**	**(0.00)**	**(100.00)**

Source: Survey Data

United effort for shaping the society and give right education to the children has been a motto in disguise for the self help groups working in Paderu agency area. Due to a multi-faceted community based monitoring the access to education facilities and services has been improving along with the maturing of the self help groups. 61.61 per cent of self help group members supported the fact of accessing better education services across the mandals of the agency area (see table-6.51). Besides, access to health services has been a critical concern for the tribals living over the ITDA. Though drastic effort has been made by the TWD, Govt of AP to open up and run the medical centers all across the ITDA, people still depend on traditional healing as a first option for treatment. The situation and scenario of trust on the modern treatment is gaining popularity and there are many people accessing the services at the PHCs and CHCs (58.71%). There are about 2 per cent of people reported of not being happy on the services provided and hence, expressed their unwillingness to access the Govt health services offered at Govt. run establishments (see table-6.52).

Table 6.51: Access to Better and Improved Education

Mandal	Same	Increased	Decreased	Deteriorated	Total
Dumbriguda	82	132	0	0	214
	(38.32)	(61.68)	(0.00)	(0.00)	(100.00)
G. Madugula	101	123	0	0	224
	(45.09)	(54.91)	(0.00)	(0.00)	(100.00)
Pedabailu	55	127	0	0	182
	(30.22)	(69.78)	(0.00)	(0.00)	(100.00)
Total	**238**	**382**	**0**	**0**	**620**
	(38.39)	**(61.61)**	**(0.00)**	**(0.00)**	**(100.00)**

Source: Survey Data

Table 6.52: Access to Health Services

Mandal	Same	Increased	Decreased	Deteriorated	Total
Dumbriguda	108	104	0	2	214
	(50.47)	(48.60)	(0.00)	(0.93)	(100.00)
G. Madugula	94	117	13	0	224
	(41.96)	(52.23)	(5.80)	(0.00)	(100.00)
Pedabailu	39	143	0	0	182
	(21.43)	(78.57)	(0.00)	(0.00)	(100.00)
Total	**241**	**364**	**13**	**2**	**620**
	(38.87)	**(58.71)**	**(2.10)**	**(0.32)**	**(100.00)**

Source: Survey Data

Immunization services provisioned in Paderu agency area have better access. There are about 72.90 self help group members have accessed the service while 27.10 per cent remain stagnant (see table-6.53). However,

Govt involving self groups across the agency area has created awareness campaigns on the benefits of immunization and SHGs have taken care to made immunization accessible to the maximum beneficiaries.

Table 6.53: Access to Immunization Programme Across Agency Area

Mandal	Same	Increased	Decreased	Deteriorated	Total
Dumbriguda	28	186	0	0	214
	(13.08)	(86.92)	(0.00)	(0.00)	(100.00)
G. Madugula	102	122	0	0	224
	(45.54)	(54.46)	(0.00)	(0.00)	(100.00)
Pedabailu	38	144	0		182
	(20.88)	(79.12)	(0.00)	(0.00)	(100.00)
Total	**168**	**452**	**0**	**0**	**620**
	(27.10)	**(72.90)**	**(0.00)**	**(0.00)**	**(100.00)**

Source: Survey Data

Throughout the agency area of Paderu, SHG members have reported for better access to sanitation facilities. In fact, the sanitation programme was executed through the IKP and implemented by the self help groups of the respective villages. There are 64.52 per cent of self help group members reported for accessing the facilities and services offered (see table-6.54) across the regions of the agency. Toilet facilities are provided in school campuses and free toilet panels are being provided for individual members.

Table 6.54: Access to Sanitation Facility Across Agency Area

Mandal	Same	Increased	Decreased	Deteriorated	Total
Dumbriguda	91	123	0	0	214
	(42.52)	(57.48)	(0.00)	(0.00)	(100.00)
G. Madugula	95	129	0	0	224
	(42.41)	(57.59)	(0.00)	(0.00)	(100.00)
Pedabailu	34	148	0	0	182
	(18.68)	(81.32)	(0.00)	(0.00)	(100.00)
Total	**220**	**400**	**0**	**0**	**620**
	(35.48)	**(64.52)**	**(0.00)**	**(0.00)**	**(100.00)**

Source: Survey Data

Rural credit and especially credit facilities in tribal areas are very complex in nature. Different modes and means of credit circulation are in adoption by different agency/ (ies), agents and agencies. SHGs in the attempt to enhance access to formal credit to the needy have surpassed the standard norms. More than average number (57.74%) of members reported to have increased access to formal credit and 36.45 per cent reported no change in access and a meager (5.81%) reported facing adverse condition to access credit (see table-6.55). The adversity for the reported few is due to huge fund requirement and SHGs are not capable of providing the desired volume at the time of need.

Table 6.55: Status of Access to Credit Sources Across Agency Area

Mandal	Same	Increased	Decreased	Deteriorated	Total
Dumbriguda	85	127	2	0	214
	(39.72)	(59.35)	(0.93)	(0.00)	(100.00)
G. Madugula	97	93	34	0	224
	(43.30)	(41.52)	(15.18)	(0.00)	(100.00)
Pedabailu	44	138	0	0	182
	(24.18)	(75.82)	(0.00)	(0.00)	(100.00)
Total	**226**	**358**	**36**	**0**	**620**
	(36.45)	**(57.74)**	**(5.81)**	**(0.00)**	**(100.00)**

Source: Survey Data

Table 6.56: Status of Asset Building

Mandal	Same	Increased	Decreased	Deteriorated	Total
Dumbriguda	54	160	0	0	214
	(25.23)	(74.77)	(0.00)	(0.00)	(100.00)
G. Madugula	142	82	0	0	224
	(63.39)	(36.61)	(0.00)	(0.00)	(100.00)
Pedabailu	35	147	0	0	182
	(19.23)	(80.77)	(0.00)	(0.00)	(100.00)
Total	**231**	**389**	**0**	**0**	**620**
	(37.26)	**(62.74)**	**(0.00)**	**(0.00)**	**(100.00)**

Source: Survey Data

The status of self help group members to create and own a good asset is amply visible. The volume of investment gone in towards asset creation is quite huge in the context of tribal area and population (see table-6.46) and a large per cent (62.74%) SHG members reported for owning better assets across the studied Mandals of the Paderu agency area (see table-6.56). Morethan 80 per cent self help group members of Pedabailu Mandal reported for enjoying loan for asset creation following Dumbriguda (74.77%) and G. Madugula (36.61%).

Table 6.57: Experience of Increased Family Income

Mandal	Same	Increased	Decreased	Deteriorated	Total
Dumbriguda	122	92	0	0	214
	(57.01)	(42.99)	(0.00)	(0.00)	100.00
G. Madugula	164	53	7	0	224
	(73.21)	(23.66)	(3.13)	(0.00)	100.00
Pedabailu	64	118	0	0	182
	(35.16)	(64.84)	(0.00)	(0.00)	(100.00)
Total	**350**	**263**	**7**	**0**	**620**
	(56.45)	**(42.42)**	**(1.13)**	**(0.00)**	**(100.00)**

Source: Survey Data

Table 6.58: Impact of SHGs on Skill Equipment

Mandal	Same	Increased	Decreased	Deteriorated	Total
Dumbriguda	62	152	0	0	214
	(28.97)	(71.03)	(0.00)	(0.00)	100.00
G. Madugula	120	98	6	0	224
	(53.57)	(43.75)	(2.68)	(0.00)	(100.00)
Pedabailu	45	137	0	0	182
	(24.73)	(75.27)	(0.00)	(0.00)	(100.00)
Total	**227**	**387**	**6**	**0**	**620**
	(36.61)	**(62.42)**	**(0.97)**	**(0.00)**	**(100.000**

Source: Survey Data

However, 42.42 per cent self help group members reported for realizing increased income and more than half of the members (56.45%) reported for no change in income level even after joining with the self help groups (see table-6.57). Self help groups have conducted various skill development programmes for its members from time-to-time. About 62.42 per cent members of SHG have accessed skill building programme across the studied regions of ITDA Paderu (see table-6.58). This scenario is confirming to the fact that there are increasing number of self help group members are getting benefitted in various forms day-by-day and in the process becoming eligible for getting themselves involved in various income generation activities.

The entire effort of self help group is to get the voice of the women and the vulnerable heard. People of the SHGs should voice their concern in public forums. The issues and difficulties should get noticed by the concerned government officials and forums and draw attention of appropriate departments/divisions/agency to resolve. About 63.06 members across the agency area reported that their issues and concerns are heard at the public platform (see table-6.59). This opportunity for them has been given by the SHG. Indeed, there are still 34.35 per cent self help group members taking a neutral stand of no change in situation and they could not ventilate their concerns yet.

Table 6.59: Impact of SHGs on Voicing the Concern at Public Forum

Mandal	Same	Increased	Decreased	Deteriorated	Total
Dumbriguda	59	153	2	0	214
	(27.57)	(71.50)	(0.93)	(0.00)	(100.000
G. Madugula	107	103	14	0	224
	(47.77)	(45.98)	(6.25)	(0.00)	(100.00)
Pedabailu	47	135	0	0	182
	(25.820	(74.18)	(0.00)	(0.00)	(100.00)
Total	**213**	**391**	**16**	**0**	**620**
	(34.35)	**(63.06)**	**(2.58)**	**(0.00)**	**(100.000**

Source: Survey Data

Addressing nutritional issues is a major concern for the SHGs. Each SHG are taking care of distribution of rations, IFA tablets and conducting awareness discussion in coordination with the Angan Wadi Workers (AWW). Nutrition and Health Day (NHD) celebration is the key responsibility of SHGs. A good per cent (58.87%) of lactating mothers and pregnant ladies have accessed the service across the agency (see table-6.60). The rate of access is more in Pedabailu Mandal (75.27%) followed by G. Madugula (60.27%) and Dumbriguda Mandal (43.46%). Better access to this service is ensuring a better dawn for the future generations to come.

Table 6.60: Impact on SHGs on Nutrition Awareness for Pregnant Ladies and Lactating Mothers

Mandal	Same	Increased	Decreased	Deteriorated	Total
Dumbriguda	119	93	2	0	214
	(55.61)	(43.46)	(0.93)	(0.00)	(100.00)
G. Madugula	89	135	0	0	224
	(39.73)	(60.27)	(0.00)	(0.00)	(100.00)
Pedabailu	45	137	0	0	182
	(24.73)	(75.27)	(0.00)	(0.00)	(100.00)
Total	**253**	**365**	**2**	**0**	**620**
	(40.81)	**(58.87)**	**(0.32)**	**(0.00)**	**(100.00)**

Source: Survey Data

Tribal in general follows the natural method of menopause. The predominant contraception method among both tribal and rural areas was female sterilization, *i.e.,* tubectomy. The contraceptive service outreach were poor among scheduled tribe population and situation was not fair in rural areas also (G.R. Varma and A. Rohini, 2008). However, issues of family planning have also been taken up by the self help groups. Awareness programme are conducted in all the SHG and VO meetings in a scheduled manner. 53.39 per cent of SHG members reported to be aware about the benefit of the programme (see table-6.61) and have accessed the service.

Table 6.61: Impact of SHGs on Family Planning Awareness

Mandal	Same	Increased	Decreased	Deteriorated	Total
Dumbriguda	137	77	0	0	214
	(64.02)	(35.98)	(0.00)	(0.00)	(100.00)
G. Madugula	110	114	0	0	224
	(49.11)	(50.89)	(0.00)	(0.00)	(100.00)
Pedabailu	42	140	0	0	182
	(23.08)	(76.92)	(0.00)	(0.00)	(100.00)
Total	289	331	0	0	620
	(46.61)	(53.39)	(0.00)	(0.00)	(100.00)

Source: Survey Data

Girl Child Development (GCD) Programme has given priority by all the agencies promoting self help groups. The IKP, NGOs and other semi-formal institution has paid attention to enhance technical skill and education of tribal girls across the agency. Hence, loan for girl child education has been given top priority by the SHGs and the Banks. There are 65.52 per cent beneficiaries aware of the programme and got benefited under the scheme (see table-6.62). Larger number of beneficiaries (80.77%) benefited under the programme are from Pedabailu Mandal followed by Dumbriguda and G. Madugula.

Table 6.62: Impact of SHGs on Girl Child Development Programmes

Mandal	Same	Increased	Decreased	Deteriorated	Total
Dumbriguda	90	124	0	0	214
	(42.06)	(57.94)	(0.00)	(0.00)	(100.00)
G. Madugula	95	129	0	0	224
	(42.41)	(57.59)	(0.00)	(0.00)	(100.00)
Pedabailu	35	147	0	0	182
	(19.23)	(80.77)	(0.00)	(0.00)	(100.00)
Total	**220**	**400**	**0**	**0**	**620**
	(35.48)	**(64.52)**	**(0.00)**	**(0.00)**	**(100.000**

Source: Survey Data

Table 6.63: Impact of SHGs on Health Awareness Across ITDA Paderu

Mandal	Same	Increased	Decreased	Deteriorated	Total
Dumbriguda	129	85	0	0	214
	(60.28)	(39.72)	(0.00)	(0.00)	(100.00)
G. Madugula	98	126	0	0	224
	(43.75)	(56.25)	(0.00)	(0.00)	(100.00)
Pedabailu	53	129	0	0	182
	(29.12)	(70.88)	(0.00)	(0.00)	(100.000
Total	**280**	**340**	**0**	**0**	**620**
	(45.16)	**(54.84)**	**(0.00)**	**(0.00)**	**(100.00)**

Source: Survey Data

Tribals of ITDA Paderu are susceptible to many seasonal diseases. And health is a major concern for the government administration every year. It is understood that 'prevention is better than cure' and hence, government in coordination with SHGs have been organizing awareness programmes across the agency using local folk media *i.e.* kalajatha and borakatha. About 54.84 per cent self help group members reported aware of the precautionary measures to be adopted before onset of the seasons (see table-6.63). The intensity of awareness programme is high in Padabailu Mandal.

Table 6.64: Impact of SHGs on Women Involvement in Decision Making on Issues of Child

Mandal	Same	Increased	Decreased	Deteriorated	Total
Dumbriguda	83	131	0	0	214
	(38.79)	(61.21)	(0.00)	(0.00)	(100.00)
G. Madugula	94	130	0	0	224
	(41.96)	(58.04)	(0.00)	(0.00)	(100.00)
Pedabailu	36	146	0	0	182
	(19.78)	(80.22)	(0.00)	(0.00)	(100.00)
Total	**213**	**407**	**0**	**0**	**620**
	(34.35)	**(65.65)**	**(0.00)**	**(0.00)**	**(100.00)**

Source: Survey Data

Table 6.65: Impact of SHGs on Women Decision Making on Spending at HHs

Mandal	Same	Increased	Decreased	Deteriorated	Total
Dumbriguda	71	143	0	0	214
	(33.18)	(66.820	(0.00)	(0.00)	(100.00)
G. Madugula	101	123	0	0	224
	(45.09)	(54.91)	(0.00)	(0.00)	(100.00)
Pedabailu	25	157	0	0	182
	(13.74)	(86.26)	(0.00)	(0.00)	(100.00)
Total	**197**	**423**	**0**	**0**	**620**
	(31.77)	**(68.23)**	**(0.00)**	**(0.00)**	**(100.00)**

Source: Survey Data

Decision making at family level in general is considered to be the subject matter of male. Let it be income spending, marriage, child future and asset accumulation the decisions are to be taken by male unilaterally. Women do not have any say and are not consulted to share their views. SHGs across regions unilaterally attempted to change the trend. As a result, women have been imparted with capacity building training, exposure visits and are also provided with a large platform to raise their voices heard at public forums as well as in matters of household decision making. The entire effort over the years could alleviate 65.65 per cent of women SHG members voice honored and opinion sought in matters relating children's future and still a better per cent (68.23%) in the matter of HHs functioning (see table-6.64 and 6.65).

The base for women participation has effectively created by being the member of the SHG. The trainings and exposures have equipped women with feathers and wisdom to judge the issues of the people and make them comfortable in dealing with issues along with men. They are participating in village development issues and taking forward the issues of community to the notice of government officials and also placing them at appropriate forums. About 62.86 per cent SHG members are involved in the activity (see table-6.66) and the process of solving larger village issues is gaining momentum. The level of participation is high in Pedabailu Mandal.

Table 6.66: Impact of SHGs on Women Participation in Development Programmes

Mandal	Same	Increased	Decreased	Deteriorated	Total
Dumbriguda	81	133	0	0	214
	(37.85)	(62.15)	(0.00)	(0.00)	(100.00)
G. Madugula	120	114	0	0	234
	(51.28)	(48.72)	(0.00)	(0.00)	(100.00)
Pedabailu	33	149	0	0	182
	(18.13)	(81.87)	(0.00)	(0.00)	(100.00)
Total	**234**	**396**	**0**	**0**	**630**
	(37.14)	**(62.86)**	**(0.00)**	**(0.00)**	**(100.00)**

Source: Survey Data

Does starvation death reported in Paderu agency? Are people getting three square meals a day? The scenario of food basket has drastically changed over the years. Subsidized ration and NREGS has redefined the living of tribal people, as also the condition of the normal people. Self help groups have been working as the point of monitoring in a given area of work. Besides, various avenues of IGAs have added significantly to the food basket of the tribal families. About 52.42 per cent self help group members reported as food secure while 47.58 per cent are living with the same condition as had been before joining the group (see table-6.67). In fact, children rations have been taken care either at schools or at AWCs.

Table 6.67: Impact of SHGs on Food Security Across Mandals

Mandal	Same	Increased	Decreased	Deteriorated	Total
Dumbriguda	140	74	0	0	214
	(65.42)	(34.58)	(0.00)	(0.00)	(100.00)
G. Madugula	109	115	0	0	224
	(48.66)	(51.34)	(0.00)	(0.00)	(100.00)
Pedabailu	46	136	0	0	182
	(25.27)	(74.73)	(0.00)	(0.00)	(100.00)
Total	295	325	0	0	620
	(47.58)	(52.42)	(0.00)	(0.00)	(100.00)

Source: Survey Data

Assured food security is the end result of effective employment of the tribal either partly or fully. Training, fund support and marketing linkages has given a moral boost to SHG members to thrive on developing enterprises on their own. The issues of employment are partially addressed. About 68.87 per cent of self help group members reported to get positive employment benefit. SHG has given a platform to enhance their family income (see table-6.68).

Table 6.68: Impact of SHGs on Provision of Employment (Partial and Full)

Mandal	Same	Increased	Decreased	Deteriorated	Total
Dumbriguda	64	150	0	0	214
	(29.91)	(70.09)	(0.00)	(0.00)	(100.00)
G. Madugula	107	117	0	0	224
	(47.77)	(52.23)	(0.00)	(0.00)	(100.00)
Pedabailu	22	160	0	0	182
	12.09)	(87.91)	(0.00)	(0.00)	(100.000
Total	**193**	**427**	**0**	**0**	**620**
	(31.13)	**(68.87)**	**(0.00)**	**(0.00)**	**(100.00)**

Source: Survey Data

Self-employment as well as timely supply of financial support to families in need has reduced dependency on moneylenders. There are many formal banks extending financial services across the mandals of Paderu agency area. One NBFI working in Araku valley Mandal is taking care of financial need of two Mandals *i.e.* Dumbriguda and Araku valley. About 61.94 per cent of self help group members across the mandals reported free from clutches of moneylender-cum-traders (see table-6.69). They are exclusively depending on formal financial institutions for all requirements. However, there is still a significant members of SHG depends on moneylender in Dumbriguda Mandal.

Table 6.69: Impact of SHGs on Dependency on Moneylender

Mandal	Same	Increased	Decreased	Deteriorated	Total
Dumbriguda	109	105	0	0	214
	(50.93)	(49.07)	(0.00)	(0.00)	(100.00)
G. Madugula	97	124	3	0	224
	(43.30)	(55.360	(1.34)	(0.00)	(100.00)
Pedabailu	27	155	0	0	182
	(14.84)	(85.16)	(0.00)	(0.000	(100.00)
Total	**233**	**384**	**3**	**0**	**620**
	(37.58)	**(61.94)**	**(0.48)**	**(0.000**	**(100.00)**

Source: Survey Data

Table 6.70: Impact of SHGs on Providing Hands on Experience on Organization Management

Mandal	Same	Increased	Decreased	Deteriorated	Total
Dumbriguda	92	122	0	0	214
	(42.99)	(57.01)	(0.00)	(0.00)	(100.00)
G. Madugula	98	126	0	0	224
	(43.75)	(56.25)	(0.00)	(0.00)	(100.00)
Pedabailu	27	155	0	0	182
	(14.84)	(85.16)	(0.00)	(0.00)	(100.00)
Total	**217**	**403**	**0**	**0**	**620**
	(35.00)	**(65.00)**	**(0.00)**	**(0.00)**	**(100.00)**

Source: Survey Data

The sole effort of educating tribal women on organization management and provision of hands on experience is a fruitful exercise. About 65 per cent of tribal women are capable of managing the organization on their own with support of IKP (see table-6.70). They learnt the art of managing self help groups. Let it be bookkeeping, collection of savings, depositing savings, prioritization of community issues, setting up of agenda for the meeting, disbursal of loans, communication in VO etc are all learnt. It is also opined that about 64.52 per cent of tribal SHGs are aware of bank transactions (see table-6.71).

Table 6.71: Impact of SHGs on Bank and Finance Management Activities

Mandal	Same	Increased	Decreased	Deteriorated	Total
Dumbriguda	91	123	0	0	214
	(42.52)	(57.48)	(0.00)	(0.00)	(100.00)
G. Madugula	109	115	0	0	224
	(48.66)	(51.34)	(0.00)	(0.00)	(100.00)
Pedabailu	20	162	0	0	182
	(10.99)	(89.01)	(0.00)	(0.00)	(100.00)
Total	**220**	**400**	**0**	**0**	**620**
	(35.48)	**(64.52)**	**(0.000**	**(0.00)**	**(100.00)**

Source: Survey Data

Transactions pertaining to deposit in the bank, updating bank passbook, withdrawal from the bank, approaching bank for funding and approaching IKP for sanctioning of matching grant are the functions independently taken up by more than average SHGs across the studied Mandals of ITDA.

Are these not symptoms of self-sufficiency in managing SHGs. This has given ability to cope with all circumstances and courage to face the situations. About 58.39 per cent members of the self help group are provided with leadership quality to manage the institutions and the community of their living (see table-6.72).

Table 6.72: Impact of SHGs on Leadership Quality

Mandal	Same	Increased	Decreased	Deteriorated	Total
Dumbriguda	89	125	0	0	214
	(41.59)	(58.41)	(0.00)	(0.00)	100.00
G. Madugula	144	80	0	0	224
	(64.29)	(35.71)	(0.00)	(0.00)	100.00
Pedabailu	25	157	0	0	182
	(13.74)	(86.26)	(0.00)	(0.00)	100.00
Total	**258**	**362**	**0**	**0**	**620**
	(41.61)	**(58.39)**	**(0.00)**	**(0.00)**	**(100.00)**

Source: Survey Data

The result of this has been visualized in participation of 64.19 per cent of women self help members in Gram Panchayat (GP) elections and victory of 68.39 per cent among them (see table-6.73 and 6.74). Better women participation in GP election is reported in Pedabailu Mandal of Paderu agency area.

Table 6.73: Influence of SHGs on Women Participation in Gram Panchayat Elections

Mandal	Same	Increased	Decreased	Deteriorated	Total
Dumbriguda	67	147	0	0	214
	(31.31)	(68.69)	(0.00)	(0.00)	(100.00)
G. Madugula	132	87	5	0	224
	(58.93)	(38.84)	(2.23)	(0.00)	(100.00)
Pedabailu	18	164	0	0	182
	(9.89)	(90.11)	(0.00)	(0.00)	(100.00)
Total	**217**	**398**	**5**	**0**	**620**
	(35.00)	**(64.19)**	**(0.81)**	**(0.00)**	**(100.00)**

Source: Survey Data

Table 6.74: Women Won in Gram Panchayat Elections

Mandal	Same	Increased	Decreased	Deteriorated	Total
Dumbriguda	41	173	0	0	214
	(19.16)	(80.84)	(0.00)	(0.00)	100.00
G. Madugula	137	87	0	0	224
	(61.16)	(38.84)	(0.00)	(0.00)	100.00
Pedabailu	18	164	0	0	182
	(9.89)	(90.11)	(0.00)	(0.00)	(100.00)
Total	**196**	**424**	**0**	**0**	**620**
	(31.61)	**(68.39)**	**(0.00)**	**(0.00)**	**(100.00)**

Source: Survey Data

Finally, it has enabled 65.81 per cent tribal women to minimize risk at household, community and society level. This is a common trend all across the studied Mandals (see table-6.75).

All SHG members are imbibed with the skill and potentiality to withstand to risk of any kind and this may be due to the versatile livelihood occupation opened up for them. There are 58.23 per cent members of self help groups reported accessing changing livelihood occupation. Hence, diversity in living has already crops in the life of tribals of Paderu. The system and procedure incorporated is self-perpetuating. The impact of this has already been proven in terms of acclaiming 'best village recognition' in some places and liquor free villages too. The tribal of Paderu are self-sufficient and would be self sufficient in the years ahead too.

Table 6.75: Impact of SHGs on Risk Management Capability

Mandal	Same	Increased	Decreased	Deteriorated	Total
Dumbriguda	66	148	0	0	214
	(30.84)	(69.16)	(0.00)	(0.00)	100.00
G. Madugula	136	88	0	0	224
	(60.71)	(39.29)	(0.00)	(0.00)	100.00
Pedabailu	10	172	0	0	182
	(5.49)	(94.51)	(0.00)	(0.00)	100.00
Total	**212**	**408**	**0**	**0**	**620**
	(34.19)	**(65.81)**	**(0.00)**	**(0.00)**	**(100.00)**

Source: Survey Data

Table 6.76: Impact of SHGs on Changing Livelihood Occupation

Mandal	Same	Increased	Decreased	Deteriorated	Total
Dumbriguda	94	120	0	0	214
	(43.93)	(56.07)	(0.00)	(0.00)	(100.00)
G. Madugula	151	73	0	0	224
	(67.41)	(32.59)	(0.00)	(0.00)	(100.00)
Pedabailu	14	168	0	0	182
	(7.69)	(92.31)	(0.00)	(0.00)	(100.00)
Total	**259**	**361**	**0**	**0**	**620**
	(41.77)	**(58.23)**	**(0.00)**	**(0.00)**	**(100.00)**

Source: Survey Data

6.8. Microfinance Vis-à-Vis Institutional Networking and Convergence

Microfinance institutions are institutions of the people, for the people and by the people. It is an institution to provide all inclusive platform for development of individual households and also the society altogether. In other words, development as well as growth of the rural economy is nesting on pace of growth of microfinance institutions operating in an area. In fact, microfinance institutions though address issues of finance of tribal's primarily, they also works in tandem with already existing institutions to resolve development issues too.

In a state like Andhra Pradesh, there are 9-10 different village level statutory institutions promoted following the Panchayat Raj principles as well as the state Government directives. All these institutions are assigned with some designed roles and responsibilities to serve the fellow community members directly or enabling Government machinery/departments to execute the specific project and programmes planned for the targeted community. The institutional upload has further been aggravated with formation of new sets of institutions by the voluntary sectors. Hence, there are broadly 'Three Major Sets of Institutions' found in each habitations *i.e.* Institutions promoted by Government, Institutions Promoted by Voluntary Organizations and the traditional Village level Institutions (Barik, 2012).

Baring a few, women participation in village level institutions is ensured because of the statutory law of the state where the participation is purely confined to putting a thumb impression and/or showing the presence on paper. However, institutions promoted exclusively with women are of paramount importance to the economic and social life of the tribal women. Hence, it is hypothesized that;

Hypothesis-VIII: SHGs in agency area are efficient, viable and well-coordinated with all other institutions of the community to deal larger community issues.

Towards the effort of understanding the institutional functioning and its relation with all other institutions in the village, an institutional mapping and ranking exercise had been under taken in the villages of tribals across the three regions. As evidenced from the ranking exercise (see table-6.77), the economic contribution of Self-help Groups (SHGs) is visible. It has been assigned 1st rank across the Mandals of the agency area.

The Education Committees formed to motivate parents sending children to school regularly stands second in rank in comparison to all other committees. The Mother Committee and Village Health and Sanitation Committee are ranked three. Almost four committees like Panchayat Raj Institutions, Forest Protection Committee (FPC), Village Development Programme (VDP) and Community Action Group (CAG) are ranked four on the basis of their contribution to the life and livelihood of tribal communities. Subsequently, the Agriculture Development Committee (ADC) and Land and Natural Resource Management Committee are ranked five and six respectably.

Self help groups interact with other institutions spelt out in table 8.77 on various issues and areas of their specialization. SHGs are therefore termed as the link-pin among the institutions. However, the 2 value is not significant at 10 per cent level of significance to accept the hypothesis that institutions working in the villages are constantly in dialogue on issues of village, community and the tribal economy at large (see table-6.78).

Table 6.77: Institution and People's Involvement

Sl. No.	Institutions	Peoples Participation	Contribution	Ranking
1	Panchayat Raj Institutions	Men and specified percentage of Women	Participation for the sake of formality	4
2	Forest Protection Committee (FPC)	Membership decided arbitrarily	No visible contribution	4
3	Village Development Committee-VDP	Men along with specified percentage of Women	Nothing to share	4
4	Mother Committee	Women Members Only	Looking after pregnant ladies and lactating mother for accessing Anganwadi services	3
5	Village Health and Sanitation Committee	Men and specified percentage of Women	Rainy season village cleanliness activities	3
6	Education Committee	Men and specified percentage of Women	Monitoring students attendance in schools	2
7	Self-help Groups	Exclusively Women with focus on POP	Regular monthly meeting for Income Generation Programme, MCP, Food Security (Grain Bank management)and other village development issues	1
8	Agriculture Development Committee	Men and specified percentage of Women	No contribution	5
9	Land and NR Management Committee	Men and specified percentage of Women	No contribution	6
10	Community Action Group-CAG	Men and Women	Overall Village Development	4

Source: Participatory Research Appraisal (Institutional Ranking, across all Mandals of ITDA, Paderu, Visakhapatnam District).

Table 6.78: Members of SHGs interact with Other Institutions

	Interact with others	Does not interact with others	Total	χ^2 value
Dumbriguda	75	139	214	
	(35.05)	(64.95)	(100.00)	0.00012
G. Madugula	90	134	224	
	(40.18)	(59.82)	(100.00)	0.00002
Pedabailu	37	145	182	
	(20.33)	(79.67)	(100.00)	0.02628
Total	**202**	**418**	**620**	
	(32.58)	**(67.42)**	**(100.00)**	**0.00032**

Source: Survey Data

χ^2 values calculated at 10 per cent level of significance

Table 6.79: Consolidation of functional issues of SHGs

Mandals	Issues
Dumbriguda	1.More educated people are required to run the SHGs
	2.Financial Development Programme to update skill
	3.All members in the group are not literate
	4.Loan repayment is poor
	5.Members are learning SHG functioning and have no prior experience
	6.No repayment in the time
	7. Seasonal employment inside the village is bleak. This is leading to repayment problems
	8. Lack of knowledge for gainful employment
	9. Lack of perfect jobs for tribal
G. Madugula	1. Non repayment of loan in time
	2. Group members have partial knowledge on SHG functioning
	3. Lack of adequate educated people within the group
	4. Book keeping is improper and often half done
	5. Lack of adequate knowledge on alternative income earning programmes by all
	6.Poor repayment
Pedabailu	1. Inadequate training and limited training provision for a few
	2. Poor participation of members in meetings during agriculture season
	3. Lack of educated people to manage the group running activity
	4. Poor loan refund

Source: Survey Data

The interaction with groups of other villages and other institutions in the same village is to gain knowledge on institutional functioning, know issues of village, status of execution of Government schemes and programmes, to make ourselves aware on issues of SHGs, systematic management of group and use of group savings etc. There are several issues still to be addressed. The benefits provided cover a good number of beneficiaries of SHG. There are some inherent issues (see table-6.79) to be dealt by SHGs in the near future.

6.9. Conclusion

Self help groups are playing all pervasive roles in agency area of Paderu, Visakhapatnam district. The catalystic role in achieving visible and tangible changes in the life and livelihood of tribal is amply visible. SHG intervention in village activities has added a new definition and meaning to their living. The impact of SHG is reflected in promotion of educational institutions and access to education, access to health services, modalities of service delivery, reduction of poverty and un-employments, imbibing tribal youths with skills and hands on experience on IGAs, promotion of women participation in HHs, community and village development, leadership promotions, ensuring food security, ensuing visible income and provisioning self-esteem to women. Hence, it can be concluded with confidence that SHG operation in agency area of ITDA is very much meaningful.

Notes

1. The coefficient of correlation is calculated using the total number of self help groups (SHGs) working in the respective mandals under study against the beneficiaries survey report of access to different sources of portable drinking water of the respective mandals. 779 SHGs of Dumbriguda is examined against 49 respondents, 871 SHGs of G. Madugula is cross examined with 33 respondents and 906 SHGs has been examined against 26 beneficiaries report.
2. India's per capita income (nominal) is $ 1219, ranked 142nd in the world, while its per capita purchasing power parity (PPP) of US $3,608 is ranked 129th. It is estimated that India's Per Capita Income will register an average growth rate of 13% during 2011-20 so as to reach $ 4,200 by 2020. India's per capita purchasing power parity (PPP) will be at $ 12,800 in the year 2020. States of India have large disparities. One of the critical problems facing India's economy is the sharp and growing regional variations among India's different states and territories in terms of per capita income, poverty, availability of infrastructure and socio-economic development.
3. Social Capital is an accumulation of various types of social, psychological, cultural, cognitive, institutional and related assets that increase the amount of mutually beneficial cooperative behavior. Social Capital refers in general to

the glue that holds groups and societies together through bonds of shared values, norms and institutions.

4. Link Workers are employees of IKP and are responsible for effective functioning of self help groups of their area of operation. They help SHGs in formation, maintenance of records and day-to-day functioning.

5. Growth Rate= $\frac{\text{Current Year-Previous Year} \times 100}{\text{Current Year}}$

6. **Employment:** The international definition of employment adopted by the 13th ICLS is formulated in terms of the labour force framework, *i.e.* with respect to a short reference period. The definition distinguishes between paid employment (employees including apprentices or trainees and members of the armed forces) and self-employment (employers, own-account workers including producers of goods for own final use, members of producers' co-operatives, and contributing family workers). It provides separate criteria for the measurement of these two types of employment. According to the definition, the "employed" comprise all persons above the age specified for measuring the economically active population

   (e.g. 15 years) who, during a specified short period of either one week or one day, were in the following categories:

   (a) paid employment:

   (a1) at work: persons who, during the reference period, performed some work (*i.e.* at least one hour) for wage or salary, in cash or in kind;

   (a2) with a job but not at work: persons who, having already worked in their present job, were temporarily not at work during the reference period and had a formal attachment to their job;

   (b) self-employment:

   (b1) at work: persons who, during the reference period, performed some work (*i.e.* at least one hour) for profit or family gain, in cash or in kind;

   (b2) with an enterprise but not at work: persons with an enterprise (which may be a business enterprise, a farm or a service undertaking) who were temporarily not at work during the reference period for any specific reason.

7. **Underemployment:** The international standards for the statistical measurement of underemployment were revised in 1998 by the 16th ICLS. The 'Resolution concerning the measurement of underemployment and inadequate employment situations' adopted by the 16th ICLS replaces the 'Resolution concerning measurement and analysis of underemployment and underutilization of manpower' of the 11th ICLS (1966) and paragraphs 14-20 and 21(5) of the 'Resolution concerning statistics of the economically active population, employment, unemployment and underemployment' of the 13th ICLS (1982).

8. **Unemployment:** The international standard definition of unemployment adopted by the 13th ICLS is based on three criteria, which have to be met

simultaneously. According to this definition, the unemployed comprise all persons above the age specified for measuring the economically active population who during the reference period were:

(a) "without work", *i.e.* were not in paid employment or self-employment as defined by the international definition of employment;

(b) "currently available for work", *i.e.* were available for paid employment or self-employment during the reference period; and

(c) "Seeking work", *i.e.* had taken specific steps in a specified recent period to seek paid employment or self-employment.

9. This is type of unemployment unusually found in overpopulated underdeveloped areas, and more particularly in agricultural sector. Due to use of family labour in agriculture (which is not paid wages in the usual sense), sometimes the number of persons working on land may be far in excess of those who are really required to produce the given volume of output. Thus, some of these workers are surplus as they do not add anything to the volume of production. Their marginal productivity, *i.e.*, the addition to production made by an additional unit of labour, is zero from the point of view of the society such person is unemployed because he does not contribute to output. Hence, his unemployment remains disguised.

10. **Seven Steps for preparation of MCPs**: ***i.*** Sensitization of members on Loan Policies and Procedures, ***ii.*** Group members discuss and identify themselves their socio-economic status of each member, ***iii.*** Members rank themselves based on the poverty status, ***iv.*** Members discuss and identify the credit needs with their family, ***v.*** Group shall collectively examine all aspects and facts of each members credit needs, ***vi.*** Consolidation of needs and setting priorities, Loan needs prioritized, finalized and appraised, Prioritization based on seasonality of the activity besides the poverty ranking, Priority ranking should take into account emergencies, and ***vii.*** Forecasting and analyzing funds flow and setting rotation-Group to finance each member based on rotation plan by utilizing internal funds first, Group may mobilize funds from internal sources, Village Organization, Bank, and Line Departments.

11. All the skill development programmes are executed from the training centres located at ITDA head quarter at Paderu. IKP has coordinated for the programme and directly run the training centres. Skill development programme is basically a residential programme and tribal youths used to be away for couple of moth for pursuing the training.

12. Who are Rural Entrepreneurs? They are (i) the rural poor, particularly in disadvantaged countries or marginal areas; (ii) all segments of the rural population in more open economies, particularly non-agricultural entrepreneurs in a situation of increasing diversification; and (iii) commodity producers, processors and traders, with a tremendous growth potential if the potential for value-added is fully mobilized.

13. Poverty is a condition where people's basic needs for food, clothing, and shelter are not being met. Poverty is generally of two types: (1) Absolute

poverty is synonymous with destitution and occurs when people cannot obtain adequate resources (measured in terms of calories or nutrition) to support a minimum level of physical health. Absolute poverty means about the same everywhere, and can be eradicated as demonstrated by some countries. (2) Relative poverty occurs when people do not enjoy a certain minimum level of living standards as determined by a government (and enjoyed by the bulk of the population) that vary from country to country, sometimes within the same country. Relative poverty occurs everywhere, is said to be increasing, and may never be eradicated. Read more: http://www.business dictionary.com/definition/poverty.html

References

1. Surender, Kumari, S., and Sehrawat, R.K. (2011): "Can self-help groups generate employment opportunity for rural poor?", European Journal of Social Sciences (EJSS).
2. "Role of Micro Finance in Economic Development", Wall Street Microfinance: Models, Policies and Experiences, http://www.wallstreetmicrofinance.org/microfinance-guide/role-of-micro-finance-in-the-economic-development.html, 2012.
3. Lewis, J. (2009): "Microfinance: Models, Policies and Experiences: An Economic Development Failure?", http://skollworldforum.org
4. "Report for Selected Countries and Subjects". Imf.org. 2006-09-14. Retrieved 2010-08-12.
5. "Economic: Adam Smith Theory of Value", Economictheories.org. Retrieved 2012-03-13.
6. Sinha Frances (2005): " Access, Use and Contribution of Microfinance: Models, Policies and Experiences in India-findings from National Study, Economic and Political Weekly, Vol. 40, No.17, (April.23-29), pp. 1714-1719.
7. Zakhir (2011): "Management of Self Help Groups", State Institute of Rural Development (SIRD), Assam, p. 48.
8. "The Impact of Microfinance: Models, Policies and Experiences on Women and economic Development- A Client Study", Women's World Banking, 2009.
9. Pantoja Enrique(2002): "Microfinance: Models, Policies and Experiences and Disaster Risk Management Experiences and Lessons Learned", Provention Consortium, The World Bank.
10. Moses, V.J.R. E (2011): "Women empowerment through SHGs: A micro study", International Referred Research Journal, Vol. II, Issue 16, January 2011, p. 25.
11. Shah, Chaitanya (2011): "Microfinance: Models, Policies and Experiences-Important Role in Economic Growth", www. Smeworld.org
12. Seibel. H.D (2007): "The Role of Microfinance: Models, Policies and Experiences in Rural Microenterprise Development", Syngenta Foundation for Sustainable Agriculture, Switzerland, www.syngentafoundation.org.

13. Bundela, A, Dubey. K and Pande. S (2013): "Role of Self Help Groups in Rural Employments of India: A Case Study of Swarn Jayanti Gram Swarojgar Yojna", IOSR Journal Of Humanities And Social Science (IOSR-JHSS) e-ISSN: 2279-0837, p-ISSN: 2279-0845.Volume 7, Issue 1 (Jan. - Feb. 2013), PP 38-41 www.Iosrjournals.Org.

14. Austinice (2010): "Impact of Micro-finance on Economic Development", www.studymode.com.

15. Varma, G.R and Rohini, A (2008): "Perceptions of Community on Post-Contraceptive Care by Primary Health Care System in Tribal and Rural Areas of Visakhapatnam District of Andhra Pradesh, Ethno-Med., 2(1): pp. 29-32

16. Barik, S.K. (2012): "Micro Credit for Empowerment of Micro People: A Study of SHGs in Paderu Integrated Tribal Development Agency (ITDA), Andhra Pradesh", Elixir Fin. Mgmt. 48 (2012) 9437-9441.

Chapter 7

The Road Ahead for MFIs

7.0 Introduction

Microfinance is the buzz word across the globe and is being looked as one of the major tools for poverty alleviation. Though there are different models of micro financing followed across the countries, the ultimate underlying principle is how to empower marginalized communities to get out of poverty. Breaking the 'vicious circle of poverty' is the main objective behind the intervention. With this pretext, self help group financing is evolving as a peoples' friendly initiative to work with people for economic upgradation of people.

Mandated to work with the women group of the rural and tribal folk, it is accepted as a better means of financing for income generation programmes in rural and semi-urban areas. Each self help group is formed with women participation of the same locality and people of same and/or similar financial background and with similar anticipation of stabilization of income base for the family. The interest is no more of the individual interest rather it is the interest of the community to supplement and complement the initiative to reach the individual target with a '*comprehensive community effort*'.

7.1 Summary of Findings of experiments

The current section summarizes and consolidates the findings of the entire research.

(i) Global poverty can be cut in half by 2015 and eliminated by 2025 if the World's richest countries including the United States, Japan

and Germany more than double aid to the poorest countries (koffi Annan).

(ii) The World Bank puts the number of poor around the world at 1.4 billion. It says this is larger than previously thought. But, it notes new estimates confirm that between 1990 and 2005, the number of people living in extreme poverty has fallen from 1.8 to 1.4 billion. This indicates the 1990 global poverty rate is likely to be halved by 2015, which would put the world on track to achieving the Millennium Development Goals.

(iii) The scenario of development is still worse in the third world countries. Riddled with extreme poverty and coupled with the political imbalances, sectoral development is lagging behind adequately to help enable a balanced development of sectors and causing for in-appropriate regional development. The debate on poverty is not for the developed and developing nations of the world, rather poverty in general.

(iv) Poverty is the prime cause for lopsided development of the world economy in general and economies of the third world countries in particular. Low level of literacy, poor health facilities, poor infrastructural facilities, declining environment and ruined natural resources, abundant supply of manual labourer, etc is thereby becoming the characteristics of the nations riddled with poverty.

(v) The provision and effort of developed nations with vigorous internal changes in governments developmental policies to look into the *in situ* development devising different need based strategies on the basis of region, gender, natural resource base, manpower etc are contributing in a way to address issues of poverty in urban and rural areas.

(vi) Development of India is *prima facie* seen in terms of the development of rural villages. The common saying of India lives in villages and "India shines" when the villages of the country prospers. Through the five years plan, special attention is being given on the development of rural villages. Time to time necessary changes had been brought up in the development strategies to deal with diversified and complex issues peeping in with the passage of time. A close analysis of the poverty alleviation programmes in India clearly falls under two categories *i.e.* (1) interventions where the poor households/persons are direct beneficiaries, and (2) area level interventions where geographical areas/locations with high incidence of unemployment and poverty are targeted for development.

(vii) The growth strategy followed in 1960s and 1970s had presumed that a higher rate of economic growth, through "Trickle Down" effect, would enhance the standard of living of the poor. But the "Trickle Down" concept has failed to precipitate. Therefore, during the 1970s the Government of India had initiated Anti-Poverty Programmes. Since the Sixth Plan (1980-85), a more distinct and direct approach was adopted.

(viii) "The 'direct' approach to poverty reduction emphasized that it is essential to directly provide the poor with adequate purchasing power, other assets or access to food grains at subsidised prices to meet their minimum consumption requirement"

(ix) The Anti-Poverty Programmes (APP henceforth) could be broadly classified into two groups: Rural Wage Employment Schemes and Rural Self- Employment Schemes. We have focussed here on Self-employment programmes only.

(x) In recent times, special emphasis is laid on social mobilization through Self-help Groups, especially DWCRA groups for creating access to development schemes and programmes and bringing together the hitherto socially marginalized, economically weaker and politically excluded sections of the rural poor. This approach is built on the fundamental principles of promoting the cause of the poor as partners in development and thereby ensuring the benefits of participation to the poor because the market oriented growth will not provide opportunities to the poor to participate in it and drive the benefits.

(xi) The microfinance movement emerged in the 1970s[1]. Pioneers of microfinance, like Grameen and Opportunity International, were driven by a view that poor people are full of inherent potential: capable of pulling themselves out of poverty, if only they could get access to capital that banks would not provide. Small loans (micro credit) were devised to meet this need. Microfinance institutions at present serve an estimated 120 million clients in the world.

(xii) The microfinance sector has grown over time with more and different types of actors coming involved, more and more geographic regions around the world being serviced, new types of products and services being developed, and with new ideas and technologies to support it.

(xiii) The global picture regarding microfinance outreach is quite impressive. From a mere 7.6 million poorest families in 1997, the

Microcredit Summit Campaign reported an outreach of more than 113 million clients by December 31, 2005. It included about 82 million poorest families when they started with the programme.

(xiv) In April 2005, the Microcredit Summit Campaign announced the extension of the Campaign up to 2015 at the Summit in Santiago, Chile. Phase II of the campaign from 2006-2015 has two goals. The first goal is to ensure that 175 million of the world's poorest families, especially the women receive credit for self-employment and other financial and business services by the end of 2015. The second goal of the campaign is to ensure that 100 million of the world's poorest families move from below US$ 1 a day adjusted for Purchasing Power Parity (PPP) to above US $ 1 a day adjusted for PPP, by the end of 2015. With an average of 5 people per family, it can be estimated that 500 million poor would rise above US$ 1 a day, which would imply the near accomplishment of the Millennium Development Goal of halving absolute poverty.

(xv) Indian history withstands to the fact of rule of kings and their kingdom. Reference of Chanakya is being quoted as a role model of management of economics and management. Through ages, it is being mentioned that poor people use to develop their own mechanism of survival and manage their crises. The multiple form of coping is seen in the formation of local level organizations/institutions to extend help and assistance in all forms to waive off the difficulties and grain banks, seed banks, etc are the symbols.

(xvi) Microfinance in India started in the early 1980s with small efforts at forming informal self-help groups (SHG) to provide access to much-needed savings and credit services. From this small beginning, the microfinance sector has grown significantly in the past decades. National bodies like the Small Industries Development Bank of India (SIDBI) and the National Bank for Agriculture and Rural Development (NABARD) are devoting significant time and financial resources to micro finance. The strength of the microfinance organizations (MFOs) in India is in the diversity of approaches and forms that have evolved over time.

(xvii) Accordingly, in February 1987, it was decided that a study team led by NABARD and comprising of representatives from various financial institutions, should be constituted to undertake the survey. The survey was undertaken in September 1987 and the

report discussed at the 18th Executive Committee Session and 10th Foundation Anniversary of APRACA held at New Delhi in November 1987. This survey report laid the foundation of the SHG-bank linkage programme in India launched as a pilot project in 1992. The pilot project was able to link 225 SHGs to the banks as against the target of 500 by the end of March 1993. The Self-help group and bank linkage programme operation in the country gave birth to three different unique models of financial delivery

(xviii) India has 400 million people spread across more than six million villages who are in need of micro-financing. The organized financial sector caters to the need of only 20 million people. Because of the lack of statistical validity of data sets, there are few reliable indications on the reach and demand of microfinance. However, the demand for microfinance in India is estimated to be around ₹ 300 billion. All this suggests that there is a huge unmet gap between demand and supply.

(xix) The wholesome picture of the country in terms of serving the women for conferring economic empowerment and self-sufficiency through extension of micro credit for undertaking entrepreneurial activities is very fragile. Moreover, more advanced states, as has been revealed from the statistics, have increased accessed to the designated fund for women empowerment of the country. The cumulative fund devolvement for women empowerment and reliability of the Rastriya Mahila Kosh of Government of India during 1993-2007 reveals that the state of Andhra Pradesh has comparatively greater access to the fund.

(xx) The genesis of SHGs has been initiated by NGOs in late 1980s in Andhra Pradesh. However, the movement for women's savings and credit in the state have really began in 1995, in Nellore district as "*podupulakshmi*" movement. This has led to large-scale promotion of SHGs in every district of AP. The *podupulakshmi* movement started with the slogan of "save a rupee a day".

(xxi) In the year 2000, DWCRA became a flagship program of Government of AP to implement poverty reduction programs. State Government introduced schemes like Deepam (subsidized gas connections for DWCRA women) and State funded Revolving Fund, in addition to the Central Government Schemes to channelize fund for poverty reduction.

(xxii) During 1996-2001, UNDP funded Poverty Alleviation Program was implemented in Mahabubnagar, Kurnool and Anantapur districts with a total outreach of 3,200 SHGs. In 1999, CARE began the implementation of a DFID funded project called Credit

and Savings for Household Enterprise (CASHE) to support the emergence of quality SHGs through NGOs and Government. Instances of competition among NGOs, DRDA and UNDP were also noticed.

(xxiii) In June 2000, World Bank funded District Poverty Initiatives Project (DPIP) was initiated in six poorest districts of AP based on the UNDP experience. In 2002, the project was scaled up to all the 22 rural districts of AP with funding support from World Bank and the project is called AP Rural Poverty Reduction Project.

(xxiv) There are 8752540 members organized into 699056 SHGs in 33700 Village Organizations (VOs) and 1083 Mandal Samakhyas (MSs). In addition, there are 126 Mandal Vikalangula Sangams, 17 Chenchu Mandal Samakhyas and 15 Fishermen Mandal Samakhyas. Zilla Samakhyas have come into existence in all the 22 districts. The SHGs in A.P. constitute around 25 per cent of all 2799850 SHGs in India.

(xxv) Following the approach of SHG-BL, ₹ 5300090 have been mobilized for 4105309 SHGs across the districts of A.P.

(xxvi) The district-wise self-help group and bank linkage scenario across the districts of the Andhra Pradesh state reveals that districts like East Godavari, West Godavari and Nallagonda records high in terms of formation of self-help groups and linking them to the banks. Districts like Visakhapatnam occupy 10th positions in ascending order both in terms of formation of self-help groups as well as establishing linkage with banks.

(xxvii) Tribals of Andhra Pradesh are primarily concentrated in 10 districts, where as very few households are found settled across the districts of the state. In order to deal with the integrated development of the tribal in the state, 10 Integrated Tribal Development Agencies have been formed in districts like Srikalulam, Vizianagaram, Visakhapatnam, East Godavari, West Godavari, Khammam, Warangal, Adilabad, Kurnool and Nellore.

(xxviii) Women's social position in different tribal communities varies. They are still the first line in the race of family work followed by men. The characteristics of low level of literacy, remote habitations, poor road and network, non-availability and dysfunctioning institutions, different dialect, ill-health is creating the visible gap of service delivery and bringing tribals to the mainstay of living. Low per capita is causing poor health, food inadequacy, and unwillingness to go for education and thereby is persistently ruining the economic condition of the tribal.

(xxix) The PoP study in the Visakha agency area shows that almost 48.49 percentage tribal households in the Paderu Agency Area falls under the category of 'Poorest of the Poor' and similarly 45.23 per cent household are Poor, and a just 6.27 per cent household are 'Above the Poverty Line'.

(xxx) With the objective of ensuring micro income, employment and food, it has been attempted by MACs as well as by Indira Kranti Patham (IKP) to ensure 100 per cent coverage of POP household as well as poor under the SHG led institutional coverage.

(xxxi) Self-help groups are generally promoted by the IKP and some of them are also formed by the Non-Governmental Organisations (NGOs) working in the agency area. The microfinance following the SHG conduits are generally executed following the hierarchy *i.e.* ZMS, MMS, VO and SHGs.

(xxxii) There are multiplicity of institutions for microfinance viz. Banks, NGOs, MACSs (AASSAV) as well as state run IKP are the players in an agency like Paderu.

(xxxiii) It is pertinent to note that all three major models of Micro Finance[22] Model-I: Self-help Groups (SHGs) formed and financed by Banks is seen operating in the state of Andhra Pradesh as one of the instrument for Poverty alleviation and women empowerment. It is also observed the NBFIs are also playing role in circulating capital for micro-enterprises. The study, therefore, intends to assess the overall visible and invisible impacts of the microfinance led poverty alleviation programmes in the agency of Visakhapatnam District.

(xxxiv) The study is particularly intending to capture the developmental impact of micro- financing operation of multiple agencies in tribals in agency areas of Andhra Pradesh and especially the tribals of Visakha Agency Area through the following research questions. Each question of the study is nothing but represents the *"specific objectives"* of the study. The questions are; a) Are Tribal Women Empowered through the targeted Micro finance programme intervention? b) How efficient Micro finance interventions in agency areas to promote institutional Development and to bring Economic Sustainability and which mode of financial operation is more effective *i.e.* NGOs Vrs. Govt., c) What do clients use Micro Credit for?, d) Are issues of Vulnerability addressed through Micro Credit and thereby brings in technical efficiency to economic interventions?, e)Does it tackle the issues of unemployed, under employed and semi-employed tribal youths?, f)Is Micro Credit

utilized for development and promotion of Micro Enterprises?, g)Does Micro Finance reduce Poverty of tribals in agency areas?, and h)How capable are the tribal women run Institutions to manage micro credit operations at different levels (SHGs, VO, MMS and GMS) adhering to the Micro Finance parameters.

(xxxv) Microfinance in India left its foot print in 1904 with the enactment of Co-operative Credit Societies Act. The primary objective of enactment of co-operative credit system was to combat the problems of usury and indebtedness of the farmers to moneylenders and to bring together people of small means for promoting thrift and mutual help for development. The Co-operative credit agencies included the primary agricultural credit society (PACS) at the village level, which was the basic unit in the short-term credit structure and were federated into the Central Co-operative Banks (CCBs) at the district level. The CCB were, in turn, federated at the state level as apex co-operative banks. The Land Development Banks (LDBs) usually meets the long-term credit needs of the farming community. Unlike -the short-term credit societies, the LDBs issued loans only against the mortgage of land and, therefore, the landless poor were excluded from the purview of these banks.

(xxxvi) After Independence, the Government of India (GOI) appointed a number of Committees to review the arrangements for institution-al credit delivery. The recommendations of these Committees, dealt more with the need for streamlining the flow of credit to the rural sectors, mainly agriculture sector, through appropriate institutional arrangements than with providing access to institutional credit for the asset less rural poor. The All-India Rural Credit Review Committee appointed in 1966 suggested that credit must be made accessible to the small farmers. As a follow up to this recommendation Small Farmers Development Agency (SFDA's) were established in 45 selected districts during the Fourth Five-Year Plan period (1969-74) to extend financial assistance to farmers with land holding size of up to 2 hectares. The Government of India had approved 40 projects under the Marginal Farmers and Agricultural Labourers (MFAL) Development Agency with a view to provide supplementary employment opportunities for marginal farmers, agricultural and landless labourers.

(xxxvii) There are different models of microfinance in operations to meet the credit need of the rural people and especially targeting running up of the institutions involving women.

(xxxviii) The pace of development and growth of Micro Finance Institutions (MFIs) is very fast. The value created in the mind and heart of the rural poor is bearing a lasting impact and more and more numbers of people are slowly attuned with the developmental contribution. Many institutions and agents are getting involved in the process. Over the passage of time the central and state Governments machinery as well as the Reserve Bank of India (RBI), SIDBI and NABARD are becoming extra cautious and more stringent control and regulatory mechanisms are evolved.

(xxxix) It is of late recognized that the local and regional level politics is influencing the functioning of the microfinance institutions in its depth and length. Caste and clan based institutions is also regulating the functioning of the MFIs at the local level. With an era of competition, and when large numbers of MFIs are working crossing the state and regional barriers, meeting the business objectives adhering to the development objective is cumbersome task. CRISIL believes that the overall asset quality of MFIs will deteriorate marginally over the short to medium term, owing to increase in large-ticket loan disbursements, multiple lending and migration of borrowers.

(xl) It is believed that poor management of operations may pose added credit risks for MFIs, especially as MFIs continue to scale up operations aggressively. MFI's risk management practices have weakened over the past couple of years, on account of a shift in focus towards business growth and network expansion. Some credit sanction and monitoring practices have been diluted.

(xli) Micro finance activities are undertaken by organizations that are registered under several legal forms. However, currently, only NBFCs are under the regulatory and supervisory purview- the NBFCs are regulated by RBI. The regulation of the sector requires more extensive consultations and public debate than is reflected in the Bill.

(xlii) It is observed that the extent and depth of MFIs operation in the country is as old as the human civilization but its operation and management is always been adhoc and sporadic in nature. In fact, regulation of MFIs is a big-gray area and there is no such systematic approaches are being taken up, either by central government or the topmost financial institutions of the country to control on the functionality.

(xliii) The Task Force (1999) made various recommendations, based on which RBI issued a number of important instructions to

microfinance practitioners. For the first time, a definition of microfinance was agreed upon. The definition is broad enough to allow for different types of institutions offering various types of financial services.

(xliv) The Task Force recommends a classification of MFIs according to the question of whether deposit facilities are offered or not and if so, what kind of facilities these are. Both the amount of savings mobilized (below or above a pre-determined cut-off limit) and the origin of savings (from clients or from the general public) determine the extent of external control by a supervisory authority. The strong reliance on self-regulation even for NGOs mobilizing deposits from the public does not go in line with more recent recommendations from other international experts, who stress the enforcement problems of a self-regulatory framework and the potential conflicts of interest of the SROs.

(xlv) Today there are nearly 32 lakh tribals, 50 lakh nomads and other backward people reside in the state of Andhra Pradesh. Their habitat spreads along the coastal and mountain strip of the Bay of Bengal from the Srikakulam district to the Khammam and Godavari districts right up to the north-eastwards to the Adilabad region. Maximum of these tribes of Andhra Pradesh have built their settlements in the hilly and forest regions of the state. In the manner of making houses also, these tribes of Andhra Pradesh have also left their marks of exuberance and artistry.

(xlvi) Tribals of Andhra Pradesh are primarily concentrated in 10 districts, where as very few households are found settled across the districts of the state and there are 10 Integrated Tribal Development Agencies have been formed in districts like Srikalulam, Vizianagaram, Visakhapatnam, East Godavari, West Godavari, Khammam, Warangal, Adilabad, Kurnool and Nellore to look at the integrated development of the people and effective management of natural resources

(xlvii) Education and especially the technical education of the state of Andhra Pradesh is recognized at the world level and that speaks of the intent, eagerness and spirit of the state as well as the education department to carry the level of education forward. Apart from focusing on the development of education for all, the state has taken a stride for provisioning access to education by tribals living in far-flung remote and hilly areas. The Tribal Welfare Department of the state of Andhra Pradesh has planned strategically to provide schools and colleges in the tribal populated districts.

(xlviii) People of rural and tribal areas are vulnerable to seasonal as well as chronic diseases and outbreak of Malaria, Jaundice, Diarrhoea and Chicken Guinea is very common. It is viewed most of the disease occur due to lack of adequate awareness on village and household sanitation and maintenance of cleanliness in the one hand and adequate availability of medical and health services on the other. The traditional food is also contributing to the deteriorated health of tribal people.

(xlix) The self-sufficiency nature of tribals often regulates their need and demand with the available resources. Hence, in order for optimal use of available nature resources and institutions, the Tribal Welfare Department (TWD) has meticulously strategized to deploy all the departments working in the Agency Areas. The intervention is aiming at reducing poverty level of tribals by way of increasing possibilities of economic gain.

(l) Agriculture is the backbone of Indian Economy and is not an exception for tribals' of Paderu ITDA. It therefore forms a significant part of the livelihood basket for the tribal. It is mostly regulated by nature and controlled agriculture in tribal areas of the state is a distant dream due to a large number of factors. Tribal farmers in particular have an intuitive understanding of natural farming techniques.

(li) The initiative for agriculture development has given emphasis on development as well as strengthening of infrastructure facilities besides distribution of quality seeds to enhance production and land productivity. 33732 beneficiaries across the Agency Areas of the state of Andhra Pradesh have been supported with an outlay of 878.64 lakhs in 2008–09, which has been multiplied in 2009–10.

(lii) Homogeneous as well as universal accessibility to the horticultural programme across the ITDA of the state has drawn criticism from all corners. In an agency like Paderu 13,882 beneficiaries reported for accessing 13,415 units of services with a financial outlay of 578.52 lakhs during the period 2008-09 stands as the highest recipient of horticultural service in agency areas of the state of Andhra Pradesh; while the accessibility of services is revealing a downward trend *i.e.,* 7418 beneficiaries in 2009-10 and 1540 beneficiaries in 2010-11.

(liii) Statistics of Animal Husbandry service across tribal agency areas of the state of Andhra Pradesh reveals that number of beneficiaries and units of service is increasing over the years. This confirms to the fact that the demand for the animal husbandry programme

has been truly accepted by the people across the agency areas of the state. Over the decade, year 2006-07 is recognized as the year for high demand and highest fund utilization.

(liv) Fishery unlike agriculture is a part of the livelihood basket of the tribals. Indigenous people were disposed of the resources that form the core of their economic and spiritual sustenance.

(lv) Sericulture is an alternative source of livelihood for rural India. Despite all efforts from all corners, the scenario of mulberry cultivation is not very encouraging so far as its coverage and recognition as a crop is concerned. With limited acceptance as a means of livelihood, its identity is retained in tribal and hilly areas and rural villages. Acceptance of sericulture within the tribals of the state of Andhra Pradesh as a fruitful means of livelihood sustenance is not welcoming. Statistics on Mulberry beneficiaries across the integrated tribal development agencies reveals that ITDAs like K.R.Puram, Bhadrachalam, Eturunagaram, and Utnoor are specializing in cultivation of cocoons. The socio-institutional benefit of the programme and economic contribution is virtually recognized in two ITDAs *i.e.* Bhadrachalam and Eturunagaram.

(lvi) The Tribal Welfare Department under the aegis of its core programme promoted construction of Minor Irrigation projects in order to restore the soil and ecology. This programme is mandated to create economies of wealth and income for the tribal families during the lean season and in turn check seasonal migration. Many MI Tanks as well as Check Dams have been constructed with people's involvement.

(lvii) ITDA Paderu has accessed and utilized the ISB scheme with greater efficiency. Per Unit of ISB is managed with an average expenditure of ₹ 9512 lakhs. In contrast, ITDA Bhadrachalam which is the biggest ITDA in the state in terms of geographic vastness but second in terms of population coverage has accessed 13457 ISB units over the decade.

(lviii) The Community Managed Grain Banks have been established in the tribal areas of Andhra Pradesh during 1990s under externally aided Andhra Pradesh Tribal Development Project (APTDP) and Andhra Pradesh Participatory Tribal Development Project (APPTDP), funded by IFAD, Rome. Grain Banks are opened up in all ITDAs. The scheme was launched in ITDA Paderu and later on spread to all other ITDAs of the state of Andhra Pradesh. The access and use of grain bank varies across the ITDAs; ITDA

Paderu access and grounded largest number of grain banks. There are about 4413 Grain Banks sanctioned and 4257 Grain Banks grounded across the agencies of the state.

(lix) Coffee cultivation has been identified as another potential activity in selected tribal pockets with cool atmosphere. Coffee loves cool atmosphere to survive and grow. Araku valley, in fact the entire ITDA of Paderu, is suitable for growing coffee. This is the only ITDA in the state of Andhra Pradesh possessing geographic as well as climatic advantage over others.

(lx) Paderu agency is the unique agency across the Agency Areas of the state in terms of its high altitude and the Primitive Tribes. The Coffee project was launched in ITDA, Paderu in the year 2002. Initially, it had been tried with 60,000 acres of land with a total outlay of ₹ 257.77 crores. Coffee program is getting momentum over the years. More and more tribal farmers are accessing as well as adopting the programme over the years and on a phased manner

(lxi) Visakhapatnam District comprises of Three Revenue Divisions divided into 43 Mandals. The District consists of two natural divisions *viz.*, the Agency and Plain areas. The Agency mainly consists of hilly regions covered by Eastern Ghats which run parallel into coast and stretches over a length of about 161 KMs., in the District from North-East to South-West lying in the interior parts of the District. It is situated between 17^0 10′ and 18^0 25′ of the Eastern Longitude.

(lxii) The Agency area consists of (11) Mandals in Paderu Revenue Division and (06) Mandals consisting of 45 villages in T.S.P. area. The Principal Hill tribes living in the Agency are Bhagatha, Kondadora, Khond, Kondakapu, Valmiki, Kammara, Gadaba, Kotias, Porja and Nookadora. They speak Telugu and of few speak tribal dialects like Gadaba and Kotiya, Oriya, which is declining. The Bhagata and Valmiki are more advanced in all walks of life than the other tribes.

(lxiii) The climate in the Hill Region is cool on account of elevation and of the green vegetation. The monsoon sets early and is along drawn in the Agency area apart from recording on average rainfall of 8412.30 MM., as against 80 MM for the District. The winter also is severe in the Agency area with the minimum temperature touching as low as 3^0 centigrade some times.

(lxiv) Half of the District area is covered by forests and a considerable portion of reserve Forest. The area is rich in Forest Produces.

The important products are Teak, Maddi, Guggilam, Kamba and Vegisa. Bomboo which is found abundance in the Forest area is mostly supplied to the paper mill at Rajahmundry. Tamarind, Ginger, Turmeric, Pippalimodi, Gantubarangi, Ranwalbia, Serpentime, Myrobalam, Adda-Leaf and Honey are the other significant items of Forest Produce collected in the Agency tracts of the District.

(lxv) The tribals are having their own culture and customs. The Scheduled Tribes who are scattering the hilly region usually assemble at the weekly shandies. It is at these places they exchange information of their welfare with conductive local conditions prevailing and at times they use these places to fix up alliances. The shandy points are also benefitting the tribals for their natural interest.

(lxvi) Tribal in general are known for their dependency on nature and more particularly on land and forest. Forest provides a major chunk of their day-to-day needs in various forms. However, with massive trees felling over the decades, dependency of tribal on land based livelihoods have been revealed through increase in cycle of land use for Podu Cultivation, Sticking to same Patch of land for several years and for generation together, and settling down near to the places of cultivation.

(lxvii) Microfinance is changing. What began many years ago as a philanthropic movement to bring finance to poor people is rapidly becoming a big business and, at the same time, a more risky business? Today, unsustainable microcredit indebtedness is commonplace across developing countries: in India; in Bangladesh; and in Peru; and also in transition countries, notably in the Balkans and especially in Bosnia and Herzegovina.

(lxviii) Home to the largest population of poor in the world, India has been a natural candidate for experimenting with microfinance as a tool for poverty alleviation. With a nationalized formal banking sector that has emphasized rural and developmental banking for several decades now, India's involvement with small credit targeted primarily at the rural poor is hardly new.

(lxix) Microfinance institutions are present in all states of India and serving the community in varied forms. Initiated with the objective of helping poorest of the poor to earn a livelihood with dignity has gradually been moved away from the prime goal. Most of these institutions have become business enterprises with an assured as well as earmarked market segment with the rural

poor and deprived section of the community in semi-urban and rural India.

(lxx) There are many institutions created and many programmes as well as schemes floated by the state and central government to increase access of fund to the poor and deprived through institutional mode to help enable a secure life and livelihood.

(lxxi) Almost 75 % microfinance institutions and the volume of operation occur in five south Indian states. The MFI data reveals that for-profit MFIs encompass nearly 90% of the total client outreach and loan portfolio.

(lxxii) A similar trend is also seen in loan portfolio- the southern states contributing 54%, Eastern states 22%, Western states 10%, Central states 7%, Northern states 5% and Northeastern states only 2%.

(lxxiii) The un-accounted MFIs are mostly working in difficult locations and against the adverse conditions and without the financial or/ and institutional support. The forms of the institutions do not fall into the organisational ambit of the so called microfinance systems and procedures. However, the numbers of microfinance institutions involved in micro-lending to the micro-people are increasing persistently over the years in terms of number as well as business volume.

(lxxiv) The other mode of microfinance operation is through the Self-help Group and Bank Linkage Programme. The programme is gaining momentum over the years and mostly identified as an established means of community reach and widely accepted by all.

(lxxv) The number of self-help groups is increasing at a faster rate. The volume of saving has increased on an average of rupees 1000 crores for the last five years.

(lxxvi) The number of Self-help Groups that had outstanding loans at the end of March 2011 had increased by about 4.9 per cent. The outstanding loans against these groups had increased by about 12.3 per cent. The number of clients added by the SBLP is about 2.9 million. The MFIs in comparison added 6.4 million clients during the year. In terms of incremental loans outstanding the SBLP added ₹ 33.6 billion. Savings by SHGs members continued to increase; 7.54 million groups had saved ₹ 69.25 billion in banks which is about 10 per cent more than the previous year level.

(lxxvii) The spread of the microfinance through SHG-BL programme is all time high in the southern region. The number of self-help groups formed across the states of the northern region is morethan 52 per cent of the total self-help groups formed in the country so far.

(lxxviii) Andhra Pradesh is the fourth largest state by area and fifth largest by population in India. The origins of microfinance in Andhra Pradesh can be traced back to government-led attempts to form "Self Help Groups (SHGs)," or groups of 12 to 20 (mostly) women which collect regular savings from members and make loans internally to members.

(lxxix) The government first formed SHGs through the Development of Women and Children in Rural Areas (DWCRA) program. Nationally, 2, 73,000 groups (2.73 lakhs) were formed under the program until its absorption into the larger Swarnajayanti Gram Swarojgar Yojana (SGSY) programme in 1999.

(lxxx) The state of Andhra Pradesh in several ways can be described as the microfinance capital of India. It accounts for 30 per cent of borrower accounts and the outstanding loan portfolio in case of MFIs. In the case of SHG linkage programme, the state had 25 per cent groups linked and 39 per cent of loans outstanding as at march 2010.

(lxxxi) The MFI industry's total exposure to Andhra Pradesh was around ₹ 7,200 crore in 2010 and they have been able to collect from the borrowers only 10 per cent of this money. They are carrying on their books ₹ 6,500 crore worth of bad assets, and when they write off this amount in March, at the end of the current fiscal, many of India's big MFIs' net worth will turn negative.

(lxxxii) Unless the Reserve Bank of India (RBI) relaxes the prudential norms that stipulate a 15% capital adequacy ratio (₹ 15 capital for every ₹ 100 worth of loans) for this set of financial intermediaries, many will have to face the sever consequence of bankruptancy.

(lxxxiii) With no fresh money in the kitty and non-payments of loans by borrowers across the of Andhra Pradesh, MFIs are forced to shrink loan books, close down offices and retrench employees. The outstanding loan book of the industry, which was ₹ 30,000 crore in October 2010, has shrunk to ₹ 15,000 crore in January, 2011 and at least 30% of 1,50,0000 employees have been retrenched.

(lxxxiv) The MFIs client base has been increased at a steady rate of about 1.3 million in 2009 and 2010. The addition of client base is none the less for SHGs. The SHG programme added an unprecedented

5.3 million members during 2008-09, 1.5 million members in 2009-10 and 4.5 millions in 2010-11.

(lxxxv) There are 525 microfinance institutions are running business over 23 districts of the state of Andhra Pradesh. Each district of the state is covered with on an average 12.17 number of MFIs registered with DRDA and 10.17 numbers of MFIs listed with MEPMA. The ratio of coverage of DRDA and MEMPA is almost equal 53: 47 for DRDA MFI's and MEMPA MFI's.

(lxxxvi) Astonishingly, 69, 12, 179 beneficiaries across the districts of the state of Andhra Pradesh are accessing financial assistance for some requirement or the other. All these beneficiaries have utilized ₹ 7895.43 crores. The money in circulation is very huge comparing to the formal institutional support extended to people to carry on business and/or activities of economic in nature for sustenance.

(lxxxvii) Recovery of loan is the key challenge for all MFIs operating in the state of Andhra Pradesh. The MFI Ordinance 2010 *i.e.* Andhra Pradesh Micro Finance Institutions (regulation of money lending) and Andhra Pradesh Micro Finance Institutions (regulation of money lending) rules 2010 is further restricting the scope of operation of MFIs and the debt recovery means adopted.

(lxxxviii) Facts corroborate that Non-Governmental Organizations (NGOs) have played a significant role in grounding and spread of the programme. NGOs are being treated as instrumental in grounding up of livelihood programmes for enhancement of economic wellbeing of the people beyond the lines of caste, creed and religion. The models followed are purely with a holistic intent and better equipped towards 'empowerment of the vulnerable and keeping away vulnerable from debt traps'.

(lxxxix) All districts of the state are home to a minimum of 30,000 self help groups. District West Godavari has the highest self help groups (80650) while Ranga Reddy is the lowest SHG concentration district with 31513 SHGs. There are 8 districts with concentration of less than 40,000 SHGs, while another 8 districts are concentrated with less than 60,000 SHGs, and there are 5 districts representing concentration of less than 50,000 SHGs.

(xc) India is a democratic country, so are the microfinance institutions. There are two major models for microfinance delivery *i.e.* the SHG Bank Linkage Programme and the MFI model in operation in the country. While the MFI model is growing rapidly the SBLP is by far the more dominant model in terms of outreach.

(xci) The MFI model uses a variety of methodologies ranging from the very popular SHG methodology to Grameen and joint liability groups, as well as individual banking arrangements. Over the years, through an ongoing process of experimentation and innovation, Indian MFIs now largely follow a mixed approach customised to their target segment and area of operation.

(xcii) A range of institutions in public sector as well as private sector offers the microfinance services in Visakhapatnam district of Andhra Pradesh.

(xciii) There are 198 microfinance institutions (MFIs) in operation at ITDA districts of the state of Andhra Pradesh. The noted MFIs among them are Asmitha Microfin Limited, Bhartiya Samruddhi Finance Limited (BSFL), Cresa Financial Services Private Limited, Fullerton India Credit Company Limited, L and T Finance Limited, Share Microfin Limited, SKS Microfinance Limited and Spandana Spoorthy Financial Limited etc.

(xciv) There are about 24 MFIs functioning in the Visakhapatnam district, with a turnover of ₹ 260 crores. Nearly one lakh people took loans from them, 65,000 of them in rural areas.

(xcv) Indira Kranti Patham (IKP) has been implementing the microfinance programme in Paderu region with involvement of community-based-CBOs namely self-help groups and with the financial assistance of World Bank. Various activities undertaken in the agency area ranges from mobilization of women into self help groups (SHGs), federating them into Village Organisation (VO), Mandal Mahila Samakhyas (MMS) and Girijana Maha Samakhya (GMS) for all 11 Mandals, providing financial access through Bank linkage (BL), Community Investment Fund (CIF), Revolving Fund (RF), Grain Banks (GB) etc. Facilitating social security through insurance and pension schemes and new pilot activities like Health and Nutrition Centers (HNC), Early Child Education Centers (ECEC) are being implemented in all the 11 Mandals and 3574 habitations of Paderu agency area.

(xcvi) Finance market is primarily and significantly controlled by the formal institution while role of informal and/or non-formal financial institutions in the agency area cannot be completely refuted. There are quite a good number of formal banks (11) operating in the agency area to cater the financial needs and requirement of tribal people and the community based organizations.

(xcvii) All banks are functioning with a targeted mandate of covering 24 self help group at the minimum to 1821. The annual mandate varies every year and so also the fund injection. Banks such as, Vijaya, DCCB and IOB have not have not supported any self help group during the financial year 2011-12.

(xcviii) There are 1331 self help groups across the district of Visakhapatnam are being supported with financial linkage of ₹ 1567.65 lakhs. This has resulted a little more than 26 per cent achievement but in terms of its turn over it has helped many self-help members to stand up their own and made them self-sustenance.

(xcix) The presence as well as contribution of Non-Banking Financial Intermediaries (NBFIs) is visible in the Paderu agency area. Tribals of the agency area have floated Mutually Aided Cooperative Society (MACS) to regulate the financial transactions of the community through an institutional mode.

(c) The MACS and popularly known as AASSAV-MACS is a four-decade old NBFC working in the Araku Mandal of Paderu agency area. It is catering to the need of 431 self-help group constituting 13.11 per cent of total fund in circulation during the financial year 2011-12.

(ci) Self help groups (SHGs) being institutions of catalystic importance are facilitating to bring the programme to people and enhance people accessibility to service delivery mechanism.

(cii) Institutions are created to regulate and exercise strategies towards goal achievement. The self-help group formation, nurturance and capacity building for a self-propelling development, is an endeavour towards this direction. Several activities are grounded in this direction and *maasostvalu* is one among them. This has not only identified the families without interest payment but also enabled formation of self-help groups of people with different abilities (PWD).

(ciii) There are 42 village Organisations included in the TFI in Paderu during 2007-08 and ₹ 15.59 crores have been mobilized. Surprisingly, the initiative of TFI for VO is not taken forward. Immediately, subsequent year had recorded TFI for 117 self help groups and a volume of ₹ 2.76 crores had been mobilized. This initiative was also for the financial year 2008-09. In totality, there are 159 institutions in Paderu agency area included the TFI and mobilized ₹ 18.35 crores during financial year 2007-09

(civ) Reasons for failure of TFI as revealed from the field survey includes limited mandate of the financial institutions for liquidation, operational stress in tribal area, poor network and communication, illiteracy and ignorance of women, credit worthiness of the SHGs/VOs, timely functioning of the SHGs in terms of credit rotation within the SHG members etc.

(cv) Formation of self help groups, nurturing and bringing them up to a level for accessing bank linkages is the preliminary challenge to the promoters let it be government and/or NGOs. Accrediting institutional status to self help groups is a milestone to be achieved through registration, structuring and provisioning legal sanctity.

(cvi) Following the principle, 583 village organizations in Paderu agency area have opted for registering themselves and 479 village organizations have successfully registered.

(cvii) Of the total number of registered village organizations, about 96.22 per cent village organizations have completed internal audit and remaining VOs have not initiated the audit process.

(cviii) Spread of Village Organizations and audited status across the mandals of Paderu agency area reveals that 14.03 per cent of Village Organizations of Munchingput, 15.21 per cent VOs of Chinatapalli, 4.76 per cent Village Organization of G.K.Veedhi and 9.61 per cent Village Originations of Koyyur Mandal have not under taken the audit as of the reporting period.

(cix) 479 village organizations have filed the return. No clearance has been accorded to any village organization as of now. About 18.35 per cent village organizations have yet to file the return for the internal audit. Of the eleven mandals, 6 mandals have filled the return fully. Rate of return in the rest of the five mandals varies from a minimum of 1.92 % to a maximum of 78.26%

(cx) UBI, UCO, IOB, DCCB and Vijaya Banks have not reported of any operational difficulty in self help group financing and its recovery from all operating branches of Paderu ITDA. The smooth functioning of all branches across the agency mandals is the strength for tribal self-help groups.

(cxi) AASSAV-the only tribal run MFI has not witnessed any operational flaws in lending finance to self help groups of Arakuvalley and Dumbriguda Mandal. It is running in two mandals viz. Arakuvalley and Dumbriguda of Paderu agency area and with staff of tribals. It is the single MFI working at par with all other formal financial institutions and catering the financial need of a handful of self help groups.

(cxii) Indira Kranti Pratham (IKP) is a World Bank supported programme to empower women of the state of Andhra Pradesh through formation of self-help groups and engaging them in IGPs of various natures. It has a special recognition in tribal areas of the state. IKP, as popularly known among the people, is the largest women empowerment microfinance programme in the world. IKP is executed with active support of the ITDAs. The staff and building have been widely engaged to popularize the programme and ensure peoples acceptance.

(cxiii) ITDA and IKP though are two institutions working in agency area. The scope of operation is very much limited for IKP. IKP takes the responsibility of physical execution of activities while technical support for execution used to be extended from the core department. IKP being perceived as an institution having deep rooted base at each and every habitation and coming up with 'community-involved-micro- credit-plans' is probably the most viable institution to establish a good linkage with community and all other government departments functioning in remote and tribal areas.

(cxiv) There are 11 NGOs involved in formation of self help groups across 11 mandals of Paderu agency area under the STEP programme. All self-help groups formed are trained on various parameters of SHG functioning, and administration.

(cxv) After the State brought in the ordinance regulating the MFIs, the sector which used to lend ₹ 7,000 crore, had not been able to lend anything now to the poor who had no access to bank loans. The MFIs had registered according to the stipulations brought in, but it was not translating into lending process.

(cxvi) The concept of SHG mode of MFI operation provides the benefits of economies of scale, cost effective alternatives for financial services, collective learning, and a democratic and participatory culture.

(cxvii) 620 SHGs out of 2556 is covered under the study *i.e.* 214 out of 779 in Dumbriguda, 224 out of 871 in G. Madugula and 182 out of 906 in Padabailu. About 24.25% of beneficiaries of SHG being interviewed

(cxviii) Women empowerment is not giving women power, women already have plenty of power, in the wealth of their knowledge and motivation, to do their jobs magnificently. Empowerment is defined as letting this power out (Blanchard, K., 1996). It

encourages people to gain the skills and knowledge that will allow them to overcome obstacles in life or work environment and ultimately, help them develop within themselves or in the society.

(cxix) The study has established that the accrued benefit of MFI for improving the household facilities and assurance of a decent living, women involvement in decision making, nature of decisions being taken in meetings, benefits of trainings on women and access and leverage of government schemes and programmes for tribal women members in self-help groups of Paderu agency area is significant. There is glaring change among the women of the agency area and more focused intervention by self-help group is the call for the day for further improvement of women potency.

(cxx) Credit plays a vital role in initiating and expanding the business, and hence, microfinance has been treated as an important tool for economic development. The SHGs offer small loans to aspiring individuals as well as group of individuals. These loans assist people in getting access to traditional financing and offer jobs to local communities. "Microfinance is about more than money. None of us in this room would accept living in a community with bad schools, little healthcare, no electricity, and filthy water – but with a great bank.

(cxxi) Microfinance is making a significant contribution to both the savings and borrowings of the poor in the country. The main use of microcredit is for direct investment. While the presence of microfinance institutions has increased the borrowing options for the poorer clients, it seems not (yet) to have significantly affected the terms and conditions of different informal credit providers.

(cxxii) Apart from the provision of assured supply of credit to all needy at a quicker time and without any collateral securities, self help groups do play a more cognitive role to shape the institutional functioning and bringing in socio-cultural changes within the tribal communities. The holistic role of MFIs and especially of self help groups as institutions of reformation is therefore not limited to finance only.

(cxxiii) Savings with self help groups are for multipurpose use. It is serving as a solution for long standing issues and also provisioning income earning support to many. Besides, the economic contribution, self help group institutions are also taking care of

developmental needs of the household and the rural communities at large.

(cxxiv) The role of tribal women self help groups in capacitating households and institutions to respond to unpredictable situations is notable and recognition worthy. The developmental role of self help groups across the mandals of ITDA Paderu would create environment for better sustainable growth in the future.

(cxxv) About 36.20 per cent members across the studied regions has been exposure to various livelihood programmes. Inter region scenario of exposure programmes varies from a minimum of 6.97 per cent in Padabailu Mandal to a maximum of 52.20 per cent for Dumbriguda Mandals. The variation in exposure program is primarily due to the locational variation of the regions in one hand the available resource potential as well as selection of livelihood baskets in particular.

(cxxvi) With a view to imbibe skill in tribal women, girls and youths, various skill development programmes[3] have been executed by self help groups from time-to-time. These includes masonry, tailoring, embroidery works, wielding, electrician, computer training, crafts making etc. About 14 per cent members across the studied regions has been exposure to above said livelihood programmes. Inter region scenario of exposure programmes varies from a minimum of 2.32 per cent in Padabailu Mandal to a maximum of 30.88 per cent for G. Madugula Mandals. Out of the total beneficiaries, 52.50 per cent of G. Madugula, 42.50 per cent of Dumbriguda and 5 per cent of Pedabailu have been covered under the programme.

(cxxvii) The direct trainings imparted to self help group members have a positive correlation (r=0.047833) with the creation of self employed professionals across the regions of ITDA Paderu.

(cxxviii) It is evident that the employment and /or engagement pattern of tribal over the years have diversified. Many self employed professional have generated, dependency on forest produce has substantially reduced and eagerness for better regulated life is created. In the process, many youths finds the definition of their engagement and living. Entire family members are no more adjusting with the age-old traditional agriculture.

(cxxix) The facts and figures presented on promotion of micro and small enterprises in ITDA of Paderu are not overwhelming but at the

same time the importance of SHGs in across the regions of agency area cannot be undermined. The efficiency of tribal self help groups in promoting Micro Enterprises (MEs) is unique. Access to loan and development of enterprises across the regions are positively correlated (r=0.889907) and so also is the scenario of correlation (r= 0.984366) livestock programmes and the investment incurred across the Mandals of Paderu agency area. The contribution of self help groups towards promotion of micro enterprises is notable, visible and enterprise promotion would have not been possible without the self help groups.

(cxxx) *'Microfinance is a means to an end but not an end itself'*. The activities taken up by the self help groups are not individual centric at all time. Issues of poor and vulnerable are heard on priority while the commitment of development of all has never been taken the back bench. The sportive spirit of dealing with micro household/ family issues to the macro level society issues and arriving at solutions have widely been recognized for 51.94 per cent of SHG members at HH level and 58.87 per cent SHGs members at the community level. The self respect for women is an accrued benefit for their dedicated effort at the platform given by SHGs.

(cxxxi) Self-employment as well as timely supply of financial support to families in need has reduced dependency on moneylenders. There are many formal banks extending financial services across the mandals of Paderu agency area.

(cxxxii) The sole effort of educating tribal women on organization management and provide hands on experience is a fruitful exercise. About 65 per cent of tribal women are capable of managing the organization of their own with support of IKP.

(cxxxiii) About 58.39 per cent members of the self help group are provided with leadership quality to management the institution and the community of their living. The result of this has been visualized in participation of 64.19 per cent of women self help members in Gram Panchayat (GP) elections and victory of 68.39 per cent among them. Better women participation in GP election is reported in Pedabailu Mandal of agency area.

(cxxxiv) It has enabled 65.81 per cent tribal women to minimize risk at household, community and society level. This is a common trend all across the studied Mandals. All SHG members are imbibed

with the skill and potentiality to withstand to risk of any kind and this may be due to the versatile livelihood occupations opened up for them. There are 58.23 per cent members of self help groups reported accessing changing livelihood occupation. Hence, diversity in living has already crops in the life of tribals of Paderu.

(cxxxv) Microfinance institutions are institutions of the people, for the people and by the people. It is an institution to provide all inclusive platform for development of individual households and also the society altogether.

(cxxxvi) Self help groups interacts with other institutions on various issues and areas of their specialization. SHGs are therefore termed as the link-pin among the institutions. However, the ^{2} value is not significant 10 per cent level of significance to accept the hypothesis that institutions working in the villages are constantly in dialogue on issues of village, community and the tribal economy at large.

(cxxxvii) Self help groups are playing all pervasive roles in agency area of Paderu, Visakhapatnam district. The catalytic role in achieving visible and tangible changes in the life and livelihood of tribal is amply visible. It has added a new definition and meaning to the livings of tribal. The impact of SHG is reflected in promotion of educational institutions and access to education, access to health services, modalities of service delivery, reduction of poverty and un-employments, imbibing tribal youths with skills and hands on experience on IGAs, promotion of women participation in HHs, community and village development, leadership promotions, ensuring food security, ensuing visible income and provisioning self-esteem to women. Hence, it can be concluded with confidence that SHG operation in agency area of ITDA Paderu is very much meaningful.

7.2. Conclusion

MFI mode of financial operation in agency area of Paderu is an efficient mechanism to address multitude needs of tribals. The self-help groups across the agency are meticulously involved in execution of group activities as well as the government schemes and programmes meant for them. In fact, SHGs are serving as a platform to give justice to execution of tribal development programme. It is not only ensuring access to schemes and programmes but also assuring community participation in execution.

Several MED programmes have been executed in the agency area of Paderu. Tribal youths are trained on several skill building programmes. Several rounds of leadership programmes have been organised to create women leaders in each and every community-not only with the view to make leaders but also to imbibe all SHGs members with the habit of understanding household and community issues and the ability to take them forward for settlement/resolution.

Food security is thoroughly addressed through opening up of grain banks. The process has adequately been supported by all NGOs working in Paderu ITDA of Visakhapatnam District. All NGOs are involved in formation of SHG, training of SHGs and are still contributing to make them sustainable in the future.

The dedicated institutional mode of MFI intervention has created a 'system of self-perpetual mechanism to address tribal issues' in the years to come. It is self-propelling because tribal women are directly taken control of management, functioning, development of action plans, maintenance of books of records and visualizing the long-term and short term goals of their operations.

It would be more stable and vibrant in the days to come because there is no other parallel efficient mechanism to address local issues with assured and undiluted fund support. *"SHGs therefore have been considered as institutions born from the villages to empower women to deal with issues of village for sustainable development of the region and the people"*.

7.3. Recommendations

Self help groups are formed with noble intention to help enable vulnerable to survive with pride and dignity. It has ensured healthy socio-economic and institutional environment. The achievements so far are visible and hence, it is termed as the largest saviour initiative for the poor/ vulnerable people in the world. However, the pace of achievement is not as had been intended. Malfunctioning is noticed in programme execution. The study has captured the hindrances reported by each region of the Paderu agency area and accordingly suggesting the following ways to overcome them for ensuring better functioning.

(i) Formation of Self help groups is a noble idea. Women of similar bent of mind are selected as members. There are hardly literate women found in SHGs. Hence, understanding the issues of society and designing plans to address the issues are difficult challenge.

Hence, educating tribal women and orienting them on SHG function is a better proposition.

(ii) PTG women are very shy in nature and hardly come out of home to speak to an outsider. This attitude has been changed to some extent but a larger initiative is required. A special drive within the tribal programme is essential to bridge the development gap between tribal and PTGs.

(iii) More handholding support to the newly formed SHG is advised. The support needs to be extended till the leaders and the members feel they are capable of managing SHGs functioning on their own.

(iv) All SHGs used to prepare MCPs and it is observed that most of the MCPs of new SHGs lack perfectness. Hence, it is recommended that a 'specific period of grooming' to be maintained to make MCPs perfect and business oriented.

(v) Opinions of Vocal members are often heard in considering RF support disbursal. The minorities' views should also be taken into consideration.

(vi) Opinion of SHG leaders is also getting squeezed in the face of bureaucratic setup. Staff of IKP is seen ruling over the institutions where ideas, ideologies and business development are imposed upon suo-motto. Hence, tribal women leaders are not getting any room to ventilate their ideas, which is more of a ground reality.

(vii) Capacity Building programmes undertaken are reported to be biased. Leaders of the groups and the active members are primarily accessing the service. This type of a community development and/or women empowerment strategy would enhance an economic gap among the members of the groups. The catalystic agents, IKP or NGOs to ensure adoption of 'fair beneficiaries' selection' for the training programme.

(viii) Redundant and monotonous trainings are conducted as a part of annual target achievement plan of IKP. Trainings need assessments have not been thoroughly organised by any agent and agency.

(ix) The livelihood activities suggested to SHGs are reported to be identified by external agents. The IKP staff and/or NGOs do recommend the livelihood activities to be taken up by self help groups. Hence, it is recommend that livelihood activities need to

be selected in consultation with SHGs and often based on their skill and potentiality and based on the available natural resources.

(x) More Micro Enterprise Development programme to be taken up in the entire agency to provide employment to tribal youths.

(xi) More and more IGA to be promoted involving all section of people. This will ensure multiple benefits of good family income and employment for the people.

(xii) The effort of bringing tribal youth to the mainstream of living through imparting technical training and provisioning employment at cities is gradually getting momentum. Many tribal youth, both boys and girls, have visualized the modern living and thereby bring changes in their living pattern. In the wake of employment and thereby changing the self-esteem, tribals are seen out of home and their culture. This has worked as a de-motivating factor for their retention. Hence, modalities need to be worked out on imparting appropriate job oriented trainings and capacity building programmes, placement in areas suitable to their cultural and religious needs and environment.

(xiii) Concept of Revolving Fund (RF) to liquidate and multiply benefit of small to all is often seen at stake. Misappropriation and extracting undue advantage of illiteracy is seen as a factor leading to under utilization of resources meant for the tribals. *Therefore, persistent effort need to be made, towards streamlining and developing procedures for effective monitoring and tracking of Revolving Fund injected to tribal economy on a regular basis.*

(xiv) Fund leveraged for Self Help Groups to take up income generation programs is steadily increasing over the years. Banks and other financial institutions are happy to extend financial leverage. However, fund liquidation of bank is subject to effective functionality and dysfunctionality of the SHGs and Village Organizations. This parameter needs to be examined very closely before recommending for Bank Linkage.

(xv) Registration of Village Organisation has taken lead. It has been tried with some VOs and there are still a large number of VOs to be registered. *"Quick registration of VOs across the agency would ensure legal sanctity to SHGs".*

(xvi) Financial audit has been introduced for VOs and SHGs to provide institutional status. This has been done for some meagre number of VOs. Hence, it is recommended to speed up the process.

(xvii) Poor Networking and Coordinated effort is reported across the studied regions of the ITDA. Institutions need to interact with each other and learn from each the modality of dealing with specific issues. Good networking environment is suggested to prevail for better functioning of self help groups across the agency area.

(xviii) Programme monitoring cell of IKP to undertake timely monitoring of the activities undertaken for self help groups.

(xix) More use of simple cash grants and Conditional Cash Transfers (CCTs), which have been shown to reduce the worst excesses of income poverty should be discouraged.

(xx) An urgent focus on the promotion of local micro-Savings, rather than microcredit, as the first step in the local accumulation of capital to be grounded.

(xxi) Pro-active local financial institutions and local industrial policies that can provide 'patient capital' and promote sustainable growth-oriented businesses, rather than 'survivalist' no-growth/high failure rate microenterprises.

(xxii) A stand alone group of poor rural women, stands a lesser chance of being effective than a confederacy of such groups. A Confederacy would offer the strength of numbers and solidarity to reach of it's constitutes. A Single stand alone group has limited visibility and importance for both the general community as well as development agency. There are certain areas where SHG's need wider support, crisis resolution, marketing, networking, mobilization for effective action or social issue etc. Thus, networking of SHGs into a federation is called for.

(xxiii) Introduction of incentive schemes for best performing SHGs across the regions would encourage SHGs to perform better. IKP may inject some competition among the self help groups.

(xxiv) Social capability building programmes should be organized from time to time to train the NGO's activists, volunteers, Panchayat representatives, members of youth clubs etc. or promote small savings and women's active and positive role in development process; ensuring their rights, entitlements and due share in developmental benefits.

(xxv) Policy makers need to recognize the potential of micro financial services to support investment and growth in key economic sectors and hence to contribute significantly to national economic growth.

(xxvi) The innovative forms of financing is imperative to supplement credit strategies for meeting the needs of the poor by combining the flexibility, sensitivity and responsiveness of the informal credit system with the technical and administrative capabilities and financial resources of formal financial institutions and also to build material trust and confidence between bankers and the rural poor and to encourage banking in a segment of population that formal financial institutions usually find difficult to reach.

(xxvii) There is urgent need to streamline the procedure for applying, seeking and releasing of credit from the banks. The procedural difficulties are one of the major impediments, which have denied women the financial benefits of the banks. Therefore, the procedure for credit access to women should be made more easy and simple.

(xxviii) In order to ensure proper utilization of the credit, there is an urgent need to introduce availability of consumption credit from the formal channel. The need is to sensitize bank staff towards the needs, constrains and inhibitions of women.

(xxix) RBI regulation of capital adequacy ratio of 15% to be liberalized for better coverage of larger number of vulnerable by MFIs. The Government and RBI to draw an MFI survival strategy before it is too late. It is required at least till such time the banks are ready to reach out the masses.

(xxx) Good coordination and liason is recommended to establish with NGOs and CBOs to realize the investment implication of fund on socio-economic and institutional development.

(xxxi) ITDA to organise a platform for IKP and NGOs to chalk out the SHG operation plant and division of roles for better grounding of the programme

(xxxii) Besides, the above recommendations, the study recorded suggestions of the SHG members and VOs through group discussions. The suggestions of the SHGs/VOs are summarized in the box-4.

Box-7.1: Consolidation of Suggestions for SHGs

Mandals	Region Specific Suggestions
Dumbriguda	1. Habit of Good Practice need to be institutionalised across SHGs
	2. To help the needy in appropriate time. This will reduce dependency on money lenders
	3. SHG is a platform to choose correct IGP for assured monthly family income
	4. Assured Marketing facilities to be provided for sale of produces produced by the self help group members
	5. Good skill building training is required for betterment of group members
G. Madugula	1. More Trainings and Exposures to be organised by Government
	2. Government i.e. IKP should ensure Marketing Facility for the already available products for disposal. The prices offered for the products are very minimal and less than the market price.
	3. Government i.e. IKP to show us better incoming earning avenues
	4. Learning is important and SHG should take up education programme for its members. This will make them understand the SHG transactions.
	5. Loan and Interest should be paid in time for common benefit of all
	6. Most Needy Person Problems to be addressed on priority
	7. Reguar meeting to be organised for future betterment
	8. Regular saving in every month to be ensured
	9. Society (SHG) gives knowledge about our village scenario and we should work for our development.
	10. There are many people with low income. It is a group norm to help them on priority basis.
Pedabailu	1. Reguar meeting to be organised for future betterment of members and SHGs functionality
	2. Regular saving collection and deposit in banks to be ensured every month
	3. Loan and interest should be paid in time. This will help better support for other members of SHGs
	4. More contextual trainings to be imparted
	5. Monthly Meeting to be conducted with regularity
	6. Permanent Meet hall is required for conduction of meetings

Contd..

Box-7.1: Contd...

Mandals	Region Specific Suggestions
	7. New people should encourage to join in the SHG group
	8. Specific Training need to be organised for specific members
	9. Every village must and should have Self help groups. It gives timely help to the needy.
	10. Society i.e. SHGs give Mirror view of our village and encourage us to work for our village development.
	11. PTG people don't know uses of their agricultural products and/or forest products. They need to be oriented with the means and ways of getting better price/revenues out of available resources with them.
	12. Some mechanism to built in to provide work all seasons
	13. SHG is the source of community Learning

Notes

1. What is microfinance, Opportunity International, Australia, 2007.
2. Model-I: Self-help Groups (SHGs) formed and financed by Banks.

   Model-II: Self-help Groups(SHGs) formed by NGOs and formal agencies, but directly financed by Banks.

   Model-III: Self-help Groups (SHGs) financed by Banks using Non-Government Organizations (NGOs) and other agencies as financial intermediaries.
3. All the skill development programmes are executed from the training centres located at ITDA head quarter at Paderu. IKP has coordinated for the programme and directly run the training centres. Skill development programme is basically a residential programme and tribal youths used to be away for couple of moth for pursuing the training.

Annexures

Annexure-2.1: Details of 241 major MFIs in the country

Sl. No.	Organisation	HQ State	No. of districts operating	No. of Poor Districts	Operation in states	Client Outreach	Loan Portfolio
1	Action for Social Advancement(ASA)	M.P.	5	3	M.P.	12876	14146520
2	ACTS Mahila Mutually Aided Cooperative Thrift Society (AMMACTS)	A.P.	3	1	A.P., T.N.., Karnatakaa	164666	84277755700
3	ADARSA	Orissa	3	3	Orissa	6160	320865400
4	Adarsha Welfare Society	A.P.	1	1	A.P.	33196	278345000
5	ADHIKAR	Orissa	12	8	Orissa	55750	205924328
6	Agradut Polly Unnayan Samity	W.B.	3	2	W.B.	3033	12133905
7	Agricultural Science Foundation	Karnatakaa	1	0	Karnatakaa	16448	36080600
8	Ajiwika	Jharkhand	8	7	Jharkhand, Bihar	4639	26282828
9	Amber Ashraye Mahila Benefit Association	Gujarat	4	2	Gujarat	8643	25620703
10	ANG Resource Ltd	W.B.	1	1	W.B.	34280	129742000
11	Annapurna Parivar Vikas Samvardhan	Maharashtra	2	0	Maharashtra	10000	60000000
12	Arman Lease and Finance Limited	Gujarat	3	0	Gujarat	4231	159844000
13	Asian Institute of Rural Regeneration	Orissa	3	2	Orissa	12307	11375200
14	Asmita Institute for Development	Maharashtra	2	2	Maharashtra	8311	89175511
15	Asmitha Microfin Ltd	A.P.	59	32	A.P., Orissa, Karnatakaa, Maharashtra	701432	3359144380

Contd...

Annexure 2.1: Contd...

Sl. No.	Organisation	HQ State	No. of districts operating	No. of Poor Districts	Operation in states	Client Outreach	Loan Portfolio
16	Bajkul Sports Association	W.B.	3	2	W.B.	15124	40207908
17	Bal Mihila Vikas Samiti-VAMA	M.P.	1	0	M.P.	2530	19192200
18	Bandhan-Konnagar	W.B.	28	20	W.B., Assam, Bihar, Jharkhand, Orissa, Meghalaya	757887	2782819910
19	Banki Anchalika Dibasi Harijan Kalyan Parisad	Orissa	3	1	Orissa	0	0
20	Batika	Bihar	1	1	Bihar	300	450000
21	Belgharia Janakalyan Samity	W.B.	1	1	W.B.	3171	8363584
22	Bhaoruka Charitable trust	Rajasthan	1	1	Rajasthan	595	1155800
23	Bharat Integrated Social Welfare Society	Orissa	60	50	Orissa, Chhattisgarh, Bihar, W.B., Jharkhand, M.P., Nagaland, Uttarakhand, U.P.	246430	1180412460
24	Bharat Sevak Samaj (BSS)	Kerala	3	1	Kerala	66794	314884375
25	Bharatha Swamukti Samsthe (BSS)	Karnataka	12	3	Karnataka	142613	814143690
26	Bharatiya Samruddhi Finance Ltd (BASIX)	A.P.	46	33	A.P., Karnataka, Chhattisgarh, Maharashtra, M.P., Rajasthan, Jharkhand, Orissa	305438	2254111
27	Bihar Development Trust	Bihar	2	0	Bihar	200	308000
28	BODHANA-Tiruvalla Social Service Society	Kerala	4	1	Kerala	8943	9844374
29	Bright Association for No.ble and Decent Human Understanding (Bandhu)	Orissa	1	0	Orissa	72550	12607672
30	Budhhanath Jubak Sangha	Orissa	0	0	Orissa	0	0
31	Bullock-Cart Workers Development Association (BWDA)	T.N.	8	4	T.N., Pondicherry	292729	851289324
32	Bureau of Obligate and Accompanier for Rural Development	Bihar	2	2	Bihar	296	820110
33	Cashpor Micro Credit (CMC)	U.P.	13	10	U.P., Bihar	302855	1472879090
34	Centre for Action and Rural Reconstruction	Orissa	2	1	Orissa	1007	2212701

Contd...

Annexure 2.1: Contd...

Sl. No.	Organisation	HQ State	No. of districts operating	No. of Poor Districts	Operation in states	Client Outreach	Loan Portfolio
35	Centre for Collective Development	A.P.	1	1	A.P.	3000	5000000
36	Centre for Development Orientation and Training	Bihar	1	1	Bihar	580	4006000
37	Centre for Rural Reconstruction through Social Action	A.P.	2	2	A.P.	26412	138686020
38	Centre for Youth and Social Development (CYSD)	Orissa	4	3	Orissa	0	0
39	Chaiduar Rural Development Centre	Assam	1	1	Assam	70	345000
40	Chaitanya Institute for Youth and Rural Development (CHINYARD)	Karnataka	2	1	Karnataka	37400	19000000
41	Chetna Mihila Vikas Kendra	Maharashtra	3	0	Maharashtra	115	496920
42	Chotanagpur Vikas Kendra	Jharkhand	1	1	Jharkhand	190	850000
43	Cohision Foundation Trust	Gujarat	4	0	Gujarat	3181	1724052
44	Community Development Centre	T.N.	5	1	T.N.	26664	105112320
45	Community Service Trust	T.N.	3	0	T.N.	5207	24301688
46	Confederation of Voluntary Association (COVA)	A.P.	1	0	A.P.	11341	27640000
47	Coperation Development Council	Orissa	2	1	Orissa	19252	89025000
48	Creation Welfare Society	Bihar	3	3	Bihar	941	940640
49	Darbar Sahitya Sansad (DSS)	Orissa	4	0	Orissa	6818	10815133
50	Dayanvikas Gramin Bahuuddeshiya Samsatha	Maharashtra	1	0	Maharashtra	376	1565000
51	Deepalaya	Delhi	5	1	New Delhi, Haryana, Uttarakhand	8627	14775000
52	Deepika Mahila MACS Ltd (Crown Social Service Society)	A.P.	1	1	A.P.	170	1000000
53	Development Initiative for Self-help and Awakening (DISHA)	Maharashtra	1	0	Maharashtra	9607	14850000

Contd...

Annexure 2.1: Contd...

Sl. No.	Organisation	HQ State	No. of districts operating	No. of Poor Districts	Operation in states	Client Outreach	Loan Portfolio
54	Development Support Team	Maharashtra	8	1	Maharashtra	12000	766097
55	Development Organisation for Village Environment (DOVE)	A.P.	7	3	A.P.	8583	81853150
56	Disha	U.P.	2	0	U.P., Uttarakhand	1831	10524962
57	Ecumenical Church Loan Fund of India (ECOLOF)	T.N.	15	4	T.N., Manipur, Nagaland, Maharashtra	10303	51259757
58	Equitas Micro Finance India Pvt. Ltd	T.N.	1	0	T.N.	16166	167380145
59	Evangelical Social Action Forum(ESAF)	Kerala	21	8	Kerala, T.N., Maharashtra, Chattisgarh	222123	706622900
60	FGC-WD	Kerala	2	1	Manipur	636	4500000
61	Forum for Rural Environment and EcoNo.mic Development (FREED)	Kerala	3	1	Kerala	53460	127182000
62	Fullerton India Credit Company Limited	Maharashtra	0	0	Maharashtra	0	0
63	Gramswaraj Seva Trust	Maharashtra	1	1	Maharashtra	232	1411805
64	Grameen Koota	Karnataka	13	3	Karnataka	139114	826461853
65	Grameen Mahila Swayam Siddha Sangh- Chaitanya	Maharashtra	1	0	Maharashtra	16125	16700000
66	Grameen Sahara	Assam	2	1	Assam	1286	6459555
67	Gramin Janakalyan Parishad	Bihar	1	1	Bihar	12062	2983435
68	Gramin Vikas Mandal, Bansarola	Maharashtra	1	0	Maharashtra	14	200000
69	Gramotkarsh & Vikas Trust	Gujarat	3	1	Gujarat	1870	650000
70	GRAM - Utthan	Orissa	6	3	Orissa	74142	372725945
71	Guidance Society for Labour Orphans and Woman (GLOW)	T.N.	2	0	T.N., Karnataka	3801	17450500
72	GUIDE	A.P.	3	2	A.P.	9845	28856088
73	Haridanga Ramkrishna Vivekananda Sangha	W.B.	1	1	W.B.	5110	19007262
74	Hindusthan Cooperative Credit Society Ltd	Maharashtra	3	1	Maharashtra	1325	29601553

Contd...

Annexure 2.1: Contd...

Sl. No.	Organisation	HQ State	No. of districts operating	No. of Poor Districts	Operation in states	Client Outreach	Loan Portfolio
75	HOPE Foundation	Kerala	9	5	Kerla, W.B., Jharkhand, T.N., Orissa	26315	46025336
76	Hope Integrated Rural Development Society	A.P.	1	1	A.P.	3170	11531260
77	Human Development Centre	W.B.	3	3	W.B.	6712	16600508
78	Humana People to People India	Rajasthan	2	1	Rajasthan	1458	7070000
79	Indian Institute of Rural Development	Rajasthan	1	1	Rajasthan	3236	19918000
80	Indure Intideepam MACS Federation Ltd	A.P.	2	2	A.P.	18645	192495859
81	Initiatives for Development Foundation	Karnataka	10	3	Karnataka	28618	78095361
82	InNo.vative Micro Finance for Poverty Allievation and Community Transformation (IMPACT)	T.N.	0	0	T.N.	0	0
83	Institute of Integrated Recourse Management(IIRM)	Assam	3	1	Assam	7400	36302523
84	Institute of Rural Credit and Entrepreneurship Development(IRCED)	Maharashtra	2	1	Maharashtra, Karnataka	10170	4085667
85	Ishara Foundation for Finance and Rural Development	Delhi	5	2	Delhi, U.P.	16798	52963063
86	Jagoo Samajik Arthik & Harit Vikas Sanghatan	U.P.	1	0	U.P.	18	70008
87	Jan Jagaran EvamPunarawas Sansthan	Bihar	2	2	Bihar	3070	2250000
88	Janalakshmi Social Services	Karnataka	2	0	Karnataka	49276	359900000
89	JaNo.daya Public Trust	Karnataka	10	4	Karnataka	0	0
90	Jhansi Rani Mahila MACS Ltd	A.P.	1	1	A.P.	1870	4200000
91	Kakatiya Mahila MACs Ltd	A.P.	1	1	A.P.	624	1600000
92	Kalanga BazarEducational Trust	Jharkhand	1	1	Jharkhand	2145	4508401
93	Kalighat Society for Development Foundation(KSDF)	W.B.	3	2	W.B.	14917	35708637

Contd...

Annexure 2.1: Contd...

Sl. No.	Organisation	HQ State	No. of districts operating	No. of Poor Districts	Operation in states	Client Outreach	Loan Portfolio
93	Kalighat Society for Development Foundation(KSDF)	W.B.	3	2	W.B.	14917	35708637
94	Karra Society for Rural Action	Jharkhand	1	1	Jharkhand	2898	4416466
95	KAS foundation	Orissa	65	39	Orissa, M.P., A.P., Chattisgarh, Maharashtra, T.N., W.B., Pondicherry	562377	1068314467
96	Khandagiri Madhyamik Mahila Samabaya Sangha Ltd	Orissa	4	1	Orissa	9547	4805629
97	Kotalipara Development Society	W.B.	7	6	W.B.	50677	129501377
98	Krishna Bhima Samrudhi Local Area Bank Ltd (KBSLAB)	A.P.	3	3	A.P.	100495	521006586
99	Krushi	A.P.	5	5	A.P.	41483	238792216
100	Krushi Vikas & Gramin Prashikshan Sanstha	Maharashtra	1	1	Maharashtra	670	1882300
101	League for Education and Development (LEAD)	T.N.	7	2	T.N.	29683	16090082
102	Liberal Association for Movement of People (LAMP)	W.B.	0	0	W.B., Jharkhand, Delhi, U.P.	50103	34411347
103	Lok Biradri Trust	M.P.	3	1	M.P.	2253	8064382
104	Mahalir Association for Literacy, Awareness and Rights (MALAR)	T.N.	1	1	T.N.	24541	23086762
105	Mahasemam	T.N.	12	4	T.N.	91756	326720062
106	Mahashakti Foundation	Orissa	6	6	Orissa	14083	48590861
107	Mahila arthin Vikas Mahamandal Ltd(MAVIM)	Maharashtra	35	18	Maharashtra	684873	1320059772
108	Mahila Kalyan Samiti	Jharkhand	1	1	Jharkhand	1061	6123500
109	Mahila Vihas Prathamika Sanchaya Samabaya Ltd	Orissa	1	0	Orissa	823	1783560
110	Manab Sewa Sangh	Assam	3	2	Assam	1310	3343000

Contd...

Annexure 2.1: Contd...

Sl. No.	Organisation	HQ State	No. of districts operating	No. of Poor Districts	Operation in states	Client Outreach	Loan Portfolio
111	Manidham Grameen Savings Cum Credit Services	T.N.	6	1	T.N., A.P.	15763	58534800
112	MANSI	Bihar	1	1	Bihar	72	72500
113	Margdarshak Community Development Support Services Pvt. Ltd	U.P.	2	1	U.P.	1253	585200
114	Mass Centre International	Bihar	5	5	Bihar, Jharkhand	1200	2575000
115	Mata Deen Mahila Manch	Bihar	1	1	Bihar	209	1500000
116	Melap Mahila Cooperative Credit Society Ltd	Gujarat	1	0	Gujarat	513	931196
117	Mission Fraternal Peace Promotion Programme(PPP)-Manipur and Nagaland	Manipur	0	0	Manipur	0	0
118	Modern Architects for Rural India (MARI)/ Sanghatitha Mahila MACS Federation Ltd	A.P.	1	1	A.P.	30085	64427968
119	Mother Teressa Mahila MACS Ltd	A.P.	1	0	A.P.	2715	3200482
120	Nanayasurabhi Development Financial Services	T.N.	0	0	T.N.	0	0
121	Nav Bharat Jagriti Kendra (NBJK)	Jharkhand	5	4	Jharkhand, Bihar	10744	49065573
122	Nav Jagriti	Bihar	3	2	Bihar	2250	8084400
123	Navachetana Foundation	Karnataka	6	2	N/A	22730	32000000
124	Navchetana	Maharashtra	1	1	Maharashtra	1067	5944000
125	Network of Entreprenurship and EcoNo.mic Development (NEED)	U.P.	0	0	U.P.	0	0
126	New Life	T.N.	5	2	T.N.	24549	7756079
127	New Modern Rural Saving Cooperative Ltd	Orissa	1	1	Orissa	470	522317
128	NIDAN	Bihar	6	5	Bihar	18508	23859201
129	Nirantara Community Services	Karnataka	1	2	Karnataka	3422	15290860

Contd...

Annexure 2.1: Contd...

Sl. No.	Organisation	HQ State	No. of districts operating	No. of Poor Districts	Operation in states	Client Outreach	Loan Portfolio
130	Nirman Bharati Samajik and Arthik Vikas Sangathan	U.P.	4	0	U.P., Uttarakhand	52040	256356660
131	OAZOANE-The Society for Development of Human Abilities and Evvironment	T.N.	2	2	T.N., Andaman	4988	79785240
132	Omalur Block Women Welfare Uplift Organisation (OBWWUO)	T.N.	1	0	T.N.	1045	9930000
133	Opportunity Microfinance India Ltd	Karnataka	40	15	Karnataka, T.N., A.P., Kerala, Orissa	16954	45607560
134	Organization for Development Coordination (ODC)	Orissa	1	0	Orissa	19554	21849740
135	Organization for Development Integrated, Social and Health Action(ODISHA)	Orissa	3	1	Orissa	14350	37881250
136	Organization for Rural Survival	Tripura	1	0	Tripura	20	80000
137	Orissa Rural Infrastructure Development Association (ORIDA)	Orissa	1	0	Orissa	231	3655000
138	OUTREACH	Karnataka	7	4	Karnataka, A.P., T.N.	2920	18742520
139	PAHAL	Uttarakhand	4	0	Uttarakhand, U.P.	533	3903165
140	PARATH Samithi	M.P.	1	1	M.P.	9231	3697392
141	People's Action for Transformation	T.N.	3	0	T.N.	13951	50044900
142	People's Action for Development and Credit Union (PACADU)	T.N.	1	0	T.N. , Kerala	632	3863434
143	Peoples Action for National Integration (PANI)	U.P.	4	2	U.P.	4220	28057398
144	People's Forum	Orissa	10	7	Orissa	32807	121515963
145	Pikepara Kamala Seva Samity	W.B.	1	1	W.B.	1730	50044900
146	Pragathi Seva Samiti (PSS)	A.P.	1	1	A.P.	42216	414657100
147	Prayas	Jharkhand	1	1	Jharkhand	1050	6350000
148	PRAYAS (Organization for Sustainable	Gujarat	4	2	Gujarat, M.P.	2943	14009457

Contd...

Annexure 2.1: Contd...

Sl. No.	Organisation	HQ State	No. of districts operating	No. of Poor Districts	Operation in states	Client Outreach	Loan Portfolio
148	PRAYAS (Organization for Sustainable Development)	Gujarat	4	2	Gujarat, M.P.	2943	14009457
149	Priyadarshani Mahila Gramin Kalyan Sanstha	Maharashtra	1	1	Maharashtra	4401	11862000
150	Priyasakhi Mahila Sangh	M.P.	1	0	M.P.	3500	5278601
151	Prochesta	Assam	14	5	Assam	39810	5763011
152	Project Concern International/India (Planned Social Concern)	Rajasthan	1	0	Rajasthan	2668	14759478
153	Purvanchal Manav Kalyan Sansthan	U.P.	1	0	U.P.	0	0
154	Pustikar Laghu Vyaparik Pratishtan Bachat and Sakh Sahakari Samiti Ltd	Rajasthan	13	7	Rajasthan, Maharashtra, A.P.	13687	46134900
155	Raghunath Pathagara	Orissa	4	4	Orissa	9220	78370000
156	Rajapur Seva Niketan	W.B.	1	0	W.B.	2870	14778945
157	Rashtriya Gramin Vikas Nidhi (RGVN)	Assam	12	5	Assam, Orissa	44772	269538134
158	Rashtriya Seva Samithi (RASS)	A.P.	2	1	A.P., T.N.	50062	339465491
159	Regional Rural Development Centre (RRDC)	Orissa	1	1	Orissa	2496	14458280
160	S E Investments Ltd	U.P.	6	0	U.P., Rajasthan, Gujarat,Delhi	83333	1210057319
161	Saadhana MicroFin Society	A.P.	4	3	A.P.	76580	324361719
162	SAATH Charitable Trust	Gujarat	1	0	Gujarat	301	2602420
163	Sahabhagi Vikas Abhiyan	Orissa	4	4	Orissa	11221	14823220
164	Sahara Utsarga Welfare Society	W.B.	7	6	W.B.	124865	340143299
165	Sahara Uttarayan	W.B.	6	6	W.B.	41762	137020943
166	Sahayata (Sri Hari Fintrade Pvt Ltd)	Rajasthan	4	2	Rajasthan	2236	11279360
167	Sakhi(An Organisation for Women)	Gujarat	3	1	Gujarat	1596	6204242
168	Sakhi Samudaya Kosh	Maharashtra	7	2	Maharashtra, Gujarat	7694	28800000

Contd...

Annexure 2.1: Contd...

Sl. No.	Organisation	HQ State	No. of districts operating	No. of Poor Districts	Operation in states	Client Outreach	Loan Portfolio
169	Samagra Gram Vikas Samstha(SAGRAS)	Maharashtra	1	1	Maharashtra	2905	4216929
170	Samajik Vikas Samsthan	Bihar	1	1	Bihar	113	340000
171	Samarthan Weekling Development Foundation	Jharkhand	3	3	Jharkhand	1804	6414929
172	Sambhav Micro Finance Institute	M.P.	7	3	M.P., U.P.	664	7026526
173	Samhita Community Development Services	M.P.	2	1	M.P.	1005	4382205
174	SamitaJan Kalyan Parisad	Bihar	1	1	Bihar	780	150000
175	Samman Foundation	Bihar	1	1	Bihar	150	200000
176	Sampada Trust	Maharashtra	12	9	Maharashtra	20894	60700000
177	Samuha	Karnataka	2	1	Karnataka	13583	50873698
178	Sangamam Women's Multipurpose Thrift and Credit Co-operative Society Ltd	T.N.	15	5	T.N., Pondicherry, Karnataka	41052	40611825
179	Sanghamithra Rural Financial Services	Karnataka	17	3	Karnataka, A.P., T.N.	21262	486068000
180	Sanginee Secondary Cooperative Ltd	Orissa	1	1	Orissa	10401	28681818
181	Sarala Women Welfare Society	W.B.	3	2	W.B.	19838	51372165
182	Sarva Jana Seva Kosh Ltd (SJSK)	T.N.	0	0	T.N.	0	0
183	Sarvodaya NaNo. Finances Ltd	T.N.	21	9	T.N., Bihar, Jharkhand, M.P., Rajasthan	523385	885260773
184	Satin Credit Care Network Ltd	Delhi	25	7	Delhi, U.P., Haryana, Rajasthan, Punjab	19648	396241029
185	SATRA	Assam	4	1	Assam	2694	5827000
186	SEARCH-KOPSA	T.N.	3	0	T.N.	25871	150097148
187	SEBA Rahara	W.B.	1	1	W.B.	1925	4706434
188	Seba Sangha	W.B.	1	1	W.B.	0	0
189	Semam Microfinance Investment Literacy and Employment Ltd (SMILE)	T.N.	7	3	T.N.	140637	739311981
190	Sevashram	Kerala	4	0	Kerala	6140	16587400

Contd...

Annexure 2.1: Contd...

Sl. No.	Organisation	HQ State	No. of districts operating	No. of Poor Districts	Operation in states	Client Outreach	Loan Portfolio
191	SEWA Bank	Gujarat	5	1	Gujarat	369837	324564682
192	Shah Shandhu Finance Co. Pvt. Ltd (Mimo Finance)	Uttarakhand	5	3	Uttarakhand, U.P., H.P	15542	57298277
193	Shalom Trust	Kerala	4	1	Kerla, Tamil Nadu	23785	150562940
194	Sharadas Women's Association for Weaker Section	A.P.	11	4	A.P.,Orissa, T.N., Maharashtra	74175	426684583
195	Share Microfin Limited	A.P.	70	47	A.P., Karnataka,Chhattisgarh, Maharashtra, M.P., Rajasthan, UP, Jharkhand, Uttara-khand, West Bengal	1289328	5952969356
196	Shikhar Development Foundation	Delhi	0	0	Delhi	0	0
197	Shramik Bharati	U.P.	2	1	U.P.	15543	43740979
198	Shree Hari Fintrade Pvt. Ltd(SAHAYATA)	Rajasthan	4	1	Rajasthan	2236	11279360
199	Shri Kshethra Dharmasthala Rural Development Project	Karnataka	5	2	Karnataka	574968	3370600000
200	SIDRIB	Bihar	1	1	Bihar	100	210000
201	Siri Micro Fin Society	A.P.	2	2	A.P.	3900	33632840
202	Social Action for Rural Community	Orissa	2	2	Orissa	393	1079500
203	Social Education and Voluntary Action (SEVA)	A.P.	1	1	A.P.	18641	97855182
204	Social Welfare Agency and Training Institute (SWATI)	Orissa	1	1	Orissa	3948	1880000
205	Societal Upliftment & Rural Action for Job and Empowerment (SURAJE)	Bihar	1	1	Bihar	262	1179984
206	Society for Empowerment and Women Advancement (SERV-SEWA)	Jharkhand	1	1	Jharkhand	1060	2669651
207	Society for Model Gram Vikash Kendra	W.B.	2	2	W.B.	7856	13167542
208	Society for Promotion of Youth and Masses (SPYM)	Delhi	1	1	Haryana	0	0
209	Sonata Finance Private Limited	U.P.	10	6	U.P., M.P.	44387	232085909
210	Spandana	A.P.	74	38	A.P., Karnataka,T.N., Maharashtra, Orissa, Chhattisgarh, M.P., Rajasthan	1188861	7284979527

Contd...

Annexure 2.1: Contd...

Sl. No.	Organisation	HQ State	No. of districts operating	No. of Poor Districts	Operation in states	Client Outreach	Loan Portfolio
210	Spandana	A.P.	74	38	A.P., Karnataka,T.N., Maharashtra, Orissa, Chhattisgarh, M.P., Rajasthan	1188861	7284979527
211	Sreema Mahila Samity	W.B.	3	3	W.B.	71786	241593423
212	Srijan Foundation	Jharkhand	1	1	Jharkhand	84	315000
213	Star Youth Association	A.P.	2	2	A.P.	24876	136824800
214	SUPPORT	Jharkhand	4	3	Jharkhand	6857	8758037
215	Swadhaar Finaccess	Maharashtra	1	0	Maharashtra	3524	11056051
216	Swayam Krishi Sangam (SKS) Microfinance Pvt Ltd.	A.P.	215	133	A.P., Karnataka, Bihar, Chhattisgarh, Delhi, Gujarat, Haryana, Jharkhand, Kerala, M.P., Orissa, U.P., Uttarakhand, West Bangal, Rajasthan	1879258	7818627031
217	Swayamshree Microcredit Services (SMCS)	Orissa	24	16	Orissa	28585	137297671
218	The Activists for Social alternatives (ASA)	T.N.	20	7	T.N., Pondichherry	263022	1153007912
219	The Luxmi Development Institute	Manipur	0	0	Manipur	0	0
220	The MAX Wealth Trust	A.P.	1	1	A.P.	30577	220868000
221	The Payakaraopta Women's Mutually Aided Co-operative Thrift and Credit Society Ltd.	A.P.	2	1	A.P.	18702	162623532
222	Tripura Rural Development Organisation	Tripura	1	1	Tripura	42	26000
223	Trust Microfin Services	Bihar	2	2	Bihar	2812	10310512
224	Ujjivan Financial Services Private Limited	Karnataka	6	0	Karnataka, Delhi, W.B.	68033	365695708
225	Ullon Social Welfare Society	W.B.	1	1	W.B.	4138	9102771
226	Unnati Mahila Sangh	M.P.	3	1	M.P.	2966	8465717
227	Upkar	Orissa	3	0	Orissa	835	1225000
228	Uttarakhand Microfinance and Livelihood promotion Cooperative Institutions(UMLPCI)	Uttarakhand	1	0	Uttarakhand	25	162400

Contd...

Annexure 2.1: Contd...

Sl. No.	Organisation	HQ State	No. of districts operating	No. of Poor Districts	Operation in states	Client Outreach	Loan Portfolio
229	Vaagdevi Mahila MACS Ltd	A.P.	1	1	A.P.	5300	7980900
230	Vardan Trust	Gujarat	2	2	Gujarat	11113	48509190
231	Vedika Credit Capital Ltd	Jharkhand	17	16	Jharkhand,Delhi	20482	134092768
232	Vikas Centre for Development	Gujarat	1	1	Gujarat	8145	25769855
233	Village Financial Services Pvt. Ltd.	W.B.	4	3	W.B.	44717	171775121
234	Village Micro Credit Services (VMCS)	W.B.	5	4	W.B.	64676	246653533
235	Village Welfare Society(VWS)	W.B.	5	4	W.B.	0	0
236	Vivekananda Seva Kendra-O-Sishu Uddyan	W.B.	1	1	W.B.	68723	83119856
237	Welfare Services Ernakulam	Kerala	4	0	Kerala	31078	68085444
238	Youth Council for Development Alternatives (YCDA)	Orissa	2	2	Orissa	614	2599000
239	Youth Volunteers Union	Manipur	4	1	Manipur	0	0
240	Yukti Samaj Sewa Society	M.P.	1	0	M.P.	3343	2516216
241	Yuva Chetna Kendra	U.P.	4	2	U.P.	6225	7260000

Source: The Bharat Micro Finance Report-Quick Data 2008

Annexure-3.1

Various Legislations on Regulation of Charging of Usurious Interest Rates

The Indian Moneylenders' Act 1918 is the base of all legislation against the charging of usurious interest rates in the country. Until the early part of the 20th century, in the absence of a proper banking network, poor farmers and petty traders were exposed to the excesses of moneylenders. Legislation to bring the charging of usurious interest rates under some control was enacted in 1918. After Independence and subsequently the enactment of the Indian Constitution, this subject was brought under the purview of state governments. Following this, various state governments have brought their own legislations and ordinances to curb the activities of moneylenders.

The Act has implications for the activities of some MFIs as well. MFIs registered as Societies and Trusts can easily be targeted under the provisions of the state-level legislations and ordinances. Since the status of microfinance as social work and an activity for the alleviation of poverty has never been under any regulatory framework, such institutions can be accused of engaging in unregulated moneylending activity. In this context, the discussion below covers some of the provisions of the Tamil Nadu Prohibition of Charging of Exorbitant Interest Ordinance, 2003 – a matter of considerable discussion in the microfinance sector at present. The explanatory statement of the ordinance says that "In order to obviate the difficulties experienced by the public at large, falling prey to any person charging exorbitant interest like daily *vatti, kandhu vatti, meter vatti* and *thandal*, the government have decided to prohibit lending money for such exorbitant interest and to provide for the stringent punishment therefore and decided to enact a new legislation for the purpose."

The above statement appears to carry all the noble intents of preventing the exploitation of poor villagers, et al. Section 3 of the ordinance says: "No person shall charge exorbitant interest on any loan advanced by him." The term exorbitant interest is defined by the Section 7 of the Tamil Nadu Moneylender's Act 1957. It states:

(a) No moneylender shall charge interest on any loan, at a rate exceeding such rate as the government may, by notification, fix from time to time. Provided that the rate of interest as may be fixed by the government shall be correlated to the current bank rates of lending as may be fixed by the Reserve Bank of India, from time to time.

(b) A moneylender may demand and take from the debtor such charges and in such cases, as may be prescribed.

(c) A moneylender shall not demand or take from the debtor any interest, profit or the sum whatever in excess of that payable under sub-section 1.

The state government has assumed the role of fixing interest rates for moneylenders. Needless to say, the interest fixed by the governments is not scientific and hardly takes into account the various costs and risks involved therein. Moreover, they are rarely followed, as is apparent from the findings of various studies of interest rates being charged by moneylenders.

In fact, the state regulations on charging of usurious interest rates vary from state to state. The Rajasthan Moneylender's Act 1963 defines interest as any sum, whatsoever name called, in excess of the principal paid or payable to a money lender in consideration of or otherwise in respect of, but does not include any sum lawfully charged by a moneylender for or on account of costs, charges or expenses in accordance with the provision of this Act or any other law for the time being in force.

The Karnataka Moneylender's Act, 1961 defines interest as the return to be made over and above what was actually lent, whether the same is charged or sought to be recovered specifically by way of interest or otherwise, but does not include any sum lawfully charged by a moneylender for or on account of costs, charges or expenses in accordance with the provisions of this Act, or any other law for the time being in force. Clearly, the government retains the right to play around with the interest rates by bringing in ordinances from time to time. While this provision appears to be in the public interest, this also leaves the possibility of other institutions becoming easy prey.

The Tamil Nadu Moneylender's Act 1957 also requires the moneylenders to obtain a license from the state government and follow some requirements for registration as a moneylender. There is a list of the requirements that needs to be followed for obtaining a license. Section 3 of the Moneylender's Act states

(a) No person shall on and after the date on which the provisions of this Act are brought into force in any area carry on or continue to carry on business as a money lender at any price in such area except under and in accordance with the terms of a license.

(b) Where a money lender has more than one shop or place of business whether in the same town or village or in different town or village or in different towns or villages, he shall obtain a separate license in respect of each such shop or place of business.

(*c*) Where a moneylender is a registered firm the license shall be obtained in the firm's name.

(*d*) Where a moneylender is an undivided Hindu family, the license shall be obtained in the name of the manager or the Karnavan or the yajaman, as the case may be described as such in the license.

(*e*) Where a moneylender is any other association of individuals, not required to be registered under the Companies Act, 1956, a separate license shall be obtained by each such individual in his name describing himself as a member of the association, provided that nothing contained in this sub-section shall affect the operation of Section 59of the Indian Partnership Act, 1932.

Similarly, Section 6 of the Bombay Moneylender's Act, 1946 states that every moneylender shall annually before such date as may be prescribed make an application in the prescribed form for the grant of a license to the Assistant Registrar of the area (within the limits of the place where he intends to carry on the business of money-lending or if he intends to carry on such business at more than one place in the area, the principal place of such business is situated). The above provisions clearly require all the MFIs (except cooperative societies, cooperative banks, NBFCs and Section 25 companies) to obtain a license for carrying out their business – if they are practicing professional money lending.

Annexure-3.2

Exemptions Available to Companies Registered under Section 25 of the Companies Act, 1956

Sl. No.	Notification number & date	Provisions dispensed with	Provisions from which exemption has been granted	
			Section	Nature of provisions
1	Sec. 263A inserted by Act No. 65 of 1960 (28.12.1960)	Provisions relating to putting a resolution to vote (provisions in the Articles of such company if any for election of directors by ballot shall prevail)	177	Voting to be by show of hands at first instance.
2	-do-	Provisions in articles of association relating to electing the directors by ballot at AGM shall prevail.	255, 256, 263	Manner of appointment of directors.
3	SO-1578 GI dated 8.7.1961	Whole provisions of Section 147 regarding publication of name and address by a company.	147	Publication of name and address of company.
4	-do-	Whole provision relating to filing with the Registrar a return stating names of members and other details.	160(1)(aa)	Annual return to accompany details.
5	-do-	Whole provisions relating to holding of AGM during business hours and on a day which is not a public holiday.	166(2)	AGM to be called during the business hours and on a day which is not a public holiday.
6	-do-	Provision relating to 21 days clear notice for calling a general meeting. (Only 14 days notice may be given for calling AGM).	171(1)	Length of notice for calling meeting.

Contd...

Annexure 3.2: Contd...

Sl. No.	Notification number & date	Provisions dispensed with	Provisions from which exemption has been granted	
			Section	Nature of provisions
7	-do-	Provisions relating to retention of books of accounts (books of accounts relating to a period of not less than four years need only be preserved).	209(4A)	Retention of books of account for a period not less than eight years.
8	-do-	Whole provisions relating to leaving a notice signifying candidature for office of director.	257	Right of persons to stand for directorship.
9	-do-	Consent for directorship not required to be filed with the company	264(1)	Consent of candidate for directorship to be filed with the company and consent to act as a director to be filed with Registrar.
10	-do-	Provision relating to holding of meeting of board of directors provided at least one meeting should be held within every six calendar months.	285	Meeting of board of directors should be held once in every three calendar months.
11	-do-	Provisions relating to holding of meeting of board of directors (quorum at a meeting of board shall be eight members of ¼ of its total strength subject to minimum of two).	287	Quorum of meetings.
12	-do-	Interest to be disclosed in the specified manner only in respect of contracts to which Sec297 (1 &3) is applicable.	299	Disclosure of interests by directors, etc.

Contd...

Annexure 3.2: Contd...

Sl. No.	Notification number & date	Provisions dispensed with	Provisions from which exemption has been granted	
			Section	Nature of provisions
13	-do-	Register of contracts to which Sec 297(1&3) is applicable only need to be maintained.	301	Maintenance of register of contracts, companies and firms in which directors are interested.
14	-do-	Provisions relating to sending an abstract terms of contract or variations of contract (such information need not be sent). Provisions relating to recording minutes. If the articles provide for confirmation of minutes by circulation, minutes may be recorded within 30 days of conclusion of every meeting.	302(2)	Abstract of terms of contract or variations of contract for appointment of managing director and interest of any director therein to be sent to the members.
15	SO-2767 dated 5.8.1964	Provisions relating to obtaining government approval for increasing number of directors (Govt. approval not required).	193	Minutes of proceedings of general meeting of board and other meetings.
16	-do-	Provisions relating to obtaining government approval for increasing number of directors (Govt. approval not required).	259	Increase in number of directors to require government sanction.
17	-do-	Provisions relating to exercising/deciding of following powers at a board meeting: a. to borrow money otherwise than on debentures; b. to invest funds of the company; c. to make loans (These powers may be exercised by circulation instead of at a meeting.)	292	Certain powers to be exercised by board only at a meeting.

Contd...

Contd...

Sl. No.	Notification number & date	Provisions dispensed with	Provisions from which exemption has been granted	
			Section	Nature of provisions
18	GSR No. 73 dated 30.12.1965	Copies of balance sheet and auditors report should be sent not less than 14 days before the date of general meeting.	219(1)	Right of members to copies of balance sheet and auditor report.
19	SO-35E dated 9.2.1976	Provisions relating to appointment of a Secretary (Person appointed to perform the duties of Secretary need not possess prescribed qualification).	2(45)	Secretary of a company to have prescribed qualification.

Source: Nabhi's Formation and Management of a Society, September 2004

Annexure-3.3

A Note on the Rating Principles and Framework of M-CRIL

Governance and strategy: The assessment depends on the quality and appropriateness of the board composition, its role and overall organizational strategy.

Management systems are rated on the quality of human resources, the strength of critical systems like accounting and management information. The performance of the MFI on staff productivity and ensuring compliance is also assessed.

Financial performance evaluation requires the rating team to construct financial statements, based on internationally accepted prudential norms, to present a fair picture of the operations. In case of organization's doing more than one activity, the costs are allocated based on a field assessment of usage. These are then used to assess the MFI's financial strength based on aspects like repayment performance, asset quality, liquidity, asset liability management (ALM) and profitability.

Annexure-6.1: SBLP Progress

Progress under Microfinance – Savings of SHGs with Banks
Region-wise / State-wise / Agency-wise position as on 31st March, 2011

(Amount ₹ lakh)

Sr. No.	Region / State	Commercial Bank		Regional Rural Bank		Cooperative Bank		Total	
		No. of SHGs	Saving Amount	No. of SHGs	Saving Amount	No. of SHGs	Saving Amount	No. of SHGs	Saving Amount
A	***Northern Region***								
1	Haryana	20397	7891.11	12919	1849.97	2003	179.37	35319	9920.45
2	Himachal Pradesh	28571	1881.90	5955	852.19	18587	974.41	53113	3708.5
3	Punjab	32156	3768.09	3483	211.18	5280	405.89	40919	4385.16
4	Jammu & Kashmir	3828	304.62	596	34.14	1145	48.38	5569	387.14
5	Rajasthan	103817	4808.50	61937	5019.74	68039	4203.46	233793	14031.7
6	New Delhi	3015	307.22	0	0.00	80	16.33	3095	323.55
7	Chandigarh	964	100.66	0	0.00	0	0.00	964	100.66
	TOTAL	**192748**	**19062.1**	**84890**	**7967.22**	**95134**	**5827.84**	**372772**	**32857.16**
B	***North Eastern Region***								
1	Assam	84597	4573.76	136642	2853.01	23881	769.83	245120	8196.6
2	Meghalaya	8484	227.39	2169	148.73	0	0.00	10653	376.12
3	Nagaland	5646	156.19	625	34.08	3595	172.72	9866	362.99
4	Tripura	10979	244.10	17644	2762.21	5689	388.99	34312	3395.3
5	Arunachal Pradesh	5822	92.03	1050	78.75	207	15.53	7079	186.31
6	Mizoram	1837	49.61	2085	106.20	670	22.30	4592	178.11
7	Manipur	5027	145.18	5279	95.05	0	0.00	10306	240.23
8	Sikkim	2212	123.70	0	0.00	599	45.24	2811	168.94
	TOTAL	**124604**	**5611.96**	**165494**	**6078.03**	**34641**	**1414.61**	**324739**	**13104.6**
C	***Eastern Region***								
1	Bihar	109840	6954.55	138357	3902.76	0	0.00	248197	10857.31
2	Jharkhand	58331	12699.95	28874	1495.81	0	0.00	87205	14195.76
3	Orissa	243965	10522.85	199679	19794.78	77508	5037.09	521152	35354.72
4	West Bengal	324200	20985.98	190403	30581.48	151711	28746.68	666314	80314.14
5	A & N Islands (UT)	1180	36.94	0	0.00	3570	78.74	4750	115.68
	TOTAL	**737516**	**51200.27**	**557313**	**55774.83**	**232789**	**33862.51**	**1527618**	**140837.61**
D	***Central Region***								
1	Chhattisgarh	45660	2265.99	50686	5050.57	21821	1112.43	118167	8428.99
2	Madhya Pradesh	81667	5116.40	57870	1891.15	14280	4666.54	153817	11674.09
3	Uttarakhand	23242	2292.19	10270	657.30	10783	1015.88	44295	3965.37
4	Uttar Pradesh	205734	25893.59	255508	10140.05	8915	235.92	470157	36269.56
	TOTAL	**356303**	**35568.17**	**374334**	**17739.07**	**55799**	**7030.77**	**786436**	**60338.01**
E	***Western Region***								
1	Goa	4985	473.66	0	0.00	2941	345.07	7926	818.73
2	Gujarat	111708	12285.44	50625	2507.22	30501	2510.47	192834	17303.13
3	Maharashtra	420259	31172.29	52685	3351.09	287217	30255.89	760161	64779.27
	TOTAL	**536952**	**43931.39**	**103310**	**5858.31**	**320659**	**33111.43**	**960921**	**82901.13**
F	***Southern Region***								
1	Andhra Pradesh	1041184	101288.99	401730	27543.00	23311	1948.23	1466225	130780.22
2	Karnataka	219667	50674.09	164100	15607.23	180778	30221.55	564545	96502.87
3	Kerala	390242	33008.78	40851	1141.09	62254	7993.71	493347	42143.58
4	Lakshadweep	164	10.36	0	0.00	0	0.00	164	10.36
5	Tamil Nadu	704598	80353.24	88789	5697.09	149711	13673.54	943098	99723.87
6	Puducherry	19495	2297.07	2586	133.80	0	0.00	22081	2430.87
	TOTAL	**2375350**	**267632.53**	**698056**	**50122.21**	**416054**	**53837.03**	**3489460**	**371591.77**
	GRAND TOTAL	**4323473**	**423006.42**	**1983397**	**143539.67**	**1155076**	**135084.19**	**7461946**	**701630.28**

Annexure-7

Govt. Order G.O.Ms .No. 83

GOVERNMENT OF ANDHRA PRADESH

ABSTRACT

Merger of Swarnajayanthi Gram Swarozgar Yojana (SGSY) and Velugu into Indira Kranthi Patham – Common guidelines – Orders – Issued.

PANCHAYAT RAJ AND RURAL DEVELOPMENT (RD.III) DEPARTMENT

G.O.Ms.No.83. **Dated.21.03.2005.**

Read:-

G.O.Ms.No.15, PR&RD (RD.III) Department, Dated.25.01.2005.

*****

O R D E R

Government in the reference read above, have issued orders for integration of SGSY and Velugu Programmes into Indira Kranthi Patham. There is need for issuing detailed guidelines for operationalisation of the Government's decision to merge SGSY and Velugu into Indira Kranthi Patham. The following guidelines may be approved for implementation of Indira Kranthi Patham.

I. Implementation arrangements

Chief Executive Officer, Society for Elimination of Rural Poverty (SERP) is so far responsible for implementation of APDPIP / APRPRP (hitherto called Velugu), through DRDAs and DPMU Velugu in the State.

1. With the merger of SGSY and Velugu into a comprehensive programme called "Indira Kranthi Patham'". The CEO, SERP assumes responsibility to implementation the following components of Indira Kranthi Patham:
   (a) AP District Poverty Initiatives Project.
   (b) AP Rural Poverty Reduction Project.
   (c) Swarnajayanthi Gram Swarozgar Yojana including Special Projects.
   (d) State Revolving Fund.
   (e) SHG-Bank Linkage.
2. In addition to Indira Kranthi Patham, Chief Executive Officer, SERP, is responsible to implement National Old Age Pension Scheme and National Family benefit Scheme.
3. Director SHGs will assist the CEO, SERP in the implementation of Indira Kranthi Patham.

P.T.O.

4. Detailed exercise for re deployment of the staff under SERP as well as DRDA will be carried out to ensure availability of uniform human resource in mandals to the extent possible.

II. Utilisation of SGSY Resources

1. In the 6 AP DPIP districts where all Mandals are covered under AP DPIP SGSY resources will be additionality to the amounts being spent under AP DPIP.
2. In the remaining districts, the SGSY resources could be utilized only in the Mandals not covered under APRPRP.
3. SGSY has the following 5 major components:-
   (a) Training.
   (b) Infrastructure.
   (c) Revolving Fund.
   (d) Individual loans.
   (e) Group loans.

i. Training

10% of the total SGSY outlay will be earmarked for trainings. Common strategy for institutional building, livelihood support and marketing support for strengthening of the SHGs will be followed in all the mandals. The additional expenditure towards capacity building of SHGs over and above 10% of the SGSY outlay could be booked to DPIP/RPRP account.

ii. Infrastructure, Revolving Fund, Group Loans

The DRDAs will combine the three components such as Infrastructure, Revolving Fund and Group Loan into a single package and make it available to the groups following the same process followed for providing Community Investment Fund under APDPIP/RPRP. In other words, the SHGs in DPIP/RPRP Mandals and the SHGs in non DPIP/RPRP Mandals will go through the same process of preparation of micro credit plan for accessing SGSY fund, just like the way CIF is accessed by the groups in DPIP / RPRP Mandals. However, bank linkage will remain a prerequisite for availing SGSY subsidy.

iii. Individual Loans

10 percentage of the total SGSY outlay in each district may go towards individual loans to most vulnerable amongst poorest- **ex:** persons with disability, commercial sex workers, victims of atrocities or natural calamities.

III. Self-Help Group - Bank Linkage Programme

Self Help Group – Bank Linkage Programme will be taken up in all the Mandals.

IV. State Revolving Fund

The State Revolving Fund will be made available to the Self Help Groups in all the Mandals following the guidelines detailed below

(i) All the Self Help Groups which have completed 6 months of savings and loaning will be eligible to access State Revolving Fund.

(ii) In each Village priority will be given to the SHGs of the poorest of the poor.

(BY ORDER AND IN THE NAME OF THE GOVERNOR OF ANDHRA PRADESH)

K. RAJU

PRINCIPAL SECRETARY TO GOVERNMENT (RD)

To

All the District Collectors.

The Chief Executive Officer, Society for Elimination of Rural Poverty, Hyderabad.

The Commissioner, Women Empowerment and Self Employment, Hyderabad.

Copy to

All Special Chief Secretaries / Principal Secretaries / Secretaries.

Principal Secretary / Special Secretary to Chief Minister.

PS to Minister, (RD,RWS & EG). / PS to Minister (PR).

PS to all Ministers.

PS to Chief Secretary to Government.

PS to Principal Secretary to Government, (RD).

PA to Additional Secretary to Government, (RD).

SF/ SC.

// FORWARDED :: BY ORDER //

Annexure-8.1

Questionnaire for SHG Groups

PADERU AGENCY AREA, VISAKHAPATNAM DISTRICT, A.P

1.0 GENERAL INFORMATION:-

A. PERSONAL DATA

1.1 Name

1.2 Address Village, ...Panchayat..........Mandal............

1.3 Age – (1) < 25 (2) 26-35 (3) 36-45 (4) 46 - >

1.4 Education – (1) Illiterate (2) Literate (3) Primary (4) Middle Class (5) High School (6) Intermediate (7) Graduation (8) Post Graduation and above (9) Any Professional Course and Technical

1.5 Marital Status (1) Married (2) Unmarried (3) Widow (4)Divorce (5) Separated

1.6 If married, no. of children -
(1) Major > 18 Minor < 18 (2) Girls Boys

2.0 FAMILY DATA

2.1 Type of Family : (1) Nuclear (2) Joint

2.2 Total members of Family

2.3 Number and Age group of children (1) 0-5 yr.(2) 6-10 yr. (3) 11-15 yr. (4) 16-20 yr. (5) 21-25 yr. (6) > 25 yr.

3.0 HUSBAND'S EMPLOYMENT DATA

3.1 Husband's Education – (1) Illiterate (2) Literate (3) Primary (4) Middle Class (5) High School (6) Intermediate (7) Graduation (8) Post Graduation and above (9) Any Professional Course and Technical

3.2 Employment - (1) Employed (2) Unemployed (3) Self employed (4) Professional

3.3 Approx. Annual Income (₹)

3.4 Subsidiary employment - (1) Agriculture(2) Animal Husbandry (3) Labour

3.5 Landholding Size -(1) < 1 Acre (2) 1-5 Acre (3) 6-10 Acre (4) 11 -> Acre (5) Landless

4.0 ABOUT YOURSELF

4.1 Employment - (1) Employed (2) Self employed(3) Professional (4) Labour

4.2 Approx. Income Monthly (₹)

4.3 How do you feel with your family (1) Fully Satisfied(2) Satisfied (3) Indifferent (4) Burdened (5) Very happy (6) Not happy

4.4 Who take the decision in your family (1) Yourself(2) Husband (3) Both

5.0 HOUSING FACILITIES

5.1 Type of House (1) Own (2) Rented (3) Others

5.2 Electricity (1) Yes (2) No

5.3 Toilet Facility (1) Open defecation (2) Toilet

5.4 Cooking Device (1) Gas (2) Kerosene Oil (3) Wood(4) Others

5.5 Drinking Water (1) Piped Water (2) Hand Pump (3) Well(4) Others

5.6 Who spend generally family income- (1) Self (2) Husband (3) Children (4) Father (5) Family member

6.0 ATTITUDES

6.1 Do you approve - (1) Dowry System (2) Inter-Caste Marriage (3) Inter-Religion marriage (4) Love Marriage (5) Widow Marriage (6) Child Marriage (7) Divorce (8) Completely Family Planning (9) Job reservation for women (10) Punishing wicked husband

2 ABOUT THE ORGANIZATION

7.0 GROUP SIZE

7.1. How many members are in your group? (1) < – 9 (2) 10 – 15 (3) 15 – 18 (4) 18 +

7.2 Group size is (1) Stable (2) Increased (3) Decreased

7.3 When you joined the group (year)?

7.4 Who suggested joining the group? (1) Self (2) Family Members (3) Friends/ Relatives (4) NGO's workers (5) Other members of the Group

7.5 What is your position in Group? (1) Ordinary Member (2) Active Member (3) Cashier/ Secretary/President

8.0 MEETINGS

8.1 Frequency of meeting- (1) Weekly (2) Fortnightly (3) Monthly

8.2 How many members attend the meeting? (1) All (2) Few (3) Some members

8.3 Who calls the meeting (1) Group Members (2) NGO (3) Projects Staff (4) Mixed

8.4 Who decides agenda of the meeting (1) Majority of members (2) Some members (3) Group members and link worker jointly (4) Link worker

8.5 Who take the decision in the meeting- (1) All members(2) President (3) IKP Staff (4) NGO Facilitator (5) Mixed

8.6 How is the decision taken?

i. By consensus
ii. By voting
iii. Group representatives
iv. Mixed

9.0 KNOWLEDGE AND AWARENESS OF SHG's ACTIVITIES

1. Meeting Calendar
2. Rules and Regulations
3. Information on Group Records
4. Cash in hand
5. Balance in Bank
6. Outstanding Loan
7. Total Capital of the Group
8. Savings of Group
9. Total loaning of Group
10. No. of Members taken loan
11. No. of Members repaid loans
12. Name of Bank
13. Income of Group
14. Objectives of Group
15. Achievements of Group
16. Constraints of Group
17. Others (Specify)

10.0 SAVING PATTERN

10.1 Your saving rate

(i) Present (₹) (ii) Initial (₹)

10.2 Main Purpose of savings (Give your order of Preferences)

Sl. No.	Purpose of Saving	Rank
1	Social Security	
2	Food Security	
3	Education	
4	Medical	
5	Marriage	
6	Festivals	
7	Emergencies	
8	Agriculture	
9	Asset Building	
10	Self Respect	
11	Others (Spe.)	

10.3 Your cumulative savings (₹)

10.4 Interest earned on savings (₹)

10.5 Your Present bank balance (₹)

11.0 TRAINING AND EXPERIENCE

11.1 Whether you received training/ orientation/ exposure related to SHG's, activities

(1) Yes

(2) No,

If Yes,

Type of Training	No. of Participations	Usefulness (Yes/ No)
Orientation		
Exposure tour		
Skill Development (masonry, tailoring, embroidery, wielding, electrician, computer operator etc.)		
Marketing Linkages		
Micro Enterprises Development		
(Turmeric processing, Rajmah value addition, medicinal plant value addition, etc)		
Others (Spe.)		

11.2 Who provided training?

(1) NGO's activists and representatives

(2) Resource Persons

(3) Govt. Officials

(4) Others

11.3 Impact of such training and exposure visits (Please give your order of preference)

Sl. No.	Impact of training	Rank
1	Confidence Building	
2	Skill Development	
3	Marketing Linkage	
4	Bank Linkages	
5	Linkages with Govt. Officials	
6	Knowledge on rights, entitlements and development programmes	
7	Managerial efficiency for Micro enterprise development	
8	Enhanced Income and earnings	
9	Active Participation in decision making in family	
10	Active Participation in Development Programmes	
11	Active Participation in Decision making outside the family	
12	Others (Spe.)	

12.0 CREDIT UTILISATION

12.1 Total No. of loans

	No. of Times	Amount (₹)
Internal Loaning		
Bank Loaning		
Cash Credit		
Repayment		
Balance		

12.2 Purpose of loaning

Sl. No.	Purpose	Required	Received	No. of Loans	Interest Rate
1	Consumption				
2	Agriculture				
3	Animal Husbandry (Goatry, Poultry etc)				
4	Income generating activities				
	i. Banana Cultivation,				
	ii. Petty-shops,				
	iii. Leaf-plate business,				
	iv. Petty-business at Sandies,				
	v. others				
5	Asset Building				
	i.Land				
	ii.Building				
	iii.Machine				
	iv.Good education				
	v.Trainings				
	vi.Getting a job				
	vii.Etc				
6	Emergencies				
7	Others				
	Total				

12.3 Whether you got benefits from government schemes

(i) Yes (ii) No

If yes,

Sl. No.	Name of Scheme/Departments	Amount/ Subsidies	Visible benefits
1	ESS		
2	Pensions		
3	AWY		
4	NREGP		
5	GCC		
6	Education		
7	Health		
8	Animal Husbandry		
9	Coffee		
10	Revenue		
11	PDS		
12	Electricity		
13	IKP		

13.0 IMPACT OF SHG's

13.1 Changes on socio-economic status after joining SHG

(1) Same (2) Increased (3) Decreased (4) Decorated

1. Increased Mobility
2. Recognition in family
3. Recognition in community
4. Interaction with outsiders
5. Access to better and improved education
6. Access to Health services
7. Access to Immunization
8. Access to sanitation facility
9. Access to credit sources
10. Asset Building
11. Increased Family Income
12. Equipped with Skills
13. Voicing your concern
14. Nutrition awareness
15. Family Planning awareness
16. Girl Child development awareness
17. Health awareness
18. Decision making related to child centered
19. Decision making related to money centered
20. Participation in Development Programmes
21. Assured Food Security
22. Got employed (Partial and full)
23. Reduced Dependency on Moneylenders
24. Hands on experience on Organization management
25. Exposure to Bank and finance management
26. Acquired Leadership quality
27. Participated in Panchayat Election
28. Own in Panchayat Election
29. Change in Livelihood/Occupation
30. Capability to manage any risk (family/community)

14.0 NETWORKING AND CONVERGENCE

14.1 Do the members want to meet and interact with other groups located in vicinity of their village?

If yes, Why? ....

If No, Why? ....

14.2 Benefits from Government Schemes.

Sl. No.	Name of the Schemes	No. of Beneficiaries	Reasons for not getting the benefit

15.0 INFLUENCE POWER OF SHG ON VILLAGE AND COMMUNITY AFFAIRS

15.1 What are the perceptions of community towards women organized in Self Help Group (SHG)?

1. Well Organized Family
2. Good relationship with their Husband
3. Check on alcoholism
4. Control over saving
5. Self confidence
6. Awareness
7. Others (Spec.)

15.2 Major community development initiatives taken by SHG at the village level?..

16.0 Your main problems related to functioning of SHGs. List them out

....

17.0 Your suggestions for improving the functioning of SHG's.

....

Annexure-8.2

Questionnaire for Self-help Group Leaders

PADERU, VISAKHAPATNAM DISTRICT

1.0 General Information

(i) Name of Village .... Panchayat:............Mandal.......

(ii) Name of the sample group: (SHG) ....

(iii) Formation Date..................

A/C opening Date ....

(iv) Group Size (Members)

(i) Initial: .... (ii) Present: ....

(v) Name of Bank: ....

Account No.:.............

2.0 Organizational Management

Group Size:

Stable Increased Decreased

If Group size has increased, how new members have been inducted: ....

If Group size has decreased, why the members have left: ....

Meeting

(i) Frequency of Meeting:- Weekly Fortnightly Monthly

(ii) Agenda of Group meetings:

Common: ....

Specific: ....

(iii) Who calls the meeting? (a) Group (b) NGO (c) IKP Staff (d) Mixed

(iv) Who decides agenda of the Meeting

(a) Majority of members

(b) Some members

(c) Group representatives

(d) Group members and IKP staff jointly

(e) IKP Staff

(f) NGOs

(vi) How the decision is taken?

(a) By consensus

(b) By voting

(c) By Group representatives

(d) Joint Decision in consultation with members

3.0 Financial Management

Thrift and Saving Management

(i) How is the saving collected?

(a) Deposited by members in group meeting on fixed date

(b) Collected by representatives from members individually out side meetings

(ii) How is the group money kept?

(a) Deposited in the bank account

(b) Distributed as a loan among the members

(c) Remains with group representatives

(d) Kept in cash box of the group

(iii) Does the group keep cash in hand for sudden requirements of members? If yes

(a) Who keeps cash..

(b) What is the Amount....................................

(c) How is the loan given in case of sudden requirement ....

(iv) Who operates Bank accounts on behalf of group?

(a) President, Treasurer and Secretary

(b) President and Treasurer

(c) President and Secretary

(d) Any two of above three

(e) Others if any

(v) Frequency to visit the bank to deposit and withdrawal

(a) Once in a month

(b) More than once in a month

(c) Occasionally

(vi) How do members save the money to deposit with the group?

(a) Curtailing their expenditure

(b) Saving from the income
(c) Provided by their family members
(d) Others (specify)

Credit Rotation

(A) Spread of loan

Sl. No.	Particulars	Numbers
1	No. of members who have taken loan	
2	No. of members who have taken loan more than once	
3	Number of non SHG persons who received loan	

(B) Credit Utilization and Repayment

Sl. No.	Loan category	Amount (in ₹)	Repayment status
1	Consumption (domestic)		
2	Agriculture		
3	Animal husbandry		
4	Income Generation Activities		
5	Asset building		
6	Emergencies		
7	Education		
8	Others		
	Total		

(C) Credit Rotation Process

(a) How do members raise their demand for loan from group?

....

(b) How does group sanction the loan?

....

(c) How does group prioritise loan demand of members if capital available with group is not enough to meet the demand of all the members?

....

....

(d) How is the loan disbursed to the members?

....

....

(e) How much time group takes in sanction and disbursement of loan to members?

....

(f) What are the documents required for withdrawal of money from the bank and

Disbursal to the members?

....

(D) Repayment

(i) What are the terms and conditions evolved by group in regards to repayments of loans?

(a) Interest rate ....

(b) Fine in case of default ....

(c) No. of defaults ....

(d) Frequency of installment ....

(e) Fixed term repayment ....

(f) Action taken by group to deal with the defaults ....

(ii) How the money generated through interest, fine and others source is utilized by the group?

(a) Added to group capital

(b) Utilized for group activities

(c) Shared as dividend among members

(d) Others (specify)

(E) Auditing of groups

(i) Frequency of group auditing?

(a) Monthly

(b) Quarterly

(c) Six monthly

(d) Annual

(e) Not Done

4.0 Knowledge and Awareness of SHG Members

4.1 About group processes and activities

Sl.No.	Activities	Majority	Some	None
1	Meeting Calendar			
2	Rules and regulations			
3	Information on Group Records			
	3.1 Cash in hand			
	3.2 Balance in Bank			
	3.3 Out standing loan			
4	Total capital of the group			
5	No. of members who have taken loan			
6	No. of members who have repaid loan regularly			
7	Name of the Bank			
8	Income of the group through interest, fine			
9	Objectives of the group			
10	Achievements of the group			
11	Cumulative Individuals saving amount			
12	Bank Transactions			
13	Constraints of the groups			
14	Others			

4.2 Social Awareness

(i) What are the local issues affecting their life as perceived and prioritized by SHG members?

....

....

(ii) Availability of basic services to group members

Sl. No.	Areas of services	Service Providers	Availability		
			Majority	Some	None
1	Maternity services				
2	Immunization of child				
3	Immunization of mothers				
4	Children going to schools				
5	Access to PDS				
6	Sanitation facility				
7	Safe drinking water				
8	Family planning				
9	Pension Scheme				
10	Aids				

(iii) What efforts are undertaken by groups to avail basic services?

....

....

5.0 Increased savings of SHGs members

(i) Saving Rate (i) Initial: .... (ii) Present: ....

(ii) What has motivated them to increase their saving rate?

....

....

(iii) Why have they chosen to save in group?

....

....

(iv) Average cumulative saving per member..

(v) Average Gains (Dividend) per member ....

(vi) Main purpose of saving as perceived by members

Sl. No.	Purpose	Code No. (Majority-1, Some-2, None-3)
1	Social security	
2	Food security	
3	Education	
4	Medical	
5	Marriage/ festivals	
6	Emergencies	
7	Agriculture	
8	Assets building	
9	Self respect	
10	IGA	
11	Micro Enterprises	

6.1 Income generation activities undertaken with project support or through internal lending

(i) Activity ....

(ii) Date of initiation ....

(iii) Members engaged ....

(iv) Details of inputs and source

6.2 Details of outputs

Sl. No.	Activities	Total Production	Consumed	Sold	Rate	Total	Income

6.3 What are the constraints/ difficulties faced by group/ members while being engaged in income generation activities?

....

....

6.4 What are appropriate solutions to overcome the constraints and difficulties as perceived by the group?

....

....

7.0 Networking and Convergence:

7.1 Do the members want to meet and interact with other groups located in vicinity of their village? If Yes, why?

....

....

7.2 How is the Village Organisation usesull to cater the need of the SHGs?

....

7.3 How can VO serve better?

....

7.4 How can a marginalized member get her voice heard at VO meeting and issues sorted out?

....

7.5 Are the members aware of on going schemes of line departments in their village?

If yes, name the Schemes?

....

7.6 Are the members availing benefits of these schemes? Give Details

Sl. No.	Name of the Schemes	No. of beneficiaries	Reasons for not availing the benefits

8.0 Influence power of SHG on village and community affairs:

8.1 What are the perceptions of community towards women organized into Self Help Group?

(a) Well organised family

(b) Good relationship with their husband

(c) Check on alcoholism

(d) Say in family affairs

(e) Control over saving

(f) Self confidence

(g) Social Awareness

Increase decision making power of SHG members in their family affairs:

Sl.No.	Areas	Before group formation			After group formation		
		Male	Female	Joint	Male	Female	Joint
1	Marriage of girls						
2	Marriage of boys						
3	Education of children						
4	Loan arrangement						
5	Purchasing						
6	Savings						
7	Expenditure on festivals						
8	Participation in meetings						
9	Interaction with outsiders						
10	Asset building						
11	IGAs						
12	Income from SHG initiated IGAs						

8.2 Major community development initiatives taken by SHG at the village level?

....

9.0 Income and Expenditure of SHG:

Sl.No.	Inflow	Amount (₹)	Outflow	Amount (₹)
1	Members contribution		Stationary	
2	Fund leveraged from Bank		Travel for staff	
3	Fund accessed through RF		Expr. on meetings	
4	Interest earnings		Others	
5	**Total**			

Bibliography

1. Ahmad, M.A., Women Empowerment: Self Help Groups, Kurukshetra, April, 1999.
2. Annual Progress Report, Sustainable Tribal Empowerment Project (STEP), CARE India, 2006, 2007.
3. Annual Progress Reports, ITDA, Paderu, Visakhapatnam District, Andhra Pradesh.
4. Annual Report: 1998-99, Department of Rural Employment and Poverty Alleviation, Ministry of Rural Development, New Delhi, 1999.
5. Antony, Valsamma (2005): "Regional Rural Banks as Champions of Rural Credit", Kurushetra, January, pp. 20-23.
6. Asian Development Bank (2000): "Finance for the Poor: Microfinance Development Strategy".
7. Austinice (2010): "Impact of Micro-finance on Economic Development", www.studymode.com.
8. Badajena, S.N. and Gundimeda, H (2009): Self Help Group Bank Linkage Model and financial Inclusion in India, www.skoch.in.
9. Bahawalpur District", 2011 International Conference on Economics and Finance Research, IPEDR vol.4 (2011)IACSIT Press, Singapore.
10. Banking with the poor Network (2009): Microfinance Industry Report: Bangladesh 2009, Banking with the Poor Network in Collaboration with SEEP.
11. Bapuji, M. (1993): "Tribal Development Administration", Kanishka Publishing House, Delhi, pp. 57-82.
12. Barik, S. (2009): "Policy of Total Financial Inclusion (TFI) and Access to Finance by the tribal Self Help-groups: Reworking on the modality of service delivery", Paper presented at the National Seminar on DYNAMICS OF BANKING SECTOR REFORMS: Emerging Issues, Challenges and Strategies, AU, July 19-21, 2009.

13. Barik, S. K (2009): Policy of Total Financial Inclusion And Access to Finance by the Tribal Self-help Groups : Re-working on the Modality of Service Delivery, paper presented at the National Seminar on Dynamics of Banking Sector Reforms: Emerging Issues, Challenges and Strategies, Centre for Banking Studies and Research, in Association with IBA Ball 19-21 July 2009.
14. Barik, S.K (1998): "NTFP and Tribal Livelihood: A study of tribal of Malkangiri and Rayagada districts of Orissa, IIFM, Bhopal.
15. Barik, S.K (2012): "Micro Credit for Empowerment of Micro People: A Study of SHGs in Paderu Integrated Tribal Development Agency (ITDA), Andhra Pradesh", Elixir Fin. Mgmt. 48 (2012) 9437-9441.
16. Barik, S.K, Rao. A, Sarath. A and Sobharani[1] (2008) in their article entitled "Are Tribal Women Empowered: Understandings from Sustainable Tribal Empowerment Project (STEP) and IKP in ITDA Paderu of Visakhapatnam Districts, Andhra Pradesh, Paper presented in Nationa Seminar in Honor of Prof. K.S Challam, A.U.
17. Basu, P and Srivastava, P (2005) "Scaling-up Microfinance for India's Rural Poor", World Bank Policy Research Working Paper 3646, June 2005.
18. Bateman, M (2011): Confronting Microfinance: Undermining Sustainable Development. Sterling VA: Kumarian Press.
19. Bateman, M and Chang, H(2012): "Microfinance and the Illusion of Development: From Hubris to Nemesis in Thirty Years", World Economic Review, Vol-I, 13-36.
20. Bhatia, N (2007): "Revisiting bank-linked self-help groups-A Study of Rajasthan State", Reserve Bank of India Occasional Papers ,Vol. 28, No. 2, Monsoon.
21. Bora, G.P (2010) "Importance of linking microfinance with livelihood promotion for tribal development", Development Matters, July 2010.
22. Bundela, A, Dubey. K and Pande. S (2013): "Role of Self Help Groups in Rural Employments of India: A Case Study of Swarn Jayanti Gram Swarojgar Yojna", IOSR Journal Of Humanities And Social Science (IOSR-JHSS) e-ISSN: 2279-0837, p-ISSN: 2279-0845.Volume 7, Issue 1 (Jan. - Feb. 2013), pp. 38-41.
23. Business maps of India.com/finance/top-finance companies –in-India.
24. Cain, P (2010): Microfinance Meltdown in Bosnia, Al Jazeera (English), 4 January.
25. Chakrabarti. R (2004): The Indian Microfinance Experience- Accomplishments and Challenges, India Development Foundation, Gurgaon.
26. Champatiray, Agrawal and Sadhu (2011): Map of Microfinnace Distribution in India, IFMR, Centre for Microfinance research.
27. Chauhan, N.M. and Kshirsagar, S.M. (2012) "Assessment of Training Needs of Members of Tribal Women SHGs or Agriculture Development", Indian Research Journal of Extension Education, Special Issue (Volume II), 2012.
28. Choudhary, M (2013): "A Detailed Study of Micro Finance as a tool for Tribal

Transformation in Areas of Madhya Pradesh", International Journal of Business and Management Invention www.ijbmi.org Volume 2 Issue 3, March. 2013, pp. 72-76.

29. Choudhary, S (2005): "Self Help Group Formation Through Frequently Asked Questions," Kurukshetra, 53 No.3, January 2005, pp.10-14.

30. Copestake.J, Greely.M, Johnson. S, Kabeer.N and Simanowitz (2006): "Money With a Mission, Volume 1: Microfinance and Poverty Reduction", Intermediate Technology publications, London.

31. Cronin. A and Ostergren, G, 2007: Tribal Watershed Management: Culture, Since, Capavity and Collaboration, American Indian Quarterly, Vol.31, No.1.

32. Dadhich. C.L (2007) "Lessons from the Micro Credit Experinece in India", http://www.apraca.th.com/Sign_Guestbook/Infomarket/Lessons_from_the_Micro_Credit_/body_lessons_from_the_micro_credit_.html.

33. Das, S.K (2011) "Women Empowerment and Self Help Group: An Analytical study of Constraints in Karbi Anglong District of Assam", Journal of North East India Studies, Vol-1, No-1.

34. Das, V (2010): Natural Farming, Tribal Farming, http//infochangeindia.org/agenda/agricultural-revival/natural-farming-tribal-farming.html.

35. Dasgupta,(2001). "An Informal Journey through Self-Help Groups", Indian Journals of Agricultural Economics, Vol.56, No.3, July-September.

36. Dieter, S and Harshkumar, R^2 (2002): "Commercial Aspects of Self-Help Group Banking in India: a Study of Bank Transaction Costs", Paper presented at a seminar on The SHG Bank Linkage Programme in India, New Delhi, 25- 26 November 2002, Published by Micro Credit Innovations Department, National Bank for Agriculture and Rural Development, Thompson Press, Mumbai 1992.

37. Economic: Adam Smith Theory of Value, Economictheories.org. Retrieved 2012-03-13.

38. Fernandez, Aloysius P (2007): "History and Spread of the Self-Help Affinity Movement in India: the Role Played by IFAD." Occasional Paper Series, IFAD, July 2007..

39. Fouillet, C and Augsburg, B (2007): "Spread of the Self-Help Groups Banking Linkage Programme in India", International Conference on Rural Finance Research: Moving Results, FAO Headquarters, Rome, Italy ,19-21 March 2007.

40. Gary Becker and Richard Posner (2006):" Microfinance and Third World Poverty and Development", http://www.becker-posner-blog.com/2006/10/microfinance-and-third-world-poverty-and-development—posner.html.

41. Gurumoorthy, T.R (2000): "Self Help Groups Empower Rural Women", Kurukshetra, Feb.

42. Haimendrof, C,V. F (1998): "Tribes in India - The Strategy for Survial", Oxford University Press, New Delhi, p. 342.

43. Harish, N (2012): "Women's empoWerment through Entrepreneurship Development With Special Reference to self help Groups in Karnataka, IJRSS, Vol-II, Issue-II.

44. Historical context of microfinance and its evolution, www.globalenvision.org.

45. http://en.wikipedia.org/wiki/Revolving_fund.

46. http://fcamin.nin.in/dfpd/Event details.asp?.

47. http://www.aptribes.gov.in/html/tcr-studies-evaluation-egbs.htm.

48. http://www.ifad.org/operations/projects/regions/pi/paper/3.pdf.

49. IFAD Evaluation Report (2003): http://www.ifad.org/english/indigenous/documents/ip_policy_e.pdf.

50. India top 50 Micro Finance Institutions, CRISIL Ratings October 2009.

51. Joseph. P (2007): www.indiareports.com/business writters/ indian microfinance sector.

52. Karmakar, K.G (2008): " Microfinance in India", Sage Publication, New Delhi.

53. Katsushi S. Imai and Arun T (2008): "Does Microfinance Reduce Poverty in India?", http://www.socialsciences.manchester.ac.uk/disciplines/economics/research/discussionpapers/pdf/EDP-0814.pdf.

54. Kevany, S (2010): Debt trends in Peru's Personal, Microfinance Sectors Spark, Warnings Wall Street Journal, 16 December.

55. Krishnaiah, P. (2003): Poverty Reduction, Self-help Group Strategy,UBI Publishers' Distributors Pvt. Ltd, pp. 77-117.

56. Krishnamurthy and Ratnaparkhi (2002): " Micro-Finance in the New Economy-India's Experience", http://theatreforum.in/static/upload/docs/MICROFINANCE.pdf.

57. Lascelles. D (2012): Microfinance-A Risky Business A Time for Strong Leadership, Centre for Financial Inclusion, Publication 17, May.

58. Latifee, H.I (2006): "The Future of Microfinance: Visioning the Who, What, When, Where, Why, and How of Microfinance Expansion Over the Next 10 Years, http://www.shatil.org.il/files/MF_the%20future%20of%20MF.pdf.

59. Laucer, K (2008): "The Pursuit of Complete Financial Inclusion -the KGFS Model in India", CGAP, No. 12, June.

60. Lauer, K(2008): Transforming NGO MFIs: Critical Ownership issues to Consider, Occasional Paper, CGAP, No.13, June.

61. Lewis, J(2009): "Microfinance: An Economic Development Failure?", http://skollworldforum.org.

62. Lokesh (2008): "Microfinance in India-Growth without Purpose", Source: www.indiamicrofinance.com.

63. Lokpal Capital (2010): Microfinance Industry in India, taken from NCAER-The great Indian market.

64. Malhotra, Rakesh,(199): Micro-Finance A Panaccea For Poverty Alleviation, Bank And Finance, June.

65. M-CRIL, 2011: Microfinance Review 2011 Anatomy of a crisis- a financial and social analysis.

66. Mishra, A (2005): "Why Microfinance needs Participatory Impact Assessment: Case Analysis of SHG-Bank Linkage programme in India", *Asia Workshop on Next Generation Participatory Monitoring and Evaluation* 9-11 November 2005, PRIA, New Delhi, India.

67. Misra. A (2006): " Microfinance in India and Millennium Development Goals: Maximizing Impact on Poverty", http://lnweb90.worldbank .org/exteu/sharepapers.nsf/7795535c333f8172c1257162003d5b69/0a0e95f1471f1d4c852571fd004dae86/$file/alok%20misra.pdf.

68. Morduch, Jonathan, (1999): "The Microfinance Promise", Journal of Economic Literature, 37 (4), December:1569 - 1614.

69. Moses, V.J.R. E (2011): "Women empowerment through SHGs: A micro study", International Referred Research Journal, Vol. II, Issue 16, January 2011, p. 25.

70. NCAER (2008): "Impact and sustainability of SHG-Bank Linkage Programme", New Delhi.

71. Nooren, S (2011): "Role of Microfinance in Empowerment of female Population.

72. Norma, S (2010): *Achieving human development objectives through microfinance institution: the case of Amanah Ikhtiar Malaysia.* Journal of Islamic Economics, Banking and Finance, 6 (2). pp. 65-78. ISSN 2070-4658.

73. Pamela Mackenzie (2006): Keeping it LOCAL, Change and Development in an Indian Tribal Community, APH, Publishing House, New Delhi.

74. Pantoja Enrique (2002): " Microfinance and Disaster Risk Management Experiences and Lessons Learned", Provention Consortium, The World Bank.

75. Patel, M.L (1984): "Planning Strategy for Tribal Development", pp. 153-159.

76. Priya Deshingkar and Johnson (2002): "Poverty Reduction in Andhra Pradesh: The Real concerns", http://www.docstoc.com/docs/23863754/poverty-reduction-in-andhra-pradesh-the-real-concerns.

77. Purushotham, S and Rao, P., (2009):Economics of Sericulture in Ananthapur District of Andhra Pradesh, Agricultural Science Digest, Vol. 29, Issue 2).

78. Rajat. W (2007): " Micro-finance in the India: The Changing face of micro-credit schemes", http://www.researchgate.net/publication/24113645_.

79. Rao, K.M (1999): "Tribal Development in Andhra Pradesh- Problems, Performance and Perspects", pp. 158-178.

80. Rao, V.M., "Women Self Help Groups: Profiles from Andhra Pradesh and Karnataka", Kurukshetra, April 2002.

81. RBI (2005): Report of the Internal Group to Examine Issues Relating to Rural Credit and Microfinance.

82. RBI (2008): Banking statistical Return Reports, RBI, Speeches Published on Financial Inclusion.

83. Reddy, C.S. and Manak, S. (2005) *Self Help Groups: A Keystone of Microfinance in India – Women Empowerment and Social Security.* [Online] Available from: http://www.aptsource.in/admin/resources/1273818040_SHGs-keystone-paper.pdf.

84. Rengarajan, V (2001): "MFI's Foray into Microinsurance", Asian Knowledge Centre for Mutual Insurance, http://www.microfinancegateway.org/.

85. Role of Micro Finance in Economic Development, Wall Street Microfinance,.

86. Sachs, D..J and Myers (2005): "The End of Poverty: Economic Possibilities for Our Time", www.carnegiecouncil.org/studio/multimedia/20050330/5132.html/_res/id=sa_File1/The_End_of_Poverty.pdf.

87. Samal, P.K (1996): "Tribal Development Options, Proceedings of National Seminar (May, 23-24), PP-363-367.

88. Samia Mahmood, (2011) "Microfinance and women entrepreneurs in Pakistan", International Journal of Gender and Entrepreneurship, Vol. 3 Iss: 3, pp.265 - 274.

89. Seibel. H.D (2007): "The Role of Microfinance in Rural Microenterprise Development", Syngenta Foundation for Sustainable Agriculture, Switzerland, www.syngentafoundation.org.

90. Shah, Chaitanya (2011): "Microfinance-Important Role in Economic Growth", www. Smeworld.org.

91. Sharma, K.C., "Micro financing through SHG's." Indian Journal of Agricultural, vol.56, no.3, (2001), p.460.

92. Shylendra, H.S (2006): "Microfinance Institutions in Andhra Pradesh", Economic and Politically Weekly, PP-1959-1963.

93. Singh Tejmani, N (2009):" Microfinance practices in India: An Overview", International Review of Business Research Papers Vol. 5 No. 5 September 2009 Pp. 131-146.

94. Sinha Anand (2012): Strengthening Microfinance Institutions (MFIs): good Governance and Strategic People Practices, Mumbai, April.

95. Sinha Frances (2005): " Access, Use and Contribution of Microfinance in India-findings from National Study, Economic and Political Weekly, Vol. 40, No.17, (April.23-29), pp.1714-1719.

96. Sinha, F (2003): "Impact Assessment of Microfinance in India", Paper prepared for the conference , Current issues in Microfinance, Misty Hills, Johannesburg–South Africa, 12-14 August.

97. Sinha.F (2009): State of Microfinance in India, Institute of microfinance (InM).

98. Srinivasan (2011): Microfinnace India, State of the Sector Report 2011, Sage Publication, New Delhi.

99. Srinivasan. N (2008) State of the Sector Report on Microfinance, published by Sage, New Delhi.
100. Srinivasan. N (2009) in his book " Microfinance India-State of The Sector Report, 2009, Sage Publication, New Delhi.
101. Sriram and Upadhyayula (2004): " The Transformation of the Microfinance Sector in India", Journal of Microfinance, Vol-6, No.2, pp. 89-112.
102. Status of Micro finance in India, 2009-10, NABARD, Mumbai.
103. Sultana and Thomson, 2007: Community Based Fisheries Management and Fisher Livelihoods: Bangladesh Case Studies, Human Ecology, Vol.35, No.5.
104. Surender, Kumari, S., and Sehrawat, R.K. (2011): **"Can self-help groups generate employment opportunity for rural poor?"**, European Journal of Social Sciences (EJSS).
105. Swain, R.B and Wallentin, F.Y (2007): "Does Microfinance Empower Women? Evidence from Self Help Groups in India", working paper-24, Department of Economics, Uppasala Universitet.
106. Thakur.B.K. (2006), "Poverty Among Tribals: A Comparative Study of Pre-reform and Post-reform Periods", Southern Economist, pp. 27-30.
107. The Financial Express (2002): Redefining the Silk Route-Andhra Sericulture Department to Set Up Model Silk Villages, statement of Commissioner and Managing Director of Sericulture Department, Hyderabad.
108. The Hindu (2007): 'Micro finance bill on the anvil', thehindu.com.
109. The Hindu (2007): "True heroines of rural development", http://www.thehindu.com/todays-paper/tp-features/tp-sundaymagazine/true-heroines-of-rural-development/article 2275003.
110. The Social Performance report (2011): "Microfinance funders and Social Performance Management", Microfinance India, ACCESS Knowledge Series, New Delhi.
111. Varma, G.R and Rohini, A (2008): "Perceptions of Community on Post-Contraceptive Care by Primary Health Care System in Tribal and Rural Areas of Visakhapatnam District of Andhra Pradesh, Ethno-Med., 2(1): pp. 29-32.
112. Vishwakarma, Y.B (1991): "Industrialization and Tribal Ecology", Rawat Publication, New Delhi, pp. 272-286.
113. Yunus, M (2003): "Banker to the Poor: Micro-Lending and the Battle Against World Poverty", Public Affairs.
114. Yunus, M. (2007): "Let us compete in poverty eradication", http://southasia.oneworld.net/archive/Article/let-us-compete-in-poverty-eradication-muhammad-yunus.
115. Zacharias, Sibi (2003): " The Micro-level impact of Tribal Development Programmes among the Kadar Tribe of Kerala", Kerala Research Programme on Local Level Development Centre for Development Studies, Thiruvananthapuram, Kerala, India.
116. Zakhir (2011): "Management of Self Help Groups", State Institute of Rural Development (SIRD), Assam, p. 48.

www.ingramcontent.com/pod-product-compliance
Ingram Content Group UK Ltd.
Pitfield, Milton Keynes, MK11 3LW, UK
UKHW041643031026
15587UKWH00001B/458

9 789351 302971